Chestnut-bellied Nuthatch

Eastern Rock Nuthatch

Beautiful Nuthatch

David Quinn

Algerian Nuthatch

THE NUTHATCHES

For Viviane, Roel and Wout

THE NUTHATCHES

by

Erik Matthysen

Illustrated by
DAVID QUINN

T & A D POYSER

London

First published 1998 by T & AD Poyser Ltd
Digital editions published 2010 by T & AD Poyser, an imprint of A&C Black
Publishers Ltd, 36 Soho Square, London W1D 3QY

ISBN (print) 978-0-8566-1101-8
ISBN (epub) 978-1-4081-2870-1
ISBN (e-pdf) 978-1-4081-4952-2

A CIP catalogue record for this book is available from the British Library

Visit www.acblack.com/naturalhistory to find out more about our authors
and their books. You will find extracts, author interviews and our blog, and
you can sign up for newsletters to be the first to hear about our
latest releases and special offers.

Printed and bound by CPI Group (UK) Ltd, Croydon CR0 4YY

MIX
Paper from
responsible sources
FSC® C013604

Contents

Preface   xvii

Acknowledgements   xx

Part I INTRODUCTION   1
Chapter 1   INTRODUCING THE NUTHATCHES   3

Part II THE EURASIAN NUTHATCH   21
Chapter 2   TAXONOMY, MORPHOLOGY AND MOULT   23
Chapter 3   HABITAT AND POPULATION DENSITIES   36
Chapter 4   FORAGING, FOOD AND HOARDING   46
Chapter 5   THE PAIR AND ITS TERRITORY   62
Chapter 6   BREEDING BIOLOGY   86
Chapter 7   FINDING A TERRITORY   112
Chapter 8   DISPERSAL AND MIGRATION   130
Chapter 9   POPULATION DYNAMICS   147
Chapter 10   NUTHATCHES IN FOREST FRAGMENTS   163

Part III NUTHATCHES OF THE WORLD   181
Chapter 11   THE MEDITERRANEAN NUTHATCHES   183
Chapter 12   THE ROCK NUTHATCHES   200
Chapter 13   ORIENTAL NUTHATCHES   217
Chapter 14   NEW WORLD NUTHATCHES   238

Appendix I   SCIENTIFIC AND COMMON NAMES OF NUTHATCHES   269
Appendix II   DIAGNOSTIC TRAITS OF THE 24 NUTHATCH SPECIES   271
Appendix III   SEX- AND AGE-RELATED MORPHOLOGICAL VARIATION
                         IN EURASIAN NUTHATCHES   272
Appendix IV   POPULATION DENSITIES OF THE EURASIAN NUTHATCH   275
Appendix V   A LIFE-TABLE FOR THE EURASIAN NUTHATCH   278
Appendix VI   SCIENTIFIC NAMES OF SPECIES MENTIONED IN
                         THE TEXT   280

References   284

Index   313

List of Photographs

1. A singing male *S. e. caesia* — 13
2. A male *S. e. caesia* in the hand — 32
3. Deciduous woodland habitat in the Peerdsbos, Belgium — 40
4. Deciduous woodland habitat near Uppsala, Sweden — 40
5. Parkland habitat in the Peerdsbos, Belgium — 41
6. Larch forest near Magadan in eastern Siberia — 43
7. Inspecting a nest cavity — 88
8. A nestbox with a half-open front, filled up with mud — 92
9. A *c.* 10-day-old nestling with colour-rings — 101
10. Aerial view of the 'fragments' study area — 167
11. Oak wood fragment of 1.5 ha — 169
12. Oak wood fragment of 0.8 ha — 169
13. Laricio pine forest near Evisa — 191
14. Western Rock Nuthatch — 202
15. Velvet-fronted Nuthatch — 234
16. Deciduous woodland inhabited by Red- and White-breasted Nuthatches — 240
17. A White-breasted Nuthatch caught in a mistnet — 244

List of Figures

1. Phylogenetic relationships between nuthatches and other passerines   6

2. Hypothetical phylogenetic tree of the genus *Sitta*   6

3. Worldwide species richness of nuthatches   8

4. Male size ranges (wing length) of all nuthatch species   9

5. 'Idealized' nuthatch showing the major diagnostic traits between species   10

6. Zigzag movement by a climbing nuthatch   15

7. Geographical distribution of the major studies on ecology and behaviour of the Eurasian Nuthatch   19

8. World distribution of the Eurasian Nuthatch and closely related species   25

9. Colour scores of brown- and white-breasted Eurasian Nuthatches in central and eastern Europe   26

10. Typical bill profiles of *europaea* and *arctica*   27

11. Postglacial expansion of the Eurasian Nuthatch across Europe   29

12. How wing moult is scored   34

13. Progression of primary flight feather moult in the Peerdsbos population in 1985   35

14. The frequency distribution of Nuthatch population densities in different habitat types   38

15. Breeding densities in different successional stages of a semi-natural oak forest in Burgundy, France   39

16. Change in breeding density with altitude in the Harz mountains, Germany   44

17. Diet of nestling Nuthatches obtained by neck collar samples   49

18. Seasonal variation in mean bill length of males and females   52

19. Foraging sites used by Nuthatches in four forests   53

20. Foraging substrate sizes in relation to the bird's own diameter   54

21. Preference for top, bottom or sides of horizontal or slanting branches by Eurasian and White-breasted Nuthatches   55

22. The number of cedar nuts taken from a feeder per hour and the proportion consumed   58

23. The study plots used in the Peerdsbos forest near Antwerp   63

24. The number of Nuthatches captured per month in the Peerdsbos from 1982 to 1987   64

25. Home-ranges of 10 pairs in the Peerdsbos   65

26. Home-ranges of four pairs in winter   67

27. Increase in the estimated size of four winter home-ranges with the number of observation periods   68

28. Movement patterns of a pair of Nuthatches in winter and spring   69

29. Threat displays of the Nuthatch   70

30. Confrontation experiments with a caged intruder   72

31. Home-ranges of a trio of Nuthatches   72

32. Daily pattern of nest provisioning and hoarding flights   75

33. Seasonal pattern of song activity in the Peerdsbos study area   77

34. Sonagrams of the five main song types of the Nuthatch   78

35. Sonagram of a transition between Trilled and Fast Ascending song   79

36. Song-matching in response to playback of different vocalizations   81

37. The geographical distribution of the song type Up-down in northern Belgium   82

38. Sonagrams of two call types   83

39. Preference of Nuthatches for different types of nestboxes   90

40. Provisioning rates of nestlings in Siberia and Germany   102

41. Summary of the fate of 400 broods in a nestbox study near Frankfurt   103

42. Clutch sizes and mean number of fledged young per nest for five habitats in Germany   106

43. Standardized fledging dates of nests with and without recruits   107

44. Breeding success in relation to population density   108

45. The proportion of male nestlings in broods of different sizes   109

46. Observations on two fledglings from the same nest that settled soon after fledging   114

47. Shifts between territories of different quality in the Peerdsbos study   116

48. One-year histories of a set of territories in the Peerdsbos study area — 116

49. Observations on a male satellite — 118

50. Successive home-ranges of a non-territorial male — 118

51. Settling date of juveniles in early summer in relation to 'breeding score' — 121

52. Changes in territory boundaries between two breeding seasons — 124

53. Autumn territories in a study area in Sweden with and without extra food — 125

54. Home-ranges of three non-territorial juveniles in Siberia — 126

55. Natal dispersal distances in the Frankfurt study — 133

56. Ringing recoveries over more than 100 km — 134

57. Decrease in the proportion of fledglings remaining on the natal territory over time — 135

58. The number of Nuthatches observed in the city of Amsterdam per month — 136

59. The number of individuals settling in the Peerdsbos study area per month — 137

60. The number of transients observed per month in the Peerdsbos study — 137

61. The occurrence of long-distance dispersal in relation to population density in the Braunschweig area — 139

62. The number of long-distance dispersers by compass quadrants — 139

63. Probable invasion routes of *S. e. asiatica* into Fennoscandia — 144

64. The relationship between male and female survival rates in six studies — 150

65. Local survival per season for different sex and age classes in the Peerdsbos study — 151

66. Mean monthly survival during winter in relation to mean winter temperature in Sweden — 153

67. Population size in the Peerdsbos study in relation to beech mast — 154

68. Autumn population sizes in two study areas in Sweden in years with and without extra food — 155

69. Changes in population size in different areas in Europe — 158

70. Past and present breeding distribution of Nuthatches in northern Belgium — 160

71. The areas used in the study on habitat fragmentation, 1990–1994   168

72. Movements by Nuthatches within the fragments study area   171

73. The rate of arrival of 'settlers' in the Peerdsbos and in forest fragments   173

74. Diagram of a simulation model of a nuthatch population in habitat fragments   177

75. The probability that a model population survives during 250 years as a function of habitat size   179

76. Changes in the estimated population size of the Algerian Nuthatch since its discovery   185

77. Geographical distribution of the Algerian Nuthatch   186

78. Geographical ranges of the Old World members of the *Sitta canadensis* superspecies   189

79. Vegetation profiles of laricio pine forest at different elevations   192

80. Sonagrams of Krueper's, Corsican and Algerian Nuthatch song   197

81. Geographical ranges of the Western and Eastern Rock Nuthatches   204

82. 'Character displacement' in the rock nuthatches   206

83. Geographical variation in eye-stripe size in Western and Eastern Rock Nuthatches   207

84. Geographical variation in bill length in Western and Eastern Rock Nuthatches   207

85. Postures and displays of rock nuthatches   213

86. Sonagrams of duetting Eastern Rock Nuthatches   215

87. Altitudinal distribution of nuthatches in different parts of Asia   219

88. Songs of the Chestnut-bellied and White-tailed Nuthatches   225

89. Magnified crown feathers of the Giant Nuthatch   231

90. Foraging niche differentiation among nuthatches in Colorado   242

91. Winter home-ranges of White-breasted Nuthatches   244

92. Vocalizations of North American Nuthatches   246

93. Anti-predator display of the White-breasted Nuthatch   248

94. The number of migrant White-breasted Nuthatches arriving and leaving per month in Baltimore   250

95. The use of foraging substrates by Red-breasted Nuthatches at temperatures above and below 0°C   253

96. Home-ranges of Red-breasted Nuthatches in winter   254

97. Occurrence of Red-breasted Nuthatch irruptions in relation
    to cone-crop size   257

98. Foraging behaviour of Brown-headed Nuthatches in relation
    to cone crops and mixed-species flocks   261

99. Schematic representation of a Pygmy Nuthatch group roosting
    in a cavity   267

100. Seasonal variation in bill length during summer and autumn   273

101. Daily variation in body weight   274

102. Seasonal variation in body weight   274

103. Lifetime survivorship curves for the Peerdsbos study population 279

List of Tables

1. Nest-building behaviour of nuthatches 17
2. Major field studies on the Eurasian Nuthatch 19
3. Diet of adult Nuthatches 48
4. Arthropod families contributing 10% or more to the nestlings' diet 50
5. Proportion of vegetable food in the diet 51
6. Proportion of foraging observations on different tree species 55
7. Sexual differences in foraging behaviour 56
8. Time budgets of Nuthatches in Sweden 74
9. Behaviour of Nuthatches after being released from capture 84
10. Breeding parameters of Nuthatches in different study areas 97
11. Mean first-egg dates of Nuthatches and several tit species 98
12. Plumage development with nestling age 102
13. Causes of breeding failure other than nest take-over or predation 105
14. Behavioural status of first-year Nuthatches when they settled in the Peerdsbos study area 117
15. Body size and condition of resident and transient juvenile Nuthatches 123
16. Main modes of establishment of young birds in four different studies 127
17. The proportion of dispersal movements to the west in dispersal studies at different scales 140
18. Annual local survival rates of adult Nuthatches 149
19. Seasonal survival of adult birds in Belgium and Sweden 152
20. Population estimates for different European countries 162
21. Effects of landscape variables on the presence/absence and breeding density of Nuthatches in forest fragments 165
22. Breeding population densities, breeding success and survival rates in different habitats 165

23. General characteristics of the Mediterranean nuthatches
     and their closest relative, the Chinese Nuthatch   187

24. The history of classification of the rock nuthatches   203

25. Probable means of ecological segregation of nuthatches in
     the mountains of central and SE Asia   218

26. Breeding densities of White-breasted Nuthatches in different
     parts of the USA   241

27. Proportion of foraging time spent on trunks by dwarf
     nuthatches in summer and winter   260

28. Sexual differences in morphometric variables   272

29. A life-table for the Nuthatch   278

Preface

In the spring of 1982 I was a third-year zoology student at the University of Antwerp, faced with the necessity of choosing a subject for my graduation dissertation. Among the more appealing proposals was a project on snail taxonomy, which, had I chosen it, would definitely have made a substantial difference to the rest of my life. Instead I chose a project on the territorial behaviour of Nuthatches in a forest in my neighbourhood. The bird and the subject proved to be inspiring as well as productive, and the project soon turned into a 5-year uninterrupted study leading to a doctoral thesis. Even though the study itself focused on territorial behaviour and population regulation, I rapidly became interested in many other aspects of Nuthatch biology such as song, nest-building, movements and moult. Studying this common but not particularly well-known bird was helping me to fulfil an old ambition from my earliest bird-watching days by collecting all the knowledge available on one particular bird, and adding to it as well. Several years (and another 5-year project on nuthatches) later, the opportunity to write this book offered itself as an irresistible chance to do a more complete literature study of the nuthatch family, and put some of my unpublished or unanalysed data to better use.

In this monograph I attempt to summarize all the information available on the natural history of nuthatches, with particular emphasis on their behaviour and ecology. The first part (Chapter 1) is a general introduction to the nuthatch family, where I discuss some features that are shared by all species (such as their climbing technique) and briefly introduce other aspects that will be covered later in more depth. Part II (Chapters 2–10) is devoted to the Eurasian Nuthatch, by far the best-studied member of the family. It is by no means intended as a summary of my own work, although some chapters (notably, 5, 7 and 10) rely on it to a large extent. I have tried to integrate information from different studies wherever possible, and to point at similarities and contradictions whenever these appear. This often leads to more general insights but, in some cases, merely demonstrates the difficulty of comparing observations between habitats, regions or time periods. Note that throughout Part II I have shortened Eurasian Nuthatch to 'Nuthatch' for convenience, except where confusion could arise. The lower-case 'nuthatch' is reserved for references to more than one particular species. The same applies to 'rock nuthatch' as a collective name for the Western and Eastern Rock Nuthatches (in Chapter 12) and 'dwarf nuthatch' for Pygmy and Brown-headed Nuthatches (Chapter 14). Scientific names of nuthatches, other animals and plants are given in

Appendices I and VI. Part III deals with the remaining 23 nuthatch species, more or less grouped by geography and/or affinities. All of these species have recently been covered by Harrap and Quinn's (1996) excellent monograph, with full details of plumages, geographical variation and distribution, including plates and maps for all species. I only summarize this information here, and refer to Harrap and Quinn for more details. However, since they admit to having followed a policy of splitting rather than lumping subspecies in any case of doubt, I have preferred to follow the more traditional views of Greenway (1967) for the designation of subspecies. Apart from plumages and geographical distribution, another aspect that has necessarily received a rather parochial treatment is the description of vocalizations. The songs and calls of the majority of nuthatches have hardly been studied at all, and, although tape-recordings are available for most of them, a proper analysis would entail a study on its own, far beyond the scope of this book. Undoubtedly such a study would be extremely interesting and could shed more light on some unsolved phylogenetic problems within the genus. Again I refer to brief descriptions in Harrap and Quinn (1996) and other sources. A few sonagrams are included, mainly to illustrate the general properties of nuthatch song and calls, and point at some differences or similarities between species.

Many people have contributed in various ways to my work on nuthatches and to the preparation of this book. Foremost among these is André Dhondt, presently at Cornell University, who has been an invaluable inspirer and supervisor for the past 15 years. He taught me not just how to practise good science, but also to become a good scientist. Many other colleagues ('dhondtons') made the Laboratory of Animal Ecology in Antwerp an inspiring, productive and sociable environment. Frank Adriaensen deserves special mention for providing the first and often highly useful comments on most of my ideas, and for endless practical and conceptual support in the field and elsewhere. Luc Lens was an excellent proofreader for an early version of the entire manuscript.

Several people have helped me to broaden my views and experiences of nuthatches. In 1986 a brief but fruitful visit to Karl-Heinz Schmidt at Frankfurt University provided the material for analyses on dispersal and breeding biology. In 1989–1990 I was able to study Red- and White-breasted Nuthatches during a 3-month stay at Ohio State University, with indispensable practical and moral support from Thomas C. Grubb Jr. and David Cimprich. Jacques Blondel and his crew at Montpellier were helpful in getting me acquainted with the Corsican Nuthatch during a short expedition in 1985, repeated in 1987. In 1989 the first of several meetings and discussions with Paul Opdam and his co-workers, Alex Schotman and Jana Verboom, provided the impetus for my later work on nuthatches in forest fragments. Finally, a continuous correspondence with Bodil Enoksson (Uppsala and Lund) has given me the benefit of her 'Swedish' view on Nuthatches on a variety of matters.

Over the years, and even more so in the final stages of writing this book, many other people have been helpful by providing me with literature, unpublished reports, tapes and sonagrams, by lending me unpublished data, or simply discussing some aspects of nuthatch biology with me. In this respect I wish to thank Russell Balda, Marcel Eens, Dave Farrow, Millicent Ficken, Lars Gabrielsen, Sandra Gaunt, Cameron Ghalambor, Judy Guinan, David Jardine, Hans Källander, Omar Kisserli, Anton Krištín, Eric Liknes, Thomas Martin, Arch McCallum, Jan-Åke Nilsson, Eugeny Panov, Vladimir Pravosudov, Kathy Purcell, Pam Rasmussen, Craig Robson, Bill Smith, Sören Svensson, Tomas Wesołowski, Wolfgang Winkel, Jay Withgott and Peter Yaukey. Mike Wilson and Vladimir Pravosudov kindly provided translations or summaries of some papers in Russian. A special thanks goes to Simon Harrap for sending me draft texts and maps of his book, as well as additional references and tape-recordings. I also want to mention the Alexander Library and librarian Linda Birch of the Edward Grey Institute for Field Ornithology at Oxford for their generous help and assistance. Sound recordings were provided by the Borror Laboratory of Bioacoustics (Ohio State University), the Cornell Laboratory of Ornithology (Ithaca) and the National Sound Archive (London). My research on Nuthatches has been supported by the Fund for Scientific Research – Flanders (FWO) (1984–present), by the Belgian Institute for promotion of Scientific Research in Industry and Agriculture (IWONL) (1983–1984) and a NATO postdoctoral fellowship while at Ohio State University (1989–1990). Finally, I want to thank Andrew Richford of T & AD Poyser for giving me the opportunity to write this book and for his encouragement, support and patience.

Artist's Acknowledgements

Thanks to Tony Parker of the Section of Birds and Mammals at the Liverpool Museum for his kind assistance during my visits to study the nuthatch specimens.

To Pete and Jane Shaw whose garden provided numerous opportunities to observe nuthatches at close quarters.

I would like to acknowledge the excellent Survival Anglia programme *Acrobat of the Woods,* filmed by Andrew Anderson, as a source of reference for some of the illustrations – specifically those of the female Eurasian Nuthatch on the nest and the 'pendulum' mating display of the male (as featured in the jacket illustration).

Part I: Introduction

Chinese Nuthatch.

CHAPTER 1

Introducing the Nuthatches

Tipsy clerics who've found
A cheery religion in browns and blacks
A family of gleaners that rattles along

Marshall (1996)

Nuthatches are among the easiest birds to describe to non-birdwatchers: if you see a bird climbing down a tree head first, it can only be a nuthatch; if it climbs upwards, it may be something else. However, they have much more to offer both to the birdwatcher and to the scientist than mere acrobatics. For one thing, they show an amazing variety of tactics for protecting their nests, from building them from mud to using resin or insect odours as predator repellents. They also possess a certain quality that many writers have described as 'agile', 'jolly' and 'good-humoured', perhaps because of their vivid plumages and simple but vigorous songs. All nuthatches are

3

typical forest birds, with the exception of the rock nuthatches – whose very name betrays their favourite habitat.

The many unique traits of the nuthatches, and their presence in almost all northern temperate forests, should make them ideal subjects for ecological and behavioural study. Nevertheless, perhaps because of their generally low population densities and reclusive lifestyle high up on tree trunks and limbs, nuthatches are certainly not the best-studied group of birds. Basic information is still lacking on some aspects of the behaviour and ecology of the majority of species, with the notable exception of the Eurasian Nuthatch. This chapter gives an overview of some of their major characteristics, with particular attention given to the evolutionary relationships within the genus and with other birds.

THE GENUS *SITTA*

All recent bird lists, with the sole exception of Wolters (1975–1982), recognize only one genus for all nuthatches, although several species have formerly been assigned to genera of their own. The name *Sitta* is derived from the ancient Greek (or even pre-Greek) σιττη, appearing first in Aristotle's *Historiae Animalarum*. The name may have originated as an imitation of the calls of the Eurasian Nuthatch, or, even more likely, the Western Rock Nuthatch (Thompson, 1936). It was introduced by Linnaeus (1758) in the well known 10th edition of his *Systema Naturae*. At that time the nuthatches were still sandwiched between the woodpeckers and the kingfishers in the order Picae, and only one species (*europaea*) was described for Europe and N America. Nowadays 24 species are recognized, the last of which was described as late as 1976. Some of the systematic relationships between the species are still unclear, and several forms have been variously treated as both full species and subspecies.

Wolters (1975–1982) divided the nuthatches into four discrete genera and several subgenera. Although these are little used, they point to the existence of several, more or less clear, species-groups (see Appendix I). The largest group (subgenus *Sitta*) includes the Eurasian Nuthatch with a few close relatives, plus the two rock nuthatches, and the Giant Nuthatch, the largest species of all. A second major group (subgenus *Micrositta*) contains six small nuthatches living throughout the Holarctic region, most of them restricted to coniferous habitat. Many were once considered conspecific (see Chapter 11). This group contains, among others, the recently discovered Algerian Nuthatch. Three minor groups have been given generic status by Wolters (*Callisitta, Poecilositta, Oenositta*) and together comprise five colourful species from subtropical Asia. The two 'dwarf nuthatches' of N America (Pygmy and Brown-headed) are not assigned to a subgenus by Wolters, and, although Glutz von Blotzheim (1993) includes them in *Mesositta* with two Asian species, their relationships remain unclear. Finally, the subgenus *Leptositta* contains a N American and central Asian species, both unique in lacking the typical eye-stripe, but otherwise probably close to *Micrositta*.

RELATIONSHIPS AND EVOLUTION OF THE SITTIDAE

Above the genus level, the nuthatches are usually separated into a sub-family or family of their own (Sittinae or Sittidae). It is generally accepted that their closest relative is the Eurasian Wallcreeper, which is often included in the Sittidae in a separate subfamily (Tichodromadinae) (Sibley and Monroe, 1990) but by others in a family of its own (e.g. Wolters, 1975–1982). Sibley and Ahlquist (1990) give a brief overview of the history of classification of the nuthatches and allies, showing that the family has traditionally been a 'scrap-basket' (Mayr and Amadon, 1951) containing a number of superficially nuthatch-like forms without clear phylogenetic affinities. The majority of these groups have recently been removed from the Sittidae. For instance, the superficially very similar Australasian sittellas were long considered close relatives until recent molecular work suggested that they are of corvid origin and are close to other Australo-Papuan lineages (Sibley and Ahlquist, 1982, 1990) (Fig. 1). This accords with some striking differences between nuthatches and sittellas, such as the spotted juvenile plumage and the use of open nests in the latter. Other nuthatch-like birds include the Malagasy Coral-billed Nuthatch, now considered a vanga with the more appropriate name of Nuthatch-Vanga; the Australo-Papuan treecreepers with still uncertain affinities, but probably belonging with other Australo-Papuan birds (Sibley and Ahlquist, 1990); and the Philippine creepers whose position is also still unclear. All of them have formerly been included in the Sittidae. Within the Holarctic region, the nearest relatives of nuthatches and the Wallcreeper have traditionally been sought with the tits and treecreepers, who share many of the ecological and behavioural characteristics of the nuthatches. Vaurie (1957) even advocated a single family for nuthatches and tits. However, whereas Sibley and Ahlquist (1990) have confirmed the close relationship between nuthatch and treecreeper lineages, they consider the tits to have branched off much earlier from the nuthatch/treecreeper group (Fig. 1).

It seems plausible that the first nuthatches evolved somewhere in SE Asia where the species diversity is still highest. However, the only fossil record is from Italy, dated to the lower Miocene (*c.* 20 million years ago): *Sitta senogalliensis* (Vielliard, 1978). According to Vielliard's phylogenetic recon-struction, the most primitive living nuthatches are probably the Chinese Nuthatch and the closely allied 'Mediterranean' nuthatches (Krueper's, Algerian, Corsican) (Fig. 2). Next in this reconstruction are the American dwarf nuthatches, the Asian/American species pair of White-cheeked and White-breasted, followed by two other small conifer-inhabiting species, the Red-breasted (N America) and Yunnan Nuthatches (Asia). The latter in particular has been regarded as an intermediate stage in the line leading to the Eurasian Nuthatch (Voous and van Marle, 1953). Vielliard's phylogeny differs somewhat from the traditional view by separating Red-breasted and Yunnan from the Mediterranean/Chinese group, which are usually placed in the same subgenus or even superspecies. Many of these 'primitive' nuthatches are thin-billed and prefer coniferous habitat. Most of them

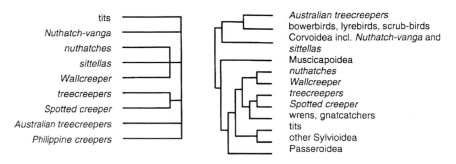

Fig. 1 *Phylogenetic relationships between nuthatches and other scansorial passerines (in italics) and some other bird groups. Left: families and subfamilies (minor branches) as listed by Greenway (1967). Right: simplified representation of the phylogenetic tree in Sibley and Ahlquist (1990). Branch lengths are drawn for graphic convenience and have no quantitative meaning.*

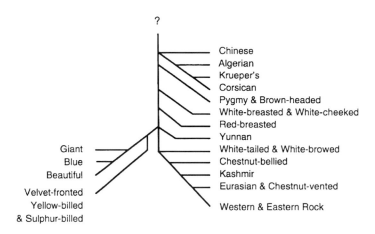

Fig. 2 *Hypothetical phylogenetic tree of the genus* Sitta *according to Vielliard (1978).*

excavate their own nest sites and none uses mud to reduce the nest entrance.

The remaining species have been tentatively assigned to two distinct lineages (Fig. 2). One fairly obvious lineage includes the Eurasian Nuthatch and a handful of morphologically quite similar species, including the rock nuthatches. They are all very similar in appearance, the main variation being in the underparts which range from creamy-white to dark chestnut. The second lineage gathers together, for convenience, several colourful tropical or subtropical species whose relationships are unresolved (Fig. 2). In particular, the largest species (Giant and Beautiful Nuthatches) show similarities to both lineages. There are very few molecular data to support this or other phylogenetic reconstructions. A study with mitochondrial DNA on 10 nuthatch species supports the monophyly of the *Sitta canadensis* superspecies (*contra* Vielliard), but remains inconclusive with respect to the other species. There is a suggestion that the Pygmy and Velvet-fronted Nuthatches would be closer to the other small species, while the Eurasian, White-breasted and White-tailed Nuthatches belong to a different main branch (Pasquet, 1998). The karyotype has been identified in only two nuthatch species. The Eurasian Nuthatch with $2n = 80$ is comparable to other passerines (Li and Bian, 1988), whereas the Western Rock Nuthatch has an unusually high number of chromosomes ($2n = 94$; Bulatova in Glutz von Blotzheim, 1993).

WORLD DISTRIBUTION

Nuthatches can be found throughout most of Asia, N America and Europe. Eighteen species live in Asia, most of them in the mountains of central and SE Asia. N America and Europe have four species each, but three of the European ones (Corsican, Krueper's and Western Rock Nuthatches) have very limited ranges within Europe. There are none in South America and Australia, and Africa has only two highly localized populations in the Atlas Mountains: a population of the Eurasian Nuthatch, and the endemic Algerian Nuthatch.

Nuthatches are a typical faunal element of the temperate to subtropical climatic zones of the Northern Hemisphere. Two species (Eurasian and Red-breasted Nuthatches) occur well into the subarctic zone, almost up to the northern fringe of the boreal forest. Only a few species live in tropical climates in southern and SE Asia, two of them reaching the equator (Velvet-fronted and Blue Nuthatches in Indonesia). Many species have small ranges, and there are few regions where more than one or two nuthatches live together (Fig. 3). The highest species diversity is found in SE Asia where up to five species may co-occur, but these are usually separated by altitude, by habitat, or both (Lack, 1971). The most complete 'tick list' of nuthatches is probably held by Dave Farrow from England (*in litt.*, 1995) who has seen all 24 species except the Beautiful Nuthatch.

GENERAL MORPHOLOGY AND SIZE

Nuthatches are unmistakable even when not climbing. They are characterized by a compact body, rounded wings, short tail, long, straight bill, relatively short legs and large feet with strong claws (Richardson, 1942). The wing has 10 primaries, the first of which is reduced. The skull and bill are rather strong and somewhat woodpecker-like, obviously adapted to hammering and excavating (Richardson, 1942; Cuisin, 1984). However, nuthatches do not excavate for wood-boring insects and lack the woodpecker's extensively protusile tongue.

The smallest species (Red-breasted and dwarf nuthatches) weigh 10–11 g

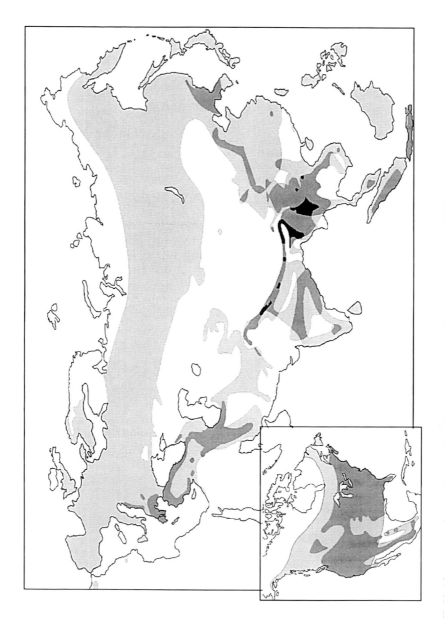

Fig. 3 *Worldwide species richness of nuthatches, based on maps in Harrap and Quinn (1996). Light grey = 1 species, dark grey = 2 or 3 species, black = 4 or more species.*

and measure about 11 cm; the largest is the Giant Nuthatch with a body length of 18–20 cm (no weight records are available). Males are often slightly larger than females. Wing lengths range from 60 to over 120 mm (Fig. 4) but because of individual and geographical variation there is a lot of overlap between species. Geographical variation may be quite complex as in the case of the two species of rock nuthatch, which have similar sizes in allopatry but are clearly distinct in sympatry (see Chapter 12). Another species with considerable geographical variation in size is the Eurasian Nuthatch, as can be seen from the wing lengths of the three major subspecies groups in Fig. 4.

One general pattern that emerges from Fig. 4 is that species in coniferous forest tend to be smaller. This tendency is found between as well as within subgenera. For instance, the largest species within the *Micrositta* subgenus is the Algerian Nuthatch, the only one that lives mainly in broadleaved forest. Also, the White-cheeked Nuthatch (in coniferous forest) is markedly smaller than the otherwise very similar White-breasted Nuthatch, which prefers broadleaved forest. A similar tendency is found within species, notably in the Eurasian Nuthatch, where the Siberian *asiatica* subspecies (in coniferous forest) is appreciably smaller than the closely related nominate subspecies in northern Russia and Scandinavia (broadleaved and coniferous forest) (see Chapter 2). This difference in body size is remarkable since in the ecologically similar tits no such trend is found in body size, though there is a difference in bill size (Snow, 1954).

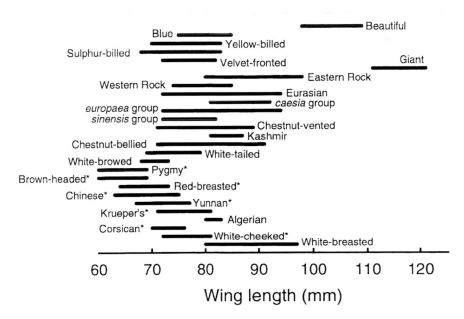

Fig. 4 *Size ranges of nuthatch species (males only) and the three subspecies groups of the Eurasian Nuthatch. Data from Harrap and Quinn (1996). Species that live mainly in coniferous forest are marked with an asterisk.*

PLUMAGES AND MOULT

Except for a few colourful tropical species, nuthatch plumages are variations on a basic theme. The underparts vary from white or grey over buff-ochraceous to rufous brown (dark chestnut in one species), while the upperparts are typically bluish-grey. Hailman (1979) suggested that this colour would have been selected to contrast with the green/brown background of the forest, implying that nuthatch plumages evolved to signal their presence, rather than to be cryptic. A few tropical species add purplish tones, most markedly in the three members of the Velvet-fronted superspecies, but also in the form of brilliant lilac streaks in the Beautiful Nuthatch and pale blue streaks in the Blue. The majority of species can be identified by a few simple traits: eye-stripe, superciliary, black (or brown) cap, spots on the tail, and bi-coloured (*vs.* uniform) undertail-coverts (Fig. 5 and Appendix II). Only a few conspicuous traits are unique to a single species, such as white wing bars (Beautiful), a rufous breast-patch (Krueper's) or a bright red bill (Velvet-fronted). A few species have a pale nuchal patch but this is present in several more species as a cryptic spot formed by whitish feather bases, and therefore its taxonomic significance is obscure. Norris (1958) proposed that the nuchal patch may have evolved as a directive mark in aggressive contexts, serving to reduce pecking in more vulnerable areas. This hypothesis is, however, untested, and does not explain variation between species, nor the vestigial presence in so many species.

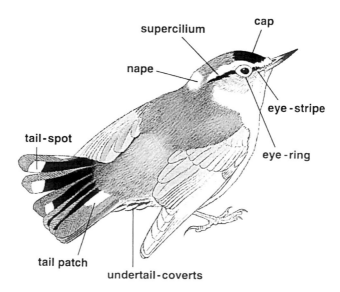

Fig. 5 *'Idealized' nuthatch showing the major diagnostic traits between species (see Appendix II)*.

Nuthatch plumages do not vary greatly with age and hardly at all with season, except for the effects of wear. First-year and adult birds are generally indistinguishable after completion of the annual summer moult, and even newly fledged juveniles closely resemble adults, though they may appear somewhat paler and duller. However, since by early summer the adults' plumage becomes faded and worn, the difference in the field is even less obvious than museum skins might suggest. Sexual dimorphism is not substantial, but in many species males and females can be discriminated, even in the field, by minor plumage differences. Males of Eurasian and related nuthatches, for instance, have dark chestnut flanks and chestnut margins to the undertail-coverts, whereas these are much paler in females. In several species, males have black caps while females have grey or dull black caps. Sexual dimorphism is most pronounced in the Chestnut-bellied Nuthatch: males have deep chestnut underparts while females show the more familiar buff-brown. Males are slightly larger in some but not all species (e.g. Wood, 1992).

As far as is known, all species have a complete annual moult after the breeding season, and young birds have a partial post-juvenile moult, excluding the flight feathers, in their first summer. However, moult has been studied in any detail in only one species (Eurasian Nuthatch, Chapter 2). A partial pre-breeding moult has been found in individuals of several species, in particular the Red-breasted, Pygmy, Blue and Velvet-fronted Nuthatches (Banks, 1978). This mainly involves the ventral feathers, and never leads to significant seasonal variation in plumage.

Velvet-fronted Nuthatch.

GENERAL ECOLOGY

Virtually all nuthatches are confined to mature forest, search for food almost exclusively on trees (rarely on shrubs or on the ground) and breed in holes in trees. Nuthatches can be found in all forest types, even in gardens and parks with enough large trees. Only the two species of rock nuthatch live on rocky slopes and cliffs, where trees are scarce or even absent. All species feed mainly on insects and other invertebrates, but seeds and nuts may be an important part of the diet, especially in winter. With their strong bills, they can open relatively large seeds such as hazelnuts and acorns, but they more often use smaller seeds such as pine, beech or sunflower seeds. Their stomachs often contain grit, probably to improve seed digestion. Fat, bread and other anthropogenic food is readily taken when available.

The forest-dwelling nuthatches spend most of their time foraging on tree trunks and large branches, though some of the smaller species also search among clusters of pine needles or on small twigs. There are several remarkable observations on the use of 'tools' in three species, all in N America. The tool is usually a piece of bark which is used to pry off other bark flakes (for details see Chapter 14). Large food items are normally broken into smaller ones. The bird wedges the food item into a crevice and, typically with the head pointing downwards, breaks off pieces with repeated strokes of the bill. This habit differentiates nuthatches from related bird families such as tits, treecreepers, corvids and sittellas, which hold food items with their feet (Noske, 1985; Löhrl, 1988). This may also explain why nuthatches are unable to pull up food from a string (Herter, 1940).

Almost all nuthatches store food items and often cover them with lichen, bark or moss. Hoarding can be elicited at almost any time of the year if seeds are provided, but insects may be stored as well (Dorka, 1980). Of 11 species studied in captivity by Löhrl (1988), the only one that did not hoard was the Velvet-fronted Nuthatch. This species took few seeds in captivity, and may live exclusively on animal food in nature. Wild-caught Chestnut-bellied Nuthatches (also subtropical) did not hoard, but their captive-bred offspring did.

POPULATION STATUS AND CONSERVATION

Many nuthatch species have small geographical ranges, and this may explain why a quarter of them are listed as threatened by the BirdLife International World Checklist (Collar *et al.*, 1994). Fortunately, none of them appears to be in immediate danger. The Algerian Nuthatch – discovered only in 1975 – which was at first thought to be restricted to a single mountain top, has now been found in several more localities with larger populations (Chapter 11). This is probably the only nuthatch to have prompted the creation of a nature reserve (the Mount Babor National Park). The White-browed Nuthatch, similarly known from a single mountain in Myanmar (Burma), had not been seen for 50 years until a 1995 expedition confirmed its presence and found no immediate threats (Chapter 13). The four remaining threatened species (Yellow-billed, Yunnan, Beautiful and Giant Nuthatches) have somewhat wider geographical ranges but are threatened by habitat loss, and their status is badly known.

BEHAVIOUR

General behaviour

Nuthatches have a rich behavioural repertoire with loud, clear calls and songs, and remarkable courtship displays. Löhrl (1964) noticed some marked behavioural differences from tits and treecreepers, not only in climbing and nest-building, but also in general display postures where the head, rather than the throat and breast, is exposed. The majority of species have an elaborate pre-copulatory display, often containing a variation on the so-called 'Pendulum' movement in which one partner, or both, swings the head slowly from side to side (Löhrl, 1988). A typical display shown at times of great excitement, often accompanying loud calling, is a rapid beating of the wings. A peculiar feature of the rock nuthatches is a Dipper- or Spotted Sandpiper-like bobbing. A typical threat display in the nest is lacking (in contrast to tits), but there are anecdotal reports of other anti-predator behaviours including direct

A singing male Sitta europaea caesia. *(Photo: Frank Adriaensen).*

attack, dropping bark flakes on the intruder, and displays (Took, 1946; Long, 1982; Löhrl, 1988). Nuthatches rarely bathe in shallow water or snow, and if they do so they seem very wary (Glutz von Blotzheim, 1993). Like many woodpeckers, they bathe much more often by exposing their feathers to drizzling rain, or in the canopy among the wet leaves after or during rain (Blackford, 1955; Slessers, 1970; Löhrl, 1988; Fiebig, 1992). Sunbathing has been described in Eurasian Nuthatches only (Löhrl, 1988) and is probably uncommon. Head-scratching is done by the indirect method (extending the leg over the outstretched wing), but, at least in Corsican Nuthatches, sometimes also by the direct method, and especially in the smaller species (but also the Eurasian) while hanging upside down (Löhrl, 1961, 1988; Glutz von Blotzheim, 1993). A curious habit, shared by all species according to Löhrl (1988; see also Kilham, 1972a), is to remain silent and motionless for several minutes without apparent cause or disturbance.

Climbing technique

Woodpeckers and treecreepers are supported by their long and rigid tail-feathers when hopping upwards on a tree trunk, but nuthatches can freely walk up or down without tail support. This habit has earned them nicknames such as 'upside-down bird' or 'devil-down-head' (Forbush, 1929), and separates them from most other climbing birds except for a handful of 'nuthatch mimics' such as the sittellas, mentioned above. Several other birds, many of them quite unrelated, climb trees without tail support, for instance the N American Black-and-White Warbler (Parulidae) and New Zealand's Rifleman (Acanthisittidae). However, I have found only one reference to head-down climbing in birds other than nuthatches, in the Spotted Creeper, a relative of the treecreepers that lacks their rigid tail-feathers (Cameron and Harrison, 1978). Nuthatches can also hang upside down from a vertical or even a horizontal surface. Only the two species of rock nuthatch rarely climb downwards, and hop rather than climb upwards (Löhrl, 1988). By analysing film of their climbing, Zippelius (1973) showed that nuthatches walk up and down, always holding one foot firmly on the bark while the other is moved, thereby confirming Law's (1929) observations. The large toes and claws are obviously adapted to this type of movement, while the tail is reduced in comparison to other trunk-climbing birds (Richardson, 1942). Upward climbing is accompanied by a typical zigzag movement (Fig. 6), which makes sense because it ensures that the bird's centre of gravity remains directly above the lower foot (Winkler and Bock, 1976). Some of the larger species, such as the Eurasian and White-breasted Nuthatches, appear to find walking on the underside of a horizontal branch difficult, and instead move from side to side on the upperside and glean from the lower surface (Löhrl, 1988). On such horizontal branches, and especially on the ground, they move by hopping with both feet. This is the terrestrial rock nuthatches' main mode of locomotion.

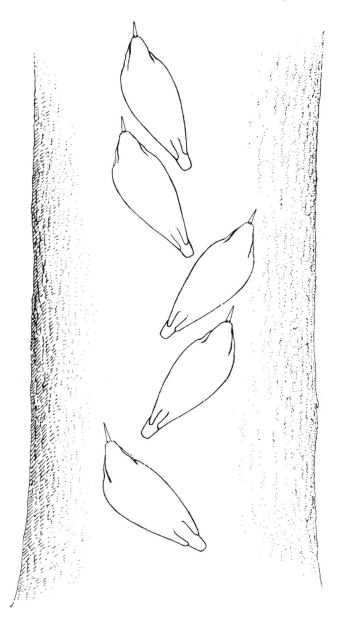

Fig. 6 *Zigzag movement by a climbing nuthatch. After Löhrl, 1958.*

Vocalizations

Nuthatch vocalizations often excel in their loudness and persistence, but rarely in their musicality or complexity. They typically consist of a series of identical notes (rarely two- or three-note combinations) that are repeated with variable speed and length. Songs and calls are often hard to distinguish and may be used in similar contexts. Norris' characterization (1958:179 ff.) of Pygmy and Brown-headed Nuthatch song probably applies to the majority of species: 'song [is] essentially a series of notes which if given non-repetitively, and/or with less volume, would amount to ordinary call notes' and 'song is distinguishable from other notes by its repetitive character and the circumstances in which given'. For the majority of species only a few song types have been described. It is noteworthy that duetting has been mentioned in four species (Pygmy, White-tailed and both rock nuthatches), but this behaviour remains to be investigated in detail.

Nest-building

Nuthatches have a diversity of nest and nest sites (Table 1); so not surprisingly, nest building is one of the better-studied elements in their natural history. The more 'primitive' small, conifer-dwelling nuthatches often excavate their own nest in soft, rotten wood, but may use existing holes as well. The larger species are completely dependent on pre-existing cavities. Many of them reduce the entrance to the nest with mud, a habit that is otherwise found only in the totally unrelated hornbills of Africa (who wall up their incubating females in the nest cavity, leaving only a small hole for food provisioning). This defence tactic is probably directed towards predators as well as nest competitors (Löhrl, 1958; Nilsson, 1984). The rock nuthatches have taken the 'plastering' a step further by building an entire dome-shaped nest of mud, attached to a rock face. Three species show entirely different adaptations to protect their nest: the Red-breasted Nuthatch smears sticky resin around the entrance, while White-breasted and White-cheeked Nuthatches do the same with crushed insects, probably as an olfactory deterrent to nest robbers or parasites. Unlike other cavity-nesters, these species make little use of fresh plant material as a parasite-repellent (Clark and Mason, 1985).

Löhrl (1988) suggested that plastering has evolved from the habit of filling up narrow cracks inside the cavity with mud or other material to improve insulation. Indeed, the Chinese Nuthatch, believed by some to be the most primitive nuthatch, uses mud inside the cavity but not around the entrance of its self-excavated hole. Pygmy and Brown-headed Nuthatches fill up cracks in the same way but with ordinary nest material instead of mud. The White-breasted Nuthatch also occasionally brings mud into the nest hole, perhaps for its anti-parasite function (Duyck *et al.*, 1991). The use of mud as a nest material is not unique to nuthatches but, for instance, is seen in Magpies (Birkhead, 1991). An unresolved question is whether there is any evolutionary connection between mud plastering and the Red-breasted and White-breasted Nuthatches' use of resin and insects, respectively. The only clue is that several species mix the mud used for plastering with berries, insects and even resin (Table 1). More information on nest-building and nest defence is given in later chapters.

TABLE 1: *Variation in nest-building behaviour among nuthatches*

	Excavation	Entrance reduced	Mud nest	Materials used		
				Mud	Resin	Insects
'PRIMITIVE' NUTHATCHES						
Chinese	+			N		
Algerian	+	(A)		N (A)		
Krueper's	+					
Corsican	+					
Pygmy	+					
Brown-headed	+					
White-cheeked						R
White-breasted				+		R
Red-breasted	+				R	
'NON-PLASTERING' NUTHATCHES						
Giant						
Beautiful		+		+		
Velvet-fronted	(A)	+		+		
'PLASTERING' NUTHATCHES						
White-tailed		+		+		+
Chestnut-bellied		+		+	+	
Kashmir		+		+		
Eurasian	(A)	+	(A)	+	(A)	
Chestnut-vented		+		+		
Western Rock			+	+	+	+
Eastern Rock		+	+	+	+	+

NOTES:
1. Species are grouped according to the hypothetical phylogenetic tree in Fig. 2 (Vielliard, 1978).
2. N = used as nest material, R = used as repellant outside the nest, (A) = anecdotal record only.
3. No information is available on five species not included in the table.

Breeding

Nuthatches generally attempt to breed when 1 year old, except for some 'helper' yearling males of Pygmy and (probably) Brown-headed Nuthatch (Chapter 14). Pair-formation and mate choice have not been studied in detail in any species. Typical clutch sizes vary from about four to eight eggs between species. Females always incubate alone but are fed by their mates. Nestlings develop relatively slowly compared with similar-sized passerines, and fledge after 18–25 days. They are fed nearly exclusively on arthropod food. Nest success is fairly high in the few species for which data are available (Chapters 6 and 14). The most common causes of nest failure are probably take-over by nest competitors, and predation by woodpeckers or squirrels. Early losses may be compensated by a repeat nest, but true second broods are exceptional in most, if not all, species.

Social behaviour and movements

The majority of nuthatches are not very social, and the main subject of this book – the Eurasian Nuthatch – proved to be the least social of all in captivity (Löhrl, 1988) in relation to both conspecifics and other species. Nuthatch pairs usually keep to their territories not only in the breeding season but also in winter, or live at most in small groups. The dwarf nuthatches of N America are more social, roosting communally and having helpers at the nest. Some of the tropical species may be moderately gregarious as well, but there are few details of their social behaviour.

There are no regular migrants among nuthatches. Red-breasted Nuthatches move south of their breeding range in large numbers every few years, and some populations of the Eurasian Nuthatch show similar but less frequent invasions (= irruptions). The other species are largely sedentary (except perhaps for the northernmost populations or those at high altitudes). There is some degree of wandering by young birds, but this is well studied only in the Eurasian Nuthatch (Chapter 8).

STUDIES ON NUTHATCHES

The undisputed pioneer of behavioural and ecological research on nuthatches is the German ornithologist Hans Löhrl. He started studying Eurasian Nuthatches when he moved to a house in the Favoritepark in Ludwigsburg in 1950, in the midst of optimal nuthatch habitat. His 12-year study not only led to several monographs on this species, but also inspired him to make comparative observations on other species. Over the next 30 years Löhrl collected behavioural observations on no fewer than 14 nuthatch species in nature, and he kept 11 of them in captivity. The results were summarized in his 1988 monograph, which remains the major reference for information on displays and vocalizations of many species. Some of his captive subjects were hand-reared and were allowed to fly at liberty around the house – sometimes hoarding seeds in the furniture or climbing upon house guests for practice.

After Löhrl, the next major field study on Eurasian Nuthatches was initiated in the 1970s by Sven Nilsson in Sweden. This was followed by a minor explosion of ecological studies (Table 2, Fig. 7). Nilsson's study remains the most thorough investigation on nuthatch population dynamics to date, and was complemented by a study on the role of territoriality in population regulation by his student Bodil Enoksson. At about the same time I started a project in the Peerdsbos forest near Antwerp (1982–1987) with a comparable, but less experimental approach. Both Enoksson and I studied marked individuals in a relatively small area (10–20 breeding pairs), and concentrated on the social mechanisms outside the breeding season. In a second project (1990–1994) I studied the effects of forest fragmentation on the ecology and behaviour of Eurasian Nuthatches. I refer to the two study areas where I worked as the 'Peerdsbos' and 'fragments' areas, respectively (see Chapters 5 and 10 for more details). The latter project was only one of

TABLE 2: *Major field studies on the Eurasian Nuthatch*

	Country	Main period	Main theme(s)
Hans Löhrl	Germany	1950–1962	Behaviour
Sven Nilsson	Sweden	1974–1982	Population dynamics
Bodil Enoksson	Sweden	1980–1986	Territoriality and food
Erik Matthysen	Belgium	1982–1987	Social organization
		1990–1994	Forest fragmentation, dispersal
Vladimir Pravosudov	Siberia	1986–1990	Breeding biology, social organization
Tomas Wesołowski	Poland	1987–1989	Breeding biology
Alex Schotman and Jana Verboom	Netherlands	1988–1994	Forest fragmentation, modelling
Lars Gabrielsen	Denmark	1994–	Forest fragmentation, dispersal

NOTE:
Localities are shown in Fig. 7.

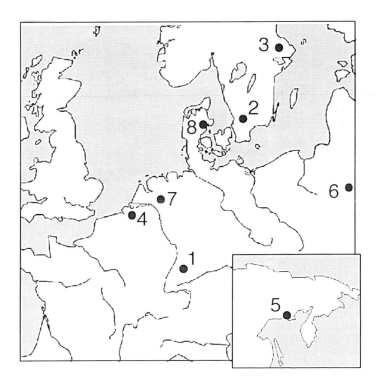

Fig. 7 *Location of the major ecological and behavioural studies on the Eurasian Nuthatch. 1. Ludwigsburg (Löhrl), 2. Stenbrohult (Nilsson), 3. Uppsala (Enoksson), 4. Antwerp (Matthysen), 5 (inset). Magadan (Pravosudov), 6. Białowieza (Wesołowski), 7. Twente (Schotman, Verboom), 8. Århus (Gabrielsen) (see Table 2 for more information).*

several parallel studies being carried out in Sweden, Denmark, the Netherlands and Belgium (Table 2). A crucial inspiration to several of these was the recognition of Eurasian Nuthatches as a suitable model system to analyse or predict the effects of landscape changes on birds (see Chapter 10). Two other studies deserving to be mentioned are those by Wesołowski and Pravosudov in Poland and Siberia, respectively.

None of the other nuthatch species has been studied to the same extent, which will be reflected in the contents of this book. Special mention must be made of the studies on the social behaviour and communal roosting of Pygmy Nuthatches by Russ Balda's team in the 1980s. The way to this study had earlier been paved by Richard Norris (1958) with a detailed comparison of the two dwarf nuthatches. A third N American species, the White-breasted Nuthatch, has been the subject of several behavioural investigations, but never of a true population study. The only nuthatches that regularly make their appearance in general ecology textbooks are the rock nuthatches, which in the 1950s provided the classical case of 'character displacement' (i.e. the divergence in characteristics of closely related species where they occur in sympatry). Another species that enjoyed brief notoriety is the Algerian Nuthatch, whose discovery in 1976 became headline news and immediately prompted some more research on this and closely related species.

Part II: The Eurasian Nuthatch

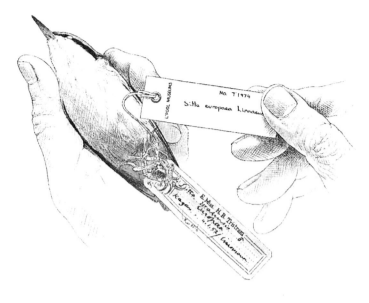

CHAPTER 2

Taxonomy, Morphology and Moult

Sitta rectricibus nigris: lateralibus quatuor infra apicem albis
('A nuthatch with black rectrices, the four outermost of which are white below the top')

Linnaeus (1758)

As the name implies, Eurasian Nuthatches occur throughout Europe and Asia, from the boreal forest zone in Siberia to higher altitudes in N Africa and the Middle East, and even into the humid subtropical zone in south-eastern China. They are absent from the steppe zone of central Asia, and in southern Asia they are replaced by a few very closely related species. Plumage and size vary considerably within this huge distributional range. To complicate the matter, variation in these traits is not consistent. For instance, both in Europe and in Asia the white-breasted populations of the north are replaced by brown-breasted birds further south and towards the fringes of the continent. European birds are in general larger than Asian ones, however, regardless of their colour.

23

Inspection of the tail-spots dictates yet another major distinction, between the *arctica* subspecies in north-eastern Siberia (with much more white on the tail) and all other forms (Kleinschmidt, 1928). Ever since Kleinschmidt, however, the colour of the underparts has been accepted as the most important trait. Geographical variation is certainly not attributable to the presence of other nuthatch species, since Eurasian Nuthatches co-occur with other species in a few small areas only, and even then are strongly separated by habitat (the rock nuthatches avoid forest, and Krueper's and Chinese Nuthatches are restricted to conifers). In the following sections the general distribution and geographical variation of Eurasian Nuthatches are discussed per major subspecies group. This is followed by a brief discussion on the relationships with three very closely related species from southern Asia, and on some more general aspects of morphology and plumages.

THE MAJOR SUBSPECIES GROUPS

Eurasian Nuthatches are characterized by blue-grey upperparts, a well-developed black eye-stripe with sometimes a faint white supercilium, subterminal white spots on the blackish outer tail-feathers, and whitish undertail-coverts with rufous or buff fringes giving a dappled appearance (Cramp and Perrins, 1993; Harrap and Quinn, 1996). Since the original description, numerous subspecies have been described, including many that were originally described as separate species. Greenway (1967) lists no fewer than 40 of these but recognizes only 14 as valid, compared to 17 in Harrap and Quinn (1996). Detailed accounts of the various subspecies and the relations between them can be found in Vaurie (1950, 1957) and Voous and van Marle (1953). Here, I shall only summarize the general patterns, especially in relation to Voous and van Marle's ideas on the postglacial expansion of Nuthatches in Europe. There are three clearly distinct subspecies groups that nevertheless interbreed freely where their distributions meet. Sibley and Monroe (1990) even attribute common names to them: 'Eurasian Nuthatch' (*europaea* group), 'Southern Nuthatch' (*caesia*) and 'Oriental Nuthatch' (*sinensis*). To avoid confusion with true species names, I will not use these. Figure 8 shows the distribution of the 14 subspecies recognized by Greenway (1967), as well as three very closely related species.

The europaea *group*

This group may be called the 'white-breasted' group, but should not be confused with the N American species bearing the same name. Most subspecies in this group indeed have whitish underparts, though some grade to

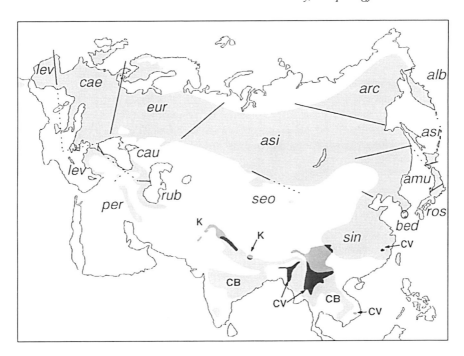

Fig. 8 *Approximate ranges of Eurasian Nuthatch subspecies (italic) and of three closely related species. Subspecies follow Greenway (1967). The 'white-breasted' group includes* albifrons, amurensis, arctica, asiatica, bedfordi, europaea, roseilia *and* seorsa. *The 'brown-breasted' group includes* caesia, caucasica, levantina, persica *and* rubiginosa. Sinensis *represents a separate group. Other species: K = Kashmir, CB = Chestnut-bellied, CV = Chestnut-vented Nuthatch.*

yellowish or buff. They occur in southern Norway and Sweden, most of Russia, and throughout the Siberian taiga up to northern China and the Pacific shore from Kamchatka to Korea (Fig. 8). The northern distribution limits are about 62° N in central Norway and Sweden (coinciding with the northern limit of the oak forest), 65° in coastal Norway and Russia, and up to 68° in parts of Siberia (Cramp and Perrins, 1993). The southern distribution limit is from northern Ukraine through the southern Ural mountains (*c.* 54°), the southern border of the forest steppe in northern Kazakhstan, the southern edge of the Altai mountains in western Mongolia – with a possibly disjunct population in the Tien Shan mountains of China – and continues through eastern Mongolia into Manchuria.

There are two intergradation zones with brown-breasted subspecies: a broad zone in Europe and a narrower one (also less well documented) in eastern Asia. The intergradation zone with the *caesia* group runs across eastern Europe from the Baltic area to the Black Sea and includes Denmark, eastern Germany, Poland, the Baltic States, Belarus, the Czech Republic, Romania, eastern Bulgaria and probably other countries as well

(Fig. 9; Voous and van Marle, 1953). Several of these intermediates have been given subspecific names, the best-known being *S. e. homeyeri* described from northern Poland (formerly East Prussia) (Voous and van Marle, 1953). Stresemann (1919), however, recognized that the variation across eastern Europe was better understood in terms of secondary hybridization than as proper subspecies or races (see Alex, 1994 for a recent description of the intergradation zone in Belarus). The contact zone between white- and brown-breasted populations follows a slightly different pattern in eastern Asia. Here there is a succession of subspecies within the *europaea* complex grading from pure white in the north to yellowish or buff farther south, especially on Japan and other islands. The intergradation zone is restricted to a relatively small area on the border of Manchuria and China, in the region of the Great Wall, where the yellowish *amurensis* grades into the smaller and brown-breasted *sinensis*. Earlier authors such as Stresemann (1919), however, considered *amurensis* as a hybrid form as well.

In Europe, the majority of white-breasted populations belong to the nominate *europaea*, the largest of all subspecies. This race has been the subject of ecological and behavioural studies by Nilsson and Enoksson in Sweden, while Wesołowski's study in Białowieza involved a population in the intergradation zone with *S. e. caesia*. Another white-breasted subspecies, *asiatica*, occurs within Europe only in the western foothills of the Ural mountains. Its main distribution is throughout Siberia up to the Pacific Coast as far north as 60°N, on Sakhalin and northern Japan. Distinctly smaller than *europaea* with a shorter bill and tail, pure white underparts with paler flanks, and often a white forehead and supercilium (see photographs

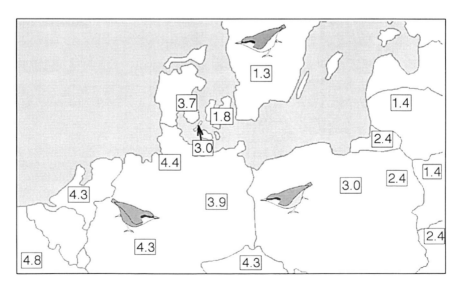

Fig. 9 *Variation in mean colour scores (from 1 = white to 5 = brown) of Eurasian Nuthatches in the intergradation zone between white- and brown-breasted populations in central Europe. Based on data from Løppenthin (1932).*

in Wahlstedt, 1965, and Klinteroth, 1978), it also has larger white spots on the tail and white tips to the greater coverts in fresh plumage (Harrap and Quinn, 1996). It is best known during irruptions in northern Europe where it is easily distinguished from resident *europaea*. However, one field study is available from its breeding range in eastern Siberia, by V. Pravosudov.

Six other white-breasted subspecies are entirely confined to Asia (Fig. 8). In Siberia, *arctica* occurs north of *asiatica*, but its exact limits are largely unknown. It is somewhat larger than *asiatica*, lighter in colour with a conspicuous white forehead, a reduced black eye-stripe, and more white on the tail and more pointed wings than other Eurasian Nuthatch populations (Harrap and Quinn, 1996). Dunajewski (1934) considered this form sufficiently distinct in bill structure (Fig. 10), and relative dimensions of wing, tail and legs, to be a separate species. The songs are also said to be markedly different from all other races except perhaps *albifrons* (Leonovich *et al.*, 1996a). There are two separate reports of *arctica* and *asiatica* coexisting in the same area: one from Olekminsk in Yakutia (Dunajewski, 1934), the other from the lower Tunguzka area some 1000 km to the west (Kleinschmidt, 1928). In short, *arctica* (perhaps together with *albifrons*, see below; Leonovich *et al.*, 1996a) might well be a candidate for the 25th nuthatch species provided that more information is available.

The subspecies *amurensis*, which has the lower half of the underparts yellowish to pale rufous, lives in north-eastern China, Korea, the Amur region of Siberia, and central Japan. In Japan it is said to overlap in range with *asiatica* which occupies the higher altitudes (Austin and Kuroda, 1953). Three more subspecies with very limited distributions are not as clearly differentiated, and not recognized at all by some handbooks (for instance Dement'ev

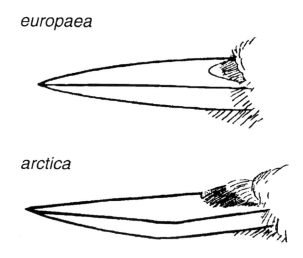

europaea

arctica

Fig. 10 *Typical bill profiles of* europaea *and* arctica *subspecies. From Dunajewski (1934).*

and Gladkov, 1954): *albifrons* on the Kamchatka peninsula, with lighter flanks and shorter bill than *arctica*; and *roseilia* and *bedfordi* in southern Japan and Cheju Island in the Korea Strait, respectively, with increasingly darker underparts contrasting with the light-coloured throat (Fig. 8). The last subspecies in this group (*seorsa*) is described from a rather isolated population in the eastern Tien Shan mountains in western China. It is slightly darker underneath than the more northerly *asiatica* (Vaurie, 1957).

The caesia *group*

This group includes the buff- or brown-breasted Eurasian Nuthatches which are familiar to most European ornithologists. They differ from the brown-breasted *sinensis* group (see below) by having whitish rather than buff cheeks and ear-coverts, and by their larger size. Originally described as *Sitta caesia* by Wolf in 1810, the conspecific status with *europaea* was probably first recognized by Taczanowski (1882) who examined individuals across the intergradation belt in eastern Europe. Populations belonging to this group occur over most of Europe except the north and east, as well as parts of the Middle East and a small stronghold in the Rif and Atlas mountains of N Africa. Within this geographical area they are notably absent from Ireland and the larger Mediterranean islands except Sicily. They are also absent from some coastal areas of the North and Mediterranean Seas, and from the steppe zone north-east of the Black and Caspian Seas. Towards the Middle East the species' distribution becomes more discontinuous.

Geographical variation within the *caesia* group is rather slight and mainly clinal, except in the Middle East (Voous and van Marle, 1953; Vaurie, 1957). Variation in size is small compared with the white-breasted group. Greenway (1967) mentions only two subspecies for most of Europe: *caesia* in western and central Europe including Great Britain and Italy, and *levantina* in most of the Mediterranean area including Spain, Morocco, Sicily and most of Turkey. This race has a more slender bill and lighter underparts than *caesia*. There are three more subspecies in the Middle East: *persica* in the Zagros mountains, *rubiginosa* in northern Iran and Azerbaijan, and *caucasica* in the Caucasus region and north-eastern Turkey. The most distinctive of these is *persica*, with a marked white frontal band and paler underparts than any other brown-breasted population (for further details see Cramp and Perrins, 1993; Harrap and Quinn, 1996). To date, *caesia* is the only well-studied brown-breasted race, except for a few limited investigations on Spanish Nuthatches.

Voous and van Marle (1953) described morphological variation across Europe in more detail, thereby recognizing seven extra subspecies, and interpreted this variation as two major clines (Fig. 11). Each cline is thought to represent a major expansion route from glacial forest refuges. One expansion supposedly started in Italy ('*cisalpina*') and moved west from the Alps towards northern France ('*hassica*') where it separated into a British ('*affinis*') and a German ('*caesia*' s.s.) branch. Along this cline the colour of the underparts (orange-buff in Italy) became less bright and more brownish, and bill size increased. A second expansion route went

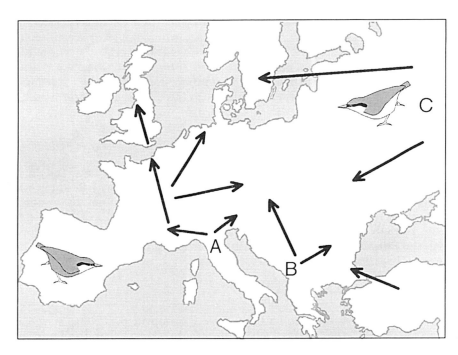

Fig. 11 *Postglacial expansion of the Eurasian Nuthatch across Europe according to Voous and van Marle (1953). A and B = origins of the first and second clines of variation in brown-breasted populations, C = invasion routes of white-breasted populations.*

northwards across the Balkans, meeting the end of the other cline in Austria and Hungary. These populations ('*harrisoni*' and '*dalmatina*') are somewhat paler than their closest neighbours in Italy, and have slightly longer bills. Vaurie (1957) has severely criticized Voous and van Marle's (1953) interpretations of the geographical variation, but I will not go into this discussion here. Voous and van Marle (1953) also recognized two more subspecies in southern Europe: '*hispaniensis*' in Iberia and '*siciliae*' in Sicily, thus restricting '*levantina*' s.s. to Asia Minor. Again the differences in coloration and morphology are slight. Finally some authors have separated the small N African population as '*atlas*' with supposedly darker underparts and a longer bill than the Iberian birds.

The sinensis *group*

The single subspecies in this group is the smallest of all Eurasian Nuthatches, and this is also the main distinction from the *caesia* group (see above). It contains the southernmost populations of the species in east—central China (roughly east of the Red Basin) and reaches down to the Northern Tropic in Taiwan (23° N). Very little is known about its ecology and behaviour. Voous and van Marle (1953) state that 'their call notes are said to have changed considerably' from the *europaea* group, but give no reference. A peculiarity of this group is the overlap in range with several other nuthatch species which replace it in coniferous forest at higher altitudes: the White-cheeked and Chinese Nuthatch in the north-western part of its range, and the closely related Chestnut-vented Nuthatch (see below) in the south-east.

THE *EUROPAEA* SUPERSPECIES

In central and southern Asia the Eurasian Nuthatch is replaced by a series of forms which were formerly treated as conspecific with it, but nowa-days are mostly regarded as true species and sometimes joined in a 'super-species' (see Chapter 1). I deal with them in more detail in Chapter 13; here I briefly consider their systematic and morphological relationships with the Eurasian Nuthatch, and some behavioural differences reported by Löhrl (1988) that support their distinctness.

The Chestnut-vented Nuthatch

This form has been allied to the *sinensis* group in particular, which it meets in two different parts of China. It is slightly larger than the Eurasian Nuthatch and has darker and bluer upperparts, but is chiefly distinguished by its greyish rather than buff underparts and highly contrasting dark flanks (brick-red in males and chestnut-rufous in females) (Harrap and Quinn, 1996). In a population that some authors distinguish as *montium* (sympatric with *sinensis*) the grey becomes visible only when the buffy edges are slightly worn (Vaurie, 1957). There is also a marked tendency, particularly in juve-niles, to have mottled pale neck patches, caused by whitish feather bases. Chestnut-vented Nuthatches are found in mountainous areas in Tibet, China, India, Myanmar (Burma) and Thailand (Fig. 8). Lack (1971) argued that it replaces *sinensis* altitudinally in two areas of China, and should therefore be considered a separate species (see also Chapter 13). As in *sinensis*, very little is known about its natural history, but its vocalizations seem to differ substantially from those of the Eurasian Nuthatch (see Chapter 13 for details).

The Kashmir Nuthatch

This nuthatch has often been considered a subspecies of the Chestnut-bellied (Greenway, 1967) or, together with the latter, of the Eurasian Nuthatch (Voous and van Marle, 1953; Vaurie, 1957; Roberts, 1992). Its main diagnostic feature is the uniform rufous colour of the undertail-coverts. The underparts are intermediate between those of the *caesia* group and the darker Chestnut-bellied Nuthatch, without the latter's white chin and cheeks. There is some sexual dimorphism (males have darker under-parts) but not as much as in the latter species. It has a limited distribution in Kashmir and adjoining areas (Fig. 8) and, where it overlaps in range with the Chestnut-bellied Nuthatch, the two are separated by habitat and altitude (Roberts, 1992). Löhrl (1988) in particular has supported its separation from both Eurasian and Chestnut-bellied Nuthatch, mainly on behavioural grounds related to alarm calls and details of nest-building and pre-copulatory behaviour (see Chapter 13). He also performed pairing experiments between the three species in captivity and managed to get viable, but apparently infertile, hybrids which did not even produce a clutch. Also, the pairing was not achieved without difficulty because of differences in voice and behaviour (in particular between Eurasian and Chestnut-bellied Nuthatch).

The Chestnut-bellied Nuthatch

The sexual dimorphism in this species is much more striking than in the previous one, with typically very dark chestnut-brown underparts in males, and cinnamon-rufous underparts in females. The sharp contrast between the whitish chin and cheeks and the darker throat is also diagnostic in males. Nevertheless, its six subspecies (Greenway, 1967) have formerly been lumped with Eurasian and Kashmir Nuthatches. The species occurs over large parts of India and throughout most of SE Asia (Fig. 8). It is separated from *sinensis* by the Himalayas (Vaurie, 1957) and from the Chestnut-vented Nuthatch in SE Asia by altitude (Lack, 1971; see Chapter 13). Playback experiments with songs of this and the Eurasian Nuthatch have revealed no response of either species to one another's song. There are also minor differences in nest-building and courtship, for instance the mixing of resin with mud for nest-building in the Chestnut-bellied Nuthatch (Löhrl, 1988).

VARIATION IN PLUMAGE WITH SEX AND AGE

In most Eurasian Nuthatch populations, males are easily distinguished from females by the dark brown colour of the flanks and the dark brown fringes to the white undertail-coverts. The dark brown is replaced by less contrasting buff-brown in females. These contrasts are a reliable criterion for sexing birds in the hand (even for nestlings from about 12 days old) and also in the field, at least in the populations studied by B. Enoksson (pers. comm.) and myself. In the white-breasted races *europaea* and *asiatica* there is an additional difference, in that the female's underparts are less whitish than the male's, tending to cream-white or pale cream-buff (Kleinschmidt, 1928; Cramp and Perrins, 1993). On the other hand, the difference in flank colour seems to be obscured in some white-breasted populations, notably in *asiatica* where some males have female-type yellow-buff flanks (Dunajewski, 1934; Cramp and Perrins, 1993). In some white-breasted birds of eastern Europe as well, flank colour is not an entirely reliable sex criterion, but always varies in accordance with the undertail-coverts (Taczanowski, 1882). Curiously, Horváth (1961) found no clear dimorphism in a series from Hungary which he attributed to *caesia*: this

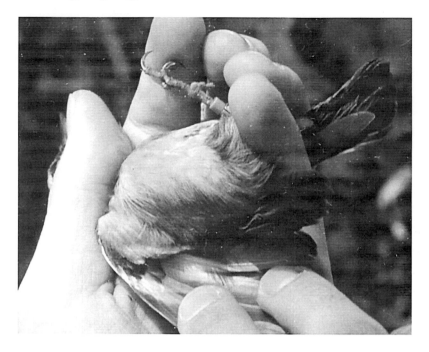

A male Sitta europaea caesia in the hand. Note the dark brown flanks, diagnostic of males. (Photo: Frank Adriaensen).

population is quite close to, or perhaps part of, the intergradation zone with white-breasted populations. Clearly, geographical variation in the degree of sexual dimorphism deserves further investigation.

Newly fledged juveniles have a slightly duller plumage than adults, have noticeably shorter bills (see below) and can also be distinguished by the absence of flight feather moult in their first summer. When moult is completed by late August or early September, they become nearly indistinguishable from adults except that moult contrasts can be seen in the median coverts, a method of ageing that I have never tried myself. Of the juveniles examined by Jenni and Winkler (1994), 93% had moulted three to six of the median coverts. The retained outer feathers are greyish-brown with at most a faint blue tinge, while the new inner ones are more bluish, similar to the renewed lesser coverts and scapulars and contrasting with the unmoulted greater coverts. Similar moult contrasts are seen in museum specimens of some other nuthatch species but the reliability as an age criterion has not been tested (Harrap and Quinn, 1996). The use of skull pneumatization for ageing is limited since adults never attain a fully pneumatized skull (Chapin, 1949; Winkler, 1979). Pneumatization scores of 1–5 indicate first-year birds, but scores of 6 may be either first-year (if caught

after November) or older (Jenni and Winkler, 1994). In Red-breasted Nuthatches as well, pneumatization is a suspect criterion after September (Yunick, 1980).

VARIATION IN SIZE

Details of morphological variation are given in Appendix III. In summary, males are, on average, slightly larger than females (about 5% in body weight and wing length, only 1% in tarsus length) but have similar bill sizes. Wing length increases by about 1 mm after the first full moult in the second summer and the wings show measurable wear in the breeding season. Juvenile bills do not approach adult size before late summer. Bill size also varies with season, being largest in October and shortest in the breeding season (see also Chapter 4). Body weight increases sharply at the end of the day, as shown by roost controls. Daytime weights are highest during summer moult and lowest at the end of the breeding season.

Few attempts have been made to compare the morphology of Nuthatches in different habitats. Some authors have found differences in male wing length between territories differing in habitat type, but this may simply reflect the fact that older males tend to live in better territories (see Chapter 7). The distinguished 19th century ornithologist C. L. Brehm identified and named separate Nuthatch 'species' in deciduous and coniferous forest (Kleinschmidt, 1928). The coniferous form '*pinetorum*' was said to have a thin, upward-pointing bill, the deciduous form '*foliorum*' a short, straight bill. As Kleinschmidt pointed out, Brehm's use of the binomial nomenclature did not at that time imply the existence of genetically distinct populations, but was used merely to catalogue variation observed in the field. Brehm's characterization would certainly make sense in the light of general trends across species and subspecies (see Chapter 1) but has not been investigated further.

MOULT

Like many other resident passerines, adult Nuthatches have a complete post-nuptial moult, i.e. a complete moult right after breeding (Fig. 12). I studied moult of flight feathers in three summers and found that moult started on average about a week before the young fledged (*c.* 31 May) and ended on average on 26 August. The mean total duration was 88 days, comparable in length to that of many other resident passerines (Matthysen, 1986a). Moult typically starts with the first primary (in the middle of the wing) progressing towards the wing tip, and – when about two primaries are replaced – a parallel progression takes place from the first secondary towards the interior. The three tertials are also replaced at an early stage,

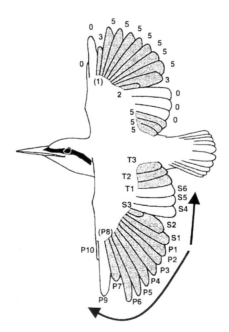

Fig. 12 *How wing moult is scored. Upper half: each feather is given a score between 0 (old feather) and 5 (fullgrown new feather) (Ginn and Melville, 1983). Lower half: sequential numbering of primaries (P1–P10), secondaries (S1–S6) and tertials (T1–T3). Note that the reduced outer primary (P10) is not scored, and P8 is hidden (score = 1). Arrows show how moult progresses in primaries and secondaries. Newly grown feathers are shaded. The individual shown on this figure has moult scores of 34, 10 and 15 for primaries, secondaries and tertials, respectively, adding up to a total score of 59 on a maximum of 90 (5 × 18 feathers).*

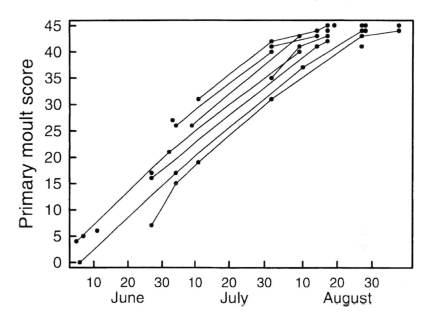

Fig. 13 *Progression of primary flight feather moult in the Peerdsbos population in 1985. Successive scores of the same individual are connected by lines.*

corresponding to primary scores between 10 and 30. Tail-feathers are moulted within the period it takes to moult the primaries. Thus, in late July all tracts (primary, secondary, tertial, tail) are moulting simultaneously.

In my study, moult rates were very similar between years and between individuals, which is illustrated by the parallel lines on Fig. 13. About 95% of all Nuthatches started their moult within a 3-week period. Individual differences in the onset of moult were not related to sex or age but appeared to depend on the breeding cycle. Early-breeding females moulted relatively early, but there was no such trend in males. As a result there was no synchronization of moult within breeding pairs. Four individuals that failed to raise a brood moulted on average 11 days before successful birds (Matthysen, 1986a).

First-year birds have an incomplete post-juvenile moult starting at about 2 months of age. The only wing-feathers moulted are the lesser wing-coverts (usually all), some or most of the median coverts (usually three to six), and very occasionally the innermost greater covert (Jenni and Winkler, 1994). According to Cramp and Perrins (1993), occasionally a few tertials may be moulted as well.

CHAPTER 3

Habitat and Population Densities

*Unter allen Umstände hält er sich an die Bäume, und nur im äußersten Notfalle
entschließt er sich, eine baumleere Strecke zu überfliegen*

*('He keeps to the trees in all circumstances, and only in the utmost necessity will
decide to fly over a treeless space')*

Brehm (1920)

In most of its range the Eurasian Nuthatch is a typical bird of mature
broadleaved forest and is only rarely found on conifers. Indeed its very presence – as betrayed by the loud calls – often indicates that some broadleaved
trees must be nearby within a largely coniferous forest. This holds for most
European populations, and probably also for the most easterly populations
(including *sinensis*), though their habitat preferences have not been particularly well studied. In between, however, is a string of (mostly white-
breasted) populations from northern Russia to Kamchatka that breed quite
commonly in purely coniferous forest, albeit at lower densities (Dement'ev

36

and Gladkov, 1954; Pravosudov, 1993a). In this chapter I present as much information as I could find on the types of forest inhabited by Nuthatches and the breeding densities in various parts of their range, primarily in Europe. First, however, some warning must be given about the interpretation of these density values.

A *CAVEAT* ON POPULATION DENSITY ESTIMATES

Population density can be estimated by a variety of methods, as is evident in the studies summarized in Fig. 14 and Appendix IV. Some studies provide estimates based on close observation of a colour-ringed population (e.g. Löhrl, 1958; Enoksson, 1990a) or at least territory mapping (e.g. Foyer, 1976; Nilsson, 1976). Others, however, rely on indirect methods where densities are extrapolated from the number of birds encountered on transects or during point counts (e.g. Ferry and Frochot, 1970; Blana, 1978). The study by Prill (1988) illustrates this point with data that were collected in the same area in different years with different methodologies. In the first year, nests were found 'on the side' during a study on woodpeckers, and total density was estimated at 4.1 pairs per 10 ha. The next year, a 'traditional' territory-mapping method resulted in 3.1 pairs per 10 ha. The third year, a thorough nest search resulted in the highest density so far, with 5.9 pairs per 10 ha. Although this variation may reflect true population fluctuations, it seems equally likely that the territory-mapping method provided a severe underestimate.

Another consideration is that the data summarized in Fig. 14 were collected in study areas that differ in many respects other than the tree species composition or the age of the stands. For instance, breeding densities may be affected by the presence of nestboxes (see below in this chapter) or by the landscape around a particular study site (Chapter 10). Also, since Nuthatch populations have increased over the last few decades in several parts of Europe (Chapter 9), old and recent records are not necessarily comparable.

GENERAL HABITAT REQUIREMENTS

Throughout the Nuthatch's range, the best habitat with the highest population densities is provided by mature forest with large, old trees that have well-developed canopies. Nuthatches are never very abundant, and areas with densities higher than 3 breeding pairs per 10 ha can be considered to be of high quality (Fig. 14). In broadleaved forest, densities typically reach their maximum in stands over 100 years old (Fig. 15). In Norway spruce stands in southern Belgium, Nuthatches were completely absent from stands younger than 70 years (Deceuninck and Baguette, 1991). The importance of tree age is also reflected in a study of Dutch woodlots where breeding density increased significantly with mean trunk diameter (van Noorden *et al.*, 1988). In a similar study in oak woodlots near Antwerp,

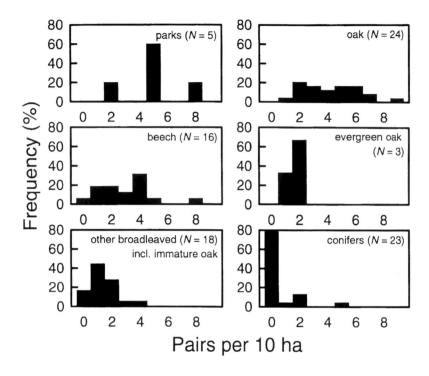

Fig. 14 *The frequency distribution of population densities recorded in different habitat types, based on the compiled data in Appendix IV. N = number of studies.*

however, I did not find such a relationship, perhaps because in this particular study area the birds failed to select the best territories (see Chapter 10). In eastern Siberia, Kharitonov (1983) also found higher densities in stands with taller trees.

Large trees are preferred because they provide a large foraging area (bark of trunk and limbs) as well as cavities for nesting. Mature ancient forests, where such trees are abundant, are nowadays scarce in Europe as they have been cut and replaced by commercial stands. In these heavily managed forests old trees are often absent. Very large trees may still be found in old parks in cities or estates in the countryside, so it is not surprising that some of the highest densities have been recorded in parks (Fig. 14). Immature stands support only low densities or may at most be visited by occasional roaming birds. A curious reference is Kozlova (1933) who observed Nuthatches in 'mountains devoid of forest vegetation' in Mongolia. For instance, a pair was observed 'among rocks, overgrown with nothing but small juniper-trees, the nearest wood being 30–40 miles distant'. This report might not be as exceptional as it seems, since terrestrial tendencies have been reported for at least two subspecies of the closely related Chestnut-bellied Nuthatch (Chapter 13). A more distant relative,

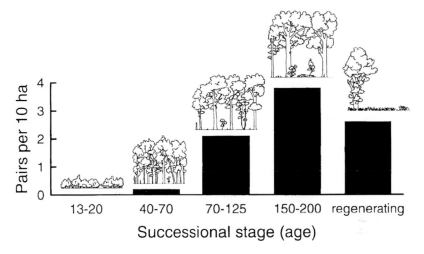

Fig. 15 *Breeding densities in different successional stages of a semi-natural oak forest in Burgundy, France. After Ferry and Frochot, 1970.*

the Red-breasted Nuthatch, is known to occur in widely different habitats during migration, including bare rocks and cliffs and even grassland with fence posts (Chapter 14). Nevertheless, the chance of finding Eurasian Nuthatches outside forest must be considered very small. In the following sections I review the major types of forest inhabited by Nuthatches. Since present-day forests over most of Europe are second-growth and often commercial plantations, these can be best described by the dominant tree species rather than by vegetation types.

OAK FOREST

Within Europe, the most important tree for Nuthatches is undoubtedly the oak, especially the common oak. In Sweden there is a close correspondence between the northern distribution limits of Nuthatches and oaks (Källander *et al.*, 1978). In mature forests dominated by common or sessile oak, densities of 3–7 pairs per 10 ha are not exceptional (Figs 14, 15). The highest densities ever recorded are 11 pairs per 10 ha in Slovakian oak–beech forest (Krištín, pers. comm.), closely followed by 10.8 pairs per ha in an oak–hornbeam forest in Germany (Pfeifer, 1955, in one year only). Studies on foraging behaviour show a clear preference for oak trees within mixed stands (Chapter 4) and, if given the choice, young birds settle preferentially in territories with oaks (Chapter 7). The reason why oaks are so attractive is probably that their rough bark provides a good grip for climbing as well as many furrows and crevices where insects and spiders may be hidden. Oak trees harbour a high diversity of arthropods (more than 400 species of insects) that surpasses most other European trees, especially

Deciduous woodland habitat in the Peerdsbos, Belgium. A stand of common oak (with bracken) is in the foreground, and beech in the background (plot T in Fig 23).

Deciduous woodland habitat near Uppsala, Sweden, with common oak trees and hazel shrub.

Parkland habitat in the Peerdsbos, Belgium (plot L in Fig. 23).

those introduced by humans (Kennedy and Southwood, 1984). The acorns, on the other hand, seem to be of little importance to the birds.

BEECH FOREST

In western Europe at least, beech trees are among the most common tree species used in long-rotation forest stands, and may reach considerable sizes. Beech nuts provide a large food resource for Nuthatches in mast years, which may affect populations noticeably (Chapter 9). However, the trees themselves seem to be less preferred as foraging sites, perhaps because of the smooth bark and the lower diversity of insects living on them (a quarter of that on oaks; Kennedy and Southwood, 1984). Population densities in beech-dominated forest are usually lower than in oak forest, generally 1–4 pairs per 10 ha (Fig. 14). For instance, Prill (1988) lists 10 censuses in different beech forests in eastern Germany, with densities between 0.9 and 3.2 pairs per 10 ha (not in Fig. 14). Another German study reported a much higher density of 8 pairs per 10 ha, but without further details (Blana, 1978). In the mixed forest of the Peerdsbos study area, territories dominated by beech trees tended to have a low level of occupation in years without beech mast (Chapters 7 and 9).

OTHER BROADLEAVED TREES

Besides oak and beech forest, Nuthatches are found in a variety of broadleaved forest types, but generally with more modest densities, around 2 pairs per 10 ha (Fig. 14). In north-temperate regions they occur in aspen groves (Durango and Durango, 1942), locust tree plantations (Székely and Juhász 1993), birch–aspen–alder stands within coniferous forest (Enoksson *et al.*, 1995) and moist alder forests, such as alder–ash–spruce forest in Poland (Wesołowski and Stawarczyk, 1991) and larch–alder and alder–poplar forests in Siberia (Dement'ev and Gladkov, 1954). In western Siberia, where the species is found mainly in coniferous forest, some Nuthatches are also found in the birch zone at the southern edge of the taiga, though only rarely in birch forest in the forest steppe zone (Johansen, 1944). Nuthatches are nearly absent from commercial poplar plantations, a type of forest that is not uncommon in western Europe. However, recently some breeding pairs were found – albeit in low densities – in lightly managed poplar stands in the reclaimed Dutch Flevo polders, typically in fairly old stands (> 20 m) with a lower layer of ash, oak or hawthorn and a fair amount of dead wood (Bijlsma, 1989). Nuthatches can also be found in evergreen oak forest (Harrison, 1955; Ceballos, 1969; Herrera, 1978) and even in olive and walnut groves (Danford, 1878; Kumerloeve, 1961; Kasparek, 1988). In the latter cases it is not clear whether the birds actually breed there, let alone whether these habitats would be capable of supporting a viable population. However, this might be true for any suboptimal habitat, where seemingly healthy populations may

persist only because of immigration from better 'source' habitats. In Chapter 10 I shall discuss such a possible situation in the context of forest fragmentation.

CONIFEROUS FOREST

In most of Europe and perhaps parts of eastern Asia as well, Nuthatches are rare inhabitants of pure coniferous forest, except for some old spruce or pine forests in the mountains of Switzerland (Glutz von Blotzheim, 1962), Greece (Peus, 1954), Spain (Carrascal, 1984a), Japan (Brazil, 1991), Korea (Fiebig, 1992) and other countries. Some of these mountain forests support fairly high densities, but more often Nuthatch densities are very low, or they are absent altogether (Fig. 14). In the Peerdsbos mixed forest, small stands with pine or spruce were used only in low-quality territories (Chapter 7). Throughout the Eurasian taiga belt (excluding Fennoscandia) from northern Russia to Kamchatka, Nuthatches are found mainly in coniferous forest (Dement'ev and Gladkov, 1954). Nevertheless, the only density figure I could find for this wide region is also very low (0.03 pairs per 10 ha, eastern Siberia; Pravosudov, 1993a). In montane coniferous forest, Nuthatches frequently occur up to the treeline at altitudes between 2000 and 2500 m, with a reported maximum of 3300 m in Taiwanese pine forest (Harrap and Quinn, 1996).

Larch forest near Magadan in eastern Siberia. (Photo: Vladimir Pravosudov).

OTHER HABITAT CHARACTERISTICS

I have found surprisingly little information on habitat characteristics other than tree species and age that might influence the presence of Nuthatches. The structure and composition of the shrub layer is probably of little direct importance, though hazel may be an exception, since hazel-nuts provide an important food resource, comparable to beech nuts. Moskát and Fuisz (1994) found no effect of thinning or bush-clearing on Nuthatch densities in Hungarian oak forest. In a study in France, Nuthatches were significantly more abundant in an oak forest where over-grazing by fallow deer had reduced the undergrowth (Lebreton *et al.*, 1991). Since Nuthatches occasionally forage on the ground, for instance in search of beech nuts, they may prefer a sparse undergrowth, which may explain Lebreton *et al.*'s result. This explanation is supported by a study on the *amurensis* subspecies in pine forest in eastern Siberia. Here, Nuthatch abundance was negatively related to herbaceous cover, and positively to the presence of tall trees and dead wood (Kharitonov, 1983). Finally, a study in the German Harz mountains shows a clear trend of decreasing density with altitude, which is only partly explained by the transition from broadleaved to coniferous habitat (Fig. 16).

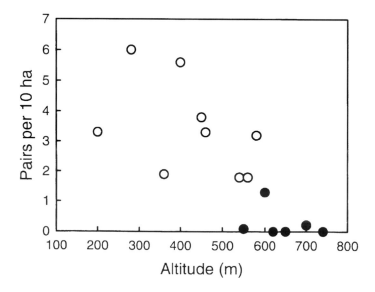

Fig. 16 *Breeding density in relation to altitude in the Harz mountains, Germany. Open dots = broadleaved forest, filled dots = spruce forest. Data from Zang (1988).*

HUMAN INFLUENCES

Nuthatches often live quite close to human habitation, for instance in city parks or residential areas. They are not particularly shy and readily visit bird tables (Thompson *et al.*, 1993) or other artificial food sources. However, since a pair needs a minimal space of about 1 ha of high-quality habitat, they are not generally found in small parks, gardens, hedgerows or small woodlots. Little is known about the effects of urbanization such as pollution, noise and disturbance. It has been suggested that the accumulation of soot on trees was the reason for the Nuthatch's disappearance from inner London earlier this century (Montier, 1977) but this hypothesis remains untested.

Nuthatches make limited use of nestboxes even when these are present at high density, probably because they prefer large cavities high up in the trees (Chapter 6). A few studies, however, suggest that the addition of nestboxes may lead to an increase in density. The two highest reported densities were found in areas with superabundant nestboxes. In German oak–hornbeam forest the density increased from 8 to 27 pairs on a 25-ha plot after 40 boxes per ha were added (Pfeifer, 1955). Unfortunately no data are given on the number of pairs that used them, therefore the Nuthatches may also have benefited from the reduced competition for natural cavities when other species also bred in these boxes. In Slovakian oak–hornbeam forest the density rose from 6 to more than 10 pairs per 10 ha after 3 boxes per ha were added, with about 70% breeding in boxes (A. Krištín, pers. comm.). The highest reported density in conifers has also been found in a plot with many nestboxes (Löhrl, cited in Glutz von Blotzheim, 1993). Nevertheless, there remains a need for proper experimental tests on the effect of nestboxes on population density.

CHAPTER 4

Foraging, Food and Hoarding

This habit has given nuthatches their name . . . They 'hatch' the nut or break its shell with the bill as with a hatchet

Forbush (1929)

Nuthatches belong to a group of forest birds that find their living mainly by searching for insects and spiders on the bare surfaces of tree trunks, branches and twigs. This group includes tits, woodpeckers, treecreepers, nuthatches and some warblers, and is commonly called the pariform guild in Europe (after the tits *Parus*) or the bark-foraging guild in N America. Within this group nuthatches are often regarded as specialized feeders, since they are so obviously adapted to an arboreal lifestyle (e.g. Carrascal *et al.*, 1990). However, they are quite versatile in their foraging behaviour in comparison to true specialists such as the treecreepers or certain woodpeckers. They eat invertebrates as well as seeds and nuts, and will consume virtually any prey species they encounter within their chosen foraging environment.

GENERAL FORAGING BEHAVIOUR

When foraging, Eurasian Nuthatches spend most of their time on tree trunks and branches, searching for adult insects, spiders, larvae, pupae and

46

eggs on or under the bark surface. They find their prey by sight or by probing with their long bill. Unlike woodpeckers they do not use their tongue for probing and rarely dig into the wood. They use their solid bill to hammer large food items into pieces before eating them; these items are not held under the foot but are wedged in a bark crevice with the bird in a head-down position. 'Scaling' or removing bits of flaky bark in search of prey underneath is also a common technique used on some tree species. In spring and summer, Nuthatches also take caterpillars or other prey from leaves or buds. Some prey, such as lycosid spiders, are picked up from the forest floor, but are typically taken to a tree to be eaten. Székely (1987) recorded the frequency of different foraging techniques and found that 'gleaning' was by far the commonest (more than 70%), followed by 'pecking' (10–20% depending on season) and 'drilling' (on average *c.* 10%). Scaling, fly-catching and food hoarding took up less than 5% of the foraging time.

Many authors have commented upon the Nuthatch's skill at catching mobile insects by surprise, even in flight. Löhrl (1958) thought they were better than a Spotted Flycatcher at catching flies. Dorka (1980), Möllersten (1985) and Källander (1993) saw Nuthatches foraging on small beetles and swarming ants in the air, and Ward (1982) and Clarke (1985) saw them hovering while taking insects from the air or from leaves. Fly-catching is known from other nuthatch species as well (e.g. Chapter 14). Another sign of their manoeuvrability is that all nuthatch species will dive down after food items that have fallen, particularly valuable food such as insects (Löhrl, 1978).

While foraging, Nuthatches move about fairly rapidly. A pair I observed in detail one winter rarely stayed longer than a minute on a particular foraging site (trunk, large branch, etc.), and hardly ever stayed in a tree for longer than 5 minutes (Matthysen, 1988). While foraging they move up- and downwards equally often (László, 1988; Matthysen, 1990a), although Székely (1987) observed *c.* two-thirds of upward movements. In the White-breasted Nuthatch, Szaro and Balda (1979) also recorded as many head-up as head-down stances.

DIET

In spring and summer, animal food clearly predominates in the adults' diet (Table 3). Beetles (Coleoptera) were the dominant prey in three out of five studies. In Spain these were mainly Curculionidae (especially *Brachyderes* and *Balaninus*, about 40%), Tenebrionidae, Coccinellidae and Elateridae (Ceballos, 1969). The importance of beetles was confirmed by Ptushenko and Inozemtsev (1968) and Nicolai (1986) (not in Table 3). Other important groups are Diptera, Hymenoptera (mainly ants), Hemiptera, Lepidoptera (also in Polivanov and Polivanova, 1986: 36 out of 46 food items, mainly adult Geometridae) and spiders. Myriapoda and Chilopoda are more rarely found. Lepidoptera are particularly important in spring, when caterpillars of species such as *Tortrix viridana* can be very abundant on newly growing leaves. Ptushenko and Inozemtsev (1968)

TABLE 3: *Diet of adult Nuthatches, expressed as proportion of food items (first two columns) and frequency of occurrence in the samples (last two columns)*

Source	Inozemtsev, 1965	Kaczmarek et al., 1981	Ceballos, 1969	Obeso, 1988
Country	Russia	Poland	Spain	Spain
Habitat	?	Mixed forest	Evergreen oak	Pine
Sample size	18	12	85	78
Coleoptera	25.0	13.3	88.9	79.2
Lepidoptera	1.8	~20.0	16.9	15.6
Diptera	24.3	–	–	17.7
Hemiptera	3.7	19.9	39.9	1.0
Hymenoptera	5.5	6.6	–	6.3
Spiders	4.9	6.6	–	21.9
Seeds and nuts	34.7	–	–	*75.0

NOTES:

~ assuming all items listed as 'herbivorous larvae' are caterpillars.

* 37% of total volume.

1. The first three columns represent data from stomachs collected in spring and summer; the last column data from faeces collected year-round on 61 different individuals.

2. Sample size = number of stomachs or faeces.

3. – No separate value provided.

found that caterpillars and spiders together comprised up to 50% of the summer diet. Maréchal (1992) and Cramp and Perrins (1993) give long lists of prey species but with no indication of their relative importance. A remarkable finding is the presence of numerous flower buds in an adult stomach collected in June by Steinfatt (1938). Obeso (1988) found that some birds' faecal sacs contained more vegetable remains than others, which probably reflects differences between territories (no relationship was found with body measurements). Nicolai (1986) also commented that usually one taxon of prey predominated in a single stomach, but provided no details.

The diet of nestlings largely resembles the adults' summer diet (Fig. 17, Table 4). In by far the largest study to date, Krištín (1994) analysed over 10,000 food items delivered to 224 nestlings in 36 nests between 3 and 23 days of age. He identified 432 arthropod species belonging to 115 families and 24 orders. Beetles alone accounted for 116 species in 23 families. Not included in Fig. 17 is Ptushenko and Inozemtsev's (1968) study where beetle larvae and caterpillars accounted for 58% of food items (164 items from 18 stomachs, collected in summer), and Diptera for 29%. While Lepidoptera (mainly caterpillars) are important in all studies, the contribution of other insect orders and spiders is more variable. Krištín's (1992a, 1994) study in beech forest also shows considerable variation between years: 7–70% for Lepidoptera, 14–33% for Heteroptera, 5–18% for spiders, 9–32% for Diptera. In Siberian larch forest similar fluctuations were found between years, in particular in the number of Hymenoptera, caterpillars

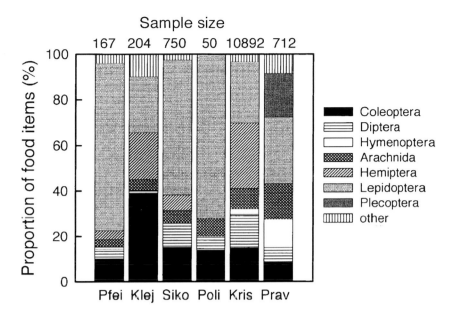

Fig. 17 *Diet of nestlings in different studies (all by means of neck collar samples). Data from Pfeifer and Keil (1959, oak–hornbeam forest, Germany), Klejnotowski (1967, oak forest, Poland, 20 nestlings from 3 nests), Sikora (1975a, mixed deciduous, Poland), Polivanova (1985, habitat unknown, Caucasus), Krištín (1994, beech, Slovakia, 224 nestlings from 36 nests) and Pravosudov et al. (1996, larch, Siberia, 19 nests). Number of nestlings or nests are not specified in some studies. Sample size = number of food items.*

and beetles, and least in the number of spiders; within-year variation between nests was less pronounced (Pravosudov *et al.*, 1996). Birds studied by Krištín (1994) took more Lepidoptera, Heteroptera and spiders than expected from their availability in the forest, and far fewer Hymenoptera. In comparison to other birds Nuthatches show less restraint in foraging on venomous or distasteful prey such as wasps, ladybirds and bugs (Krištín, 1992b). Some anecdotal records show that nestlings may occasionally be fed with a variety of other foods, particularly in northern populations: pine seeds, snails and small fish in Siberia (Pravosudov *et al.*, 1996), bread, fat and sunflower seeds in Sweden (Durango and Durango, 1942; Blomgren *et al.*, 1979) and bee larvae stolen from a Honey Buzzard's nest in Germany (Busch, 1959).

The winter diet of Nuthatches is much less well known. They probably rely to a higher degree on hidden eggs, larvae and pupae of insects. Spiders are also probably more important in winter, as in other bird species (Askenmo *et al.*, 1977). Plant food (seeds and nuts) becomes very important in winter, especially in northern localities, and this is probably a general rule among nuthatches; however, there is a lot of variation within as well as

TABLE 4: *Arthropod families that contributed 10% or more to the nestlings' diet in any particular study (or year of study by Krištín or Pravosudov)*

Order/family	Frequency (%)	Source
ARACHNIDA		
Phalangiidae	13	Krištín (1992a)
LEPIDOPTERA		
Noctuidae	39	Klejnotowski (1967)
Geometridae	32	Pfeifer and Keil (1959)
Tortricidae	29	Pfeifer and Keil (1959)
COLEOPTERA		
Tenebrionidae	13	Krištín (1992a)
HEMIPTERA		
Miridae	23	Pfeifer and Keil (1959)
Aphidinea	14	Krištín (1992a)
DIPTERA		
Syrphidae	23	Krištín (1992a)
Tipulidae	10	Krištín (1992a)
HYMENOPTERA		
~Symphyta	10	Krištín (1992a)
Tenthredinidae	13	Krištín (1992a)
PLECOPTERA		
Perlodidae	21	Pravosudov *et al.* (1996)

NOTES:
~ suborder (family not specified).
See Fig. 17 for more information on sources.

between species, perhaps because of fluctuating seed crops or weather-dependent availability of arthropod prey (Table 5).

Nuthatches eat various seeds and nuts whenever they are available. The seeds are wedged in a crevice and then hammered open. The grit that is found in the stomachs of adults as well as nestlings (Glutz von Blotzheim, 1993) may aid the digestion of seeds and hard insect parts. At least in Europe, hazel and beech are probably the most important seeds, while others such as acorns, hornbeam, sycamore and ash seeds are less often consumed. Seeds are mostly taken on the ground, and in winter Nuthatches may concentrate on snowfree patches to look for beech nuts among or under the leaves (Kneis and Görner, 1986). Beech nuts may also be taken from the tree as soon as the cupulae open, and sometimes the birds hover in the air in order to extract the nuts (Källander, 1993). Seeds are also extracted from open cones of pine or spruce. Fruits are taken much less frequently (see lists in Schuster, 1930 and Creutz, 1953). Turček (1961) lists no fewer than 55 tree species that are used for food in various ways. There are several reports of Nuthatches drinking sap from trees in spring (Mylne, 1959; Glutz von Blotzheim, 1962; Bardin, 1987; Cramp and Perrins, 1993)

TABLE 5: *Proportion of plant food in the diet of Eurasian Nuthatches and other species in spring/summer (March–August) and autumn/winter (September–February)*

Species	Habitat	Spring/ summer	Autumn/ winter	Source
Eurasian	?	35	–	Inozemtsev, 1965
	Evergreen	0	–	Ceballos, 1969
	Oak–birch	0	25	Rivera, 1985
	Coniferous	~24	40	Steinfatt, 1938
	Coniferous?	–	100	Ptushenko and Inozemtsev, 1968
	?	0	–	Polivanov and Polivanova, 1986
White-breasted	Oak	0	17	Anderson, 1976
	Pine	0	8	Anderson, 1976
	?	21	74	Beal, cited by Forbush, 1929
	Oak	–	72	Williams and Batzli, 1979c
	Deciduous	*13	67	Sanderson, 1898
	?	24	48	Martin et al., 1951, cited by Pravosudov and Grubb, 1993
Red-breasted	Fir	0	17	Anderson, 1976
	Pine	0	12	Anderson, 1976
Pygmy	Pine	0	4	Anderson, 1976
	Pine	55	76	Norris, 1958
Brown-headed	Pine	26	88	Norris, 1958
	Pine	–	79	Morse, 1970
	Pine	45	–	Withgott and Smith, 1988

NOTES:
~ based on three stomachs only.
* all from April.
All records are from stomach contents, except for observational data by Rivera and by Withgott and Smith. Data are expressed as % of food items (usually seeds and nuts) except for Anderson (volume %), Beal, Norris and Sanderson (unspecified).

and feeding on flower buds (Löhrl, 1958; De Vries, 1985). Particularly in winter, Nuthatches will take several anthropogenic sources of fat such as lard, butter, margarine, cheese and bread, and even slaughterhouse offal (Eriksson, 1970). They are enthusiastic visitors to any bird tables close to or within their territory, especially in cold weather (Glutz von Blotzheim, 1993), and are usually able to locate new bird feeders within days (e.g. Nilsson *et al.*, 1993). Nuthatches have even been seen to feed on the spawn and fry of sock-eye salmon in Kamchatka (Ladygin, 1991).

In general, the various studies on diet suggest a great flexibility in foraging behaviour within the constraints of the habitat and the foraging techniques to which the Nuthatches are adapted. This is illustrated by Krištín's (1990) study where nearby nests were more similar in diet than distant ones, suggesting spatial variation in food that is opportunistically used by the birds. In a Slovakian beech forest, the diet was most similar to that of Wren, Robin and Collared Flycatcher, even though these species forage on different sites and use different techniques. The overlap with other species, including tits and treecreepers, was much lower. The tits' diet contained a much higher proportion of caterpillars, while the Treecreeper foraged more often on Diptera and spiders, and less on beetles and caterpillars compared to the Nuthatch (Krištín, 1992b).

Diet and bill morphology

As the Nuthatch's diet changes with the season, so does the bill. These seasonal changes amount to 1 mm or more in individual birds in the course of a year (Fig. 18). Bills generally shorten from October to April and then lengthen again throughout the summer, with a slight temporary reduction at the end of the breeding season. The most likely explanation lies in the continuous growth at the base, and wear at the tip, of the *rhamphotheca*, the horny outer structure overlaying the mandibles. In winter when the birds often feed on seeds or poke for hidden insects in crevices, this probably results in net abrasion, while in summer when they live mainly by gleaning soft animal food from the surface there may be net growth (Matthysen, 1989a). Male and female bills change mostly in parallel. Female bills are slightly shorter, except that they temporarily exceed male bill length in spring. I hypothesized that this is because of the male's greater share in foraging when the female incubates eggs and broods small young (Matthysen, 1989a). As already mentioned in Chapter 2, young birds have shorter bills for the first few months of life. Whether this affects their foraging behaviour is an unresolved question, but there may be an effect on their ability to handle seeds. Juvenile birds visiting feeders in summer took on average fewer seeds in their bill than adults, whereas in autumn there was no difference (Enoksson, 1988a).

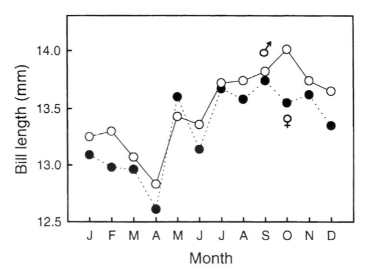

Fig. 18 *Monthly variation in mean bill length of male and female Nuthatches. Measurements of first-year birds are excluded from June to October.*

Seasonal variation in bill length is something that should be accounted for in biometric studies using data collected in different seasons. For instance, Eriksson (1970) used bill length to infer the origin of invading *asiatica* Nuthatches in Finland, but did not mention the dates when the reference specimens were collected (Chapter 8). Also, bill dimensions are quite often mentioned in descriptions of subspecific variation, but always without reference to season. Similar seasonal patterns may occur in other nuthatch species but have never been studied. There is only the anecdotal record of a White-breasted male caught in November with an unusually long bill (in fact, only the upper mandible) which was subsequently reduced by over 2 mm by next April (Brackbill, 1969a).

FORAGING SITES

Since the days of MacArthur (1958) there have been countless studies attempting to address the question of food competition between insectivorous birds by looking at the sites they use while searching for food. The underlying idea is that birds may be more specialized in the use of foraging sites than in the selection of prey on these sites. The Nuthatch's foraging niche is basically the same in different forest types, consisting mainly of trunks and large branches (Fig. 19). To put it another way, they spend most

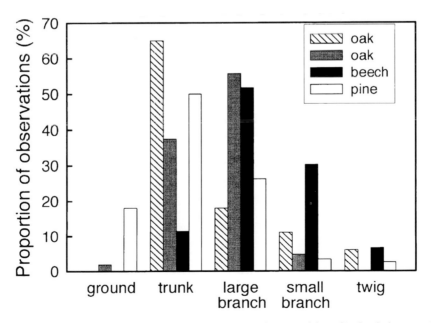

Fig. 19 *Foraging sites used in mixed deciduous forest (dominated by oak), beech forest and pine forest. 'Twig' includes cones and needle clusters. Data from László (1988; hatched), Bilcke et al. (1986; grey and black) and Moreno (1981; open).*

of their time on branches thicker than their own body diameter (Fig. 20). Although Fig. 19 shows data only from winter studies, a similar pattern was found in summer observations by Edington and Edington (1972) and Carrascal (1984b), in dense deciduous forest and pine forest, respectively. Small branches and twigs appear to be used more often on beech than on oak or pine. One reason may be that the smooth trunk of beech harbours fewer arthropods than the rough, fissured bark of oak and flaky bark of pine, and that this difference is less pronounced on the smaller branches. Alternatively, beeches may have proportionally more thin branches and twigs than the other trees. Foraging on the ground is rather common in pine and beech forest, perhaps in search of fallen seeds, but nearly absent in oak forest (Fig. 19). In a holm oak forest nearly all foraging was done on trunks (85%) and none on the ground (Tellería and Santos, 1995, based on a relatively small sample). In the Peerdsbos study area equal use was made of vertical, horizontal or slanting branches but with a noticeable preference for the upper sides (75%) (Fig. 21). The same tendency was found in White-breasted Nuthatches in an aviary study (Fig. 21).

Common oaks are clearly preferred as foraging sites to other tree species (Table 6; see also Rivera, 1985 and Nicolai, 1986). In Hungary, László (1988) found that Nuthatches showed little preference for tree species, except that sessile oak was used somewhat more, and turkey oak somewhat less, frequently than expected. Mattes (1988, cited by Glutz von Blotzheim, 1993) concluded that larch trees were preferred to spruce and other conifers.

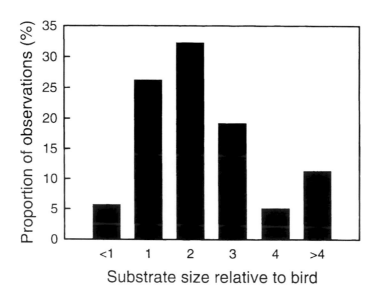

Fig. 20 *Foraging substrate sizes expressed in relation to the bird's own diameter. Data from a mixed forest (Peerdsbos, 1988–1989).*

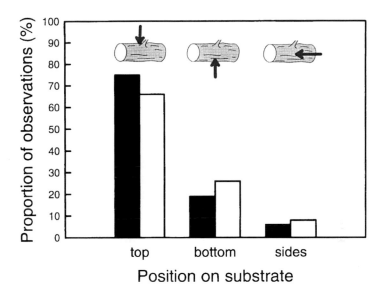

Fig. 21 *Preference for top, bottom or sides of horizontal or slanting branches by Eurasian Nuthatches (black; data from Peerdsbos 1988–1989) and White-breasted Nuthatches (white; data from aviary study by Pierce and Grubb, 1981).*

TABLE 6: *Proportion of foraging observations on different tree species compared to their availability within the study area (data from Peerdsbos study, 1988–1989)*

Tree species	% observations	% in study area
Common oak	63.7	46.1
Red oak	12.7	6.7
Pine	10.0	12.7
Beech	6.8	9.8
Birch	2.6	10.5
Other	4.2	14.2

Nuthatches tend to forage somewhat lower on colder days, but with no obvious difference in the type of substrate used (Bilcke *et al.*, 1986). In the same forest as Bilcke *et al.*, I found that Nuthatches foraged somewhat higher in winter than in autumn or spring, and that small branches were used increasingly from autumn to the breeding season. Foraging on oak trees increased slightly over the same period at the expense of beech (Matthysen, 1988). It is not clear to what extent these seasonal shifts represent shifts in food abundance or increasing specialization as food becomes more difficult to find. The increasing use of small branches in spring may be explained by the increasing availability of active insects relative to hidden prey. Short-term changes in foraging behaviour have been studied in more detail in the

N American White-breasted Nuthatch, which also tends to forage in less exposed sites, and at a slower pace, in harsher conditions (Chapter 14). Nuthatches appear relatively insensitive to weather conditions compared with other small passerines, perhaps because they are partly sheltered from the wind on the large substrates they use for foraging (Grubb, 1982).

Sex differences

Male and female Nuthatches differ little in morphology, and in fact are more similar in bill length and depth than expected from their general body sizes (see Appendix III for details). Nevertheless, differences in foraging behaviour may arise through behavioural specialization alone, as has been shown for other species (e.g. Peters and Grubb, 1983). In a pairwise comparison of males and females foraging together, I found no differences in foraging height, foraging substrate diameter, substrate orientation (vertical/horizontal/slanting) or position relative to the substrate (top/bottom/sides) (Matthysen, unpubl.). In fact, pair members used similar foraging sites more often than expected by chance, which was also found in the White-breasted Nuthatch (Grubb, 1982). The sexes did differ in one aspect, though, which is that males foraged significantly more often on oak trees. In cases where one bird foraged on an oak and its partner did not, this was the male in 14 out of 16 cases. This may suggest that females are forced to use less favoured sites in order to avoid aggression from males (for a comparable effect in Downy Woodpeckers, see Matthysen *et al.*, 1991). Even so, the overall use of oaks differed little between the sexes (75% in males, 64% in females) since very often both foraged on oaks at the same time. Sexual variation in foraging behaviour has been demonstrated more clearly in some other nuthatches (Table 7) but the pattern of variation across studies is not well explained. There is a suggestion that sex-specific niches are more clearly expressed in coniferous habitat, especially

TABLE 7: *Sexual niche differentiation in various nuthatch species (see also Chapters 11 and 14)*

Species	Habitat	Difference in:	Source
Eurasian	Deciduous	Use of oaks	Matthysen, unpubl.
White-breasted	Deciduous	None	Grubb (1982)
	Aviary	~None	Pierce and Grubb (1981)
	Conifer	Substrate/height	McEllin (1979)
Corsican	Conifer	*Substrate?	Matthysen and Adriaensen (1989a)
Pygmy	Conifer	†None	Norris (1958)
Brown-headed	Conifer	Height	Norris (1958)
Red-breasted	Deciduous	None	Matthysen, unpubl.

NOTES:
~ following the figures, not the text (*fide* T.C. Grubb Jr., *in litt.*).
* not statistically significant.
† no details given.
When there is a difference, males forage more often on oaks (Eurasian), more on trunks and at lower heights (White-breasted, Brown-headed) or less often on trunks (Corsican Nuthatch).

if we disregard the inconclusive evidence on Pygmy and Corsican Nuthatches (see also Chapters 11 and 14).

The general lack of difference in foraging niche of male and female Eurasian Nuthatches, and some other species as well (Table 7), is somewhat puzzling, since such differentiation seems to be the rule rather than the exception in many bark-foraging birds (Peters and Grubb, 1983; Ekman and Askenmo, 1984; Noske, 1986; Gosler, 1987). Whether this means that paired nuthatches do not compete for food is an unresolved question. Besides the male's preferential access to oaks, there is other evidence that the male's presence affects the female's behaviour, since she makes shorter visits to bird tables and removes fewer seeds than the male, perhaps to avoid harassment (Enoksson, 1988a). In the White-breasted Nuthatch, females spent more time looking around when their mate was present, probably for the same reason (Waite, 1987) and females seem to avoid males when hoarding food (Chapter 14 and later in this chapter). However, the overall low frequency of intra-pair conflicts, and the tendency of the pair to forage in close contact (Chapter 5), point to a limited role of competition and interference.

The foraging niche in the guild

Despite their unique locomotory skills, Nuthatches are not particularly regarded as ecological specialists within the bark-foraging guild. The relatively wide variety of foraging sites they use is reflected in an intermediate value for niche breadth (Carrascal *et al.*, 1990; see also Bilcke *et al.*, 1986). Their overall niche use does separate them, however, from the majority of other birds except for the treecreepers. The niche overlap between Nuthatches and the Short-toed Treecreeper is among the highest in the guild (Moreno, 1981; Bilcke *et al.*, 1986). The two genera probably avoid competition by specializing on different prey, and of course treecreepers do not consume seeds. Also, Short-toed Treecreepers bring a smaller variety of prey, and on average smaller prey, to the nest (Krištín, 1990, 1992b). The spatial niches of Nuthatches and treecreepers are both similar to that of woodpeckers. For instance, Székely (1987) showed that the Nuthatch overlaps as much with the Middle Spotted Woodpecker as with the Treecreeper. However, competition with woodpeckers is probably even less than with treecreepers because of differences in feeding techniques and body size.

FOOD HOARDING

Nuthatches store food throughout the year, but especially so in late summer and autumn. For instance, Enoksson (1990a) saw Nuthatches visiting her bird tables 4–13 times per hour in July–August, and 30–50 times per hour in September–October. However, few data are available on the hoarding intensity of naturally available food items. Bromley *et al.* (1974) saw Nuthatches storing up to 30 cedar nuts daily in autumn, and suggested that this habit could be put to use in cedar reafforestation plans. The stored food is mainly nuts and seeds, but occasionally insects as well (Dorka, 1980; Källander, 1993). Hoarding activity generally decreases during the day while the consumption of seeds or nuts may even increase (Fig. 22; see also Löhrl, 1958). In an aviary study on White-breasted Nuthatches, hoarding rate also decreased during the day while consumption did not (Waite and Grubb, 1988a). In another aviary study on this species, K. Petersen and D. Sherry (unpubl.) found a 50% decrease in hoarding rate from winter to spring, but no effects of temperature or time of day. Although the hippocampus region of the brain – the site for spatial memory – also decreases in size from winter to spring in this species, no difference in food retrieval accuracy was found (K. Petersen and D. F. Sherry, unpubl.).

Hoarding sites

Food items are stored one at a time which makes Eurasian Nuthatches 'scatterhoarders' (in contrast to 'larder hoarders' which concentrate hidden food in stores). Food is stored in bark furrows and cracks where it is normally (in at least 80% of all cases) covered with bits of bark, lichen or moss (Löhrl, 1958; Dorka, 1980; Källander, 1993). Caching on average takes half a minute to a minute, depending on the kind of food, whether it is already shelled, and where it is cached (Carrascal and Moreno, 1993; Källander, 1993). Hoarding sites in an oak–beech forest in southern Sweden were described in detail by Källander (1993). Preferred hoarding sites were in trunks and large branches in oaks, corresponding to the

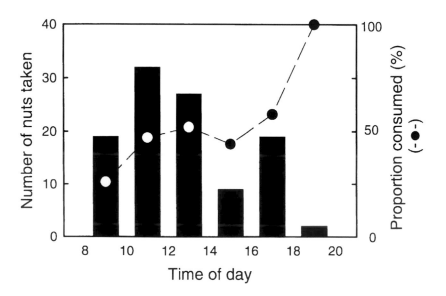

Fig. 22 *The number of cedar nuts taken from a feeder per hour (bars) and the proportion consumed (dots). Data from Bromley et al. (1974).*

preferred foraging niche. In these sites, items were stored mainly behind loose bark (26%), in rotten wood (21%), in rough bark (19%), in small holes and cracks (18%) and in moss (6%). However, about 30% of all caches were placed in the ground, between roots of trees or in fallen logs (see also Richards, 1949; Boyle, 1955). It may be that rotten wood, partly on the ground, was used more often in southern Sweden than in other areas because lichens were rare. Hazelnuts were stored more frequently in the ground (60%) than beech nuts (20%) or sunflower seeds (5%). The most unusual storing place recorded from nuthatches is undoubtedly human hair (Blomgren *et al.*, 1979).

Beech nuts were often shelled in Källander's (1993) study, but mainly so if they were stored in trees; the shelled seeds would probably rot in the ground or in fallen logs. Sunflower seeds were usually (71%) shelled before storage in Spain, but only rarely (6%) in Sweden, perhaps because in Spain seeds were more often stored for short-term retrieval, with a much smaller chance of pilfering, so that investing in shelling before storage is more profitable (Moreno and Carrascal, 1995).

Nuthatches often remove two or even three seeds on a single visit to a feeder (Enoksson, 1988a; Matthysen, pers.obs.; but see Moreno *et al.*, 1981, and Moreno and Carrascal, 1995). The seeds are then cached near each other, but in separate sites (Källander, 1993). Occasionally a stored item is retrieved immediately and stored elsewhere, which may happen up to four or five times before the bird moves on, for reasons that are not entirely clear. Stored items may also be re-hoarded if they are found later; in the aviary study by K. Petersen and D. F. Sherry (unpubl.), White-breasted Nuthatches re-hoarded about half of the previously stored seeds they found. One pair that was studied in detail by Moreno *et al.* (1981) hoarded the majority of its seeds close to the food source (up to 40 m away), much closer than the Willow Tits visiting the same feeder, which might be explained by interspecific dominance but also by differences in territory size. When sunflower seeds are shelled before storing, the birds take more time to store them and do so further from the food source (on average, 32 *vs.* 15 m away). The probable explanation is that unshelled seeds are more valuable than shelled seeds, since the latter require extra work when they are retrieved (Moreno and Carrascal, 1995).

Food theft

The obvious purpose of food hoarding is to keep a store of food to use later in the year when food is less available, and at the same time to make it unavailable for competing individuals or species. Great Spotted Woodpeckers are well known to rob Nuthatches of hazelnuts (Källander, 1983; Goodwin, 1991) but not always with success: for instance, in Källander's (1983) study, Nuthatches were followed on nine of 40 occasions, but in only two cases was the nut actually stolen. Chaffinches are also reported as pilferers of stored food (Svensson, 1975). Carrascal and Moreno (1993) made some observations on food theft from Nuthatches attempting to hoard sunflower seeds. Nuthatches were persecuted by another bird in 24% of their attempts to store food, but only three out of 27 followers stole the food immediately after hoarding (one Nuthatch, one Crested Tit, one Great Spotted Woodpecker). Nuthatches were apparently disturbed by the presence of potential robbers, and lost their seeds more frequently when they were followed (15%) than if they were not (4%). When they were followed they also spent more time looking for a hoarding site (5.5 *vs.* 3.6 seconds), probably to discourage followers, but hoarding itself did not take more time. In another experiment when bits of peanuts were offered, Nuthatches ate them immediately rather than hoard them when 'robbers' were present (Carrascal and Moreno, 1993).

Retrieval of stored food

When Nuthatches retrieve stored food their behaviour is noticeably different from normal foraging (Källander, 1993). They suddenly fly directly to a particular site, for instance a rotten branch, start hammering and extract a food item which is then consumed partly or in its entirety. This suggests that they memorize specific storage sites, though this remains difficult to prove. In Sweden, Nuthatches retrieved between one and five nuts per hour, depending on food availability in the preceding autumn (Källander, 1993). Some food is retrieved after only a few hours (Dorka, 1980) but this is probably unusual (but see Moreno and Carrascal, 1995). Nilsson *et al.* (1993) found that Nuthatches used sunflower seeds up to 98 days after storage, and that the retrieval rate of seeds stored in autumn did not decrease in the course of the winter. The birds apparently used their stored food in a prudent way, saving it for days with low temperatures when energy demand was high and food less available. Experiments by Härdling *et al.* (1995, 1997) provide some support for individuals remembering their own storage sites. In one experiment, free-living males and females were supplied with a different variety of sunflower seeds for each sex. Only 5% of the retrieved seeds were found by the other pair member, even though the sexes did not differ in the choice of particular storage sites. In another experiment, hoarders proved to be significantly more efficient at retrieving their own stored seeds (after 8 days) than naive individuals.

The importance of stored food in winter has been documented by means of 'ptilochronology' (Nilsson *et al.*, 1993; Nilsson, 1994). This technique uses the regrowth of a tail-feather (after pulling the original one) as an index of the birds' condition over the period of regrowth. Regrown feathers were longer and more symmetrical (left *vs.* right) in birds that had access to sunflower seeds in autumn. Nevertheless, they remained 5–8% shorter than the original feather and 12–20% lighter, and less symmetrical as well, suggesting that even with extra food the Nuthatches had limited energy or nutrients available for feather growth. Birds with highly symmetrical original tail-feathers were not more likely to have symmetrical regrown feathers. However, the side that was longest in summer remained so after feather regrowth, which suggests consistent differences in follicle 'quality' (Nilsson, 1994). Previous studies in N American White-breasted Nuthatches have also shown that feather growth improved with extra food present, but was not reduced by extreme cold or a toxin (Herbert *et al.*, 1989; Grubb and Cimprich, 1990; Zuberbier and Grubb, 1992) which further documents the usefulness of the technique as an index of 'nutritional' status.

CHAPTER 5

The Pair and its Territory

L'observation de cet oiseau donne beaucoup de plaisir (. . .) à lui seul il anime un bois

('The observation of this bird gives a lot of pleasure (. . .) it animates a forest on its own')

Géroudet (1963)

Many animal species defend territories against other individuals of their own kind. The nesting territories defended by birds in spring, often with conspicuous song or displays, are almost proverbial, and the Nuthatch is no exception to this habit. What is more peculiar to the Nuthatch's social system, however, is that pairs defend their territories year-round, and that young birds settle in territories of their own very early in life. Such permanent pair territories are not all that uncommon in larger birds such as swans, corvids and birds of prey, and also in various tropical birds. In small birds of temperate latitudes, however, they are quite rarely found (Matthysen, 1993). In this chapter I describe some aspects of the daily behaviour of a pair of Nuthatches in their territory, including the size of the

62

territory, its defence, the daily movements, the behaviour of mates to one another, their time budgets and vocalizations. Since I draw to a considerable extent on my own study in the Peerdsbos forest from 1982 to 1989, I first give a brief introduction to the study area and the methods involved. Some more advanced aspects of the social system are discussed in Chapter 7, notably the ways in which young birds acquire a territory and the implications of territoriality for the population in general.

THE PEERDSBOS STUDY

From August 1982 until June 1987 I closely observed a fully colour-ringed Nuthatch population in the Peerdsbos forest near Antwerp in northern Belgium. The forest itself is 150 ha in size but is surrounded by several hundred hectares of parks and residential areas which are also suitable breeding habitat. The study area contained two rather different parts or subplots (Fig. 23). Plot L is a 9-ha old park area with large (up to 30 m and more) beeches, oaks, sycamores, sweet chestnuts and yews. The adjacent plot T is 31 ha and contains somewhat younger stands (20–25 m high) of oak, beech, Scots pine and Corsican pine, and some mixed stands as well, separated from one another by lanes with two or four rows of large oaks (common oak or red oak).

In this area all individuals were ringed as soon as possible after they settled in the area, and their presence and social status were checked at least twice a month, often more regularly. Each individual received one aluminium ring and one to three coloured rings. Some individuals were painted on the breast in order to facilitate easy observation for a short period. Since most new birds appeared in summer, August was the peak month for captures with about one-quarter of the total number (Fig. 24). In

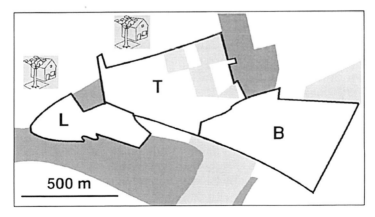

Fig. 23 *The study plots used in the Peerdsbos forest near Antwerp. White = broadleaved forest or (upper left) residential areas with dense tree cover, grey = conifer forest, dark grey = open areas. Plots L and T were used from 1982 to 1987, plot B from 1991 to 1994, and parts of T and B in 1988–1989.*

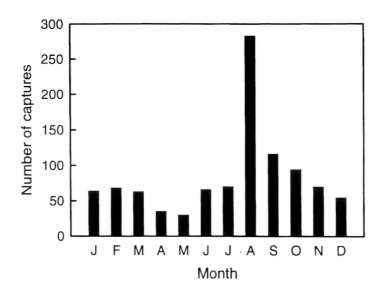

Fig. 24 *The total number of captures of Nuthatches per month in the Peerdsbos from 1982 to 1987.*

plot L, which was inaccessible to the public, I operated a line of 15–20 automatic traps at fixed locations, which were regularly baited with small amounts of sunflower seeds. With these I could easily capture up to 10 Nuthatches per day. Several individuals were trapped more than 20 times during the study, the record-holder (male 'Red – Metal Green') as many as 62 times in 4 years. In plot T birds were also trapped at feeders but with mistnets. In both subplots young birds were also lured into nets by playback of calls and songs in summer. Since the emphasis of this study was on the social system, only few data were collected on the breeding biology.

In the winter of 1988–1989 I collected some extra data on foraging behaviour, home-range use and participation in mixed flocks, partly in plot T and partly in the adjacent plot B which contained mainly oaks (Fig. 23). In this study I followed a much more rigorous protocol for data collection, by visiting 10 pairs in a pre-determined order and recording observations for 10 minutes in a standardized way. When territories are approximately known it is fairly easy to locate a pair of Nuthatches and keep track of them, thanks to their constant vocalizations. Usually the best strategy is to start by searching for mixed-species flocks which they often join.

THE TERRITORY

A *territory* is best defined as the space that is defended by an individual animal against a specific class of other individuals, and through this act is

reserved for the exclusive use (or nearly so) of the defending individual
or owner (Emlen, 1957). By this definition a territory can be owned by
an individual, a pair or a group of birds. The *home-range* of an individual
(or pair, or group) is much more loosely defined as the area used by
that individual for all of its activities. If an animal regularly strays beyond
the area it defends, the home-range may be larger than the territory, and
home-ranges of neighbours may overlap even if their territories do not. I
found it generally unnecessary to discriminate between home-ranges and
territories, since each pair of Nuthatches typically keeps to its own exclusive area throughout the year. This is illustrated by the winter home-
ranges for 10 pairs in Fig. 25. There is good evidence that these exclusive
areas are generally maintained by aggressive defence, and not just by
neighbours avoiding each other. This evidence includes observations of
border conflicts, advertisement by calls and song, and aggressive
responses by owners to real or simulated intruders (see below in this
chapter). In some cases I observed owners in areas that were normally
used by another pair, but the behaviour of such intruders was recogniz-
ably different; for instance, if an intruder was trapped in another pair's
territory it did not call when released, which owners almost invariably did
(Matthysen and Dhondt, 1988). Intruders were somewhat more common
in summer when territories were more closely packed, and they were
usually young birds. Löhrl (1958) also claimed that females would tres-
pass more often than males, but my own observations on trespassing do
not support this (23 and 17 observations of males and females, respectively, over 5 years; Matthysen, 1988).

Winter territories as illustrated in Fig. 25 do not fill up the entire area. A
similar situation can be seen in published territory maps by Nilsson (1976,
winter and spring) and Wesołowski and Stawarczyk (1991, spring), where it
also seems as if the habitat is not saturated. In such cases territorial borders

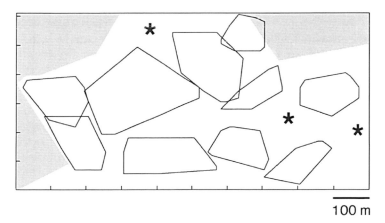

100 m

Fig. 25 *Home-ranges of 10 pairs in the Peerdsbos study area (November–December 1988),
based on 10 observations each lasting 10 minutes. Asterisks indicate additional unmapped
home-ranges. Shaded area = non-forest.*

may not be sharply defined. In two other studies, however, breeding territories were drawn with contiguous borders and virtually no empty space (Löhrl, 1958; Pravosudov, 1993a). The behavioural observations in these papers confirm that these habitats were completely covered by territories (see Chapter 7). In such a situation, especially if the territories are small, borders can be very sharply defined, and sometimes parts of a single tree can belong to different territories (Löhrl, 1958).

Territorial borders may be stable over several years even when individual owners change (Löhrl, 1958; Matthysen, 1990b). In the Peerdsbos study I was able to divide the study area into a number of permanent territories that at any moment could support a pair or remain unoccupied. It is not entirely clear why these territories are so stable, apart from mere tradition. Of course, some territories have little room for changes since they are surrounded by open areas or unsuitable forest on two or more sides. However, others seem to lack any natural boundaries. The most likely explanation is the mosaic nature of the study area with relatively small patches of variable quality, including lanes. The majority of territories were probably centred around one or a few patches of good quality which remained constant over time.

Territory size

Since territories do not overlap, there is an upper limit to their size which is determined by population density. At the highest known densities of 1 pair per ha (see Chapter 3) mean territory size must be equal to or smaller than 1 ha. At more typical densities (3–5 pairs per 10 ha) average sizes may reach 2 or 3 ha. One very small territory in Löhrl's (1958) study contained only 50 trees on less than 1 ha (27 oaks, 12 limes, 6 sweet chestnuts and 5 hornbeams). On the other hand, the largest reported territories were about 30 ha, in Siberia (Pravosudov, 1993a). However, territory sizes have rarely been measured. The size of a territory or home-range is typically measured as the polygon connecting the outermost observations of a single individual or pair. There are more refined techniques as well, which, for instance, estimate one or several core areas within the home-range that together include a certain proportion of all observations (Kenward, 1990). Figure 26 shows the 100%, 90% and 75% polygons for four Nuthatch pairs in winter. The 100% polygons vary in size from 2.8 to 4.7 ha. Of course, the estimated territory size increases as more observations are collected and the picture of the territory becomes more complete. This increase is shown in Fig. 27, which suggests that the asymptote is not yet reached by the 20th observation, and that with only 10 observations, at most 80% of the territory has been mapped. Thus, the polygons in Fig. 25 have a mean surface of 2.4 ha (range: 1.2–5.6 ha) but the true size is probably about 20% higher or close to 3 ha. Nilsson (1976) measured breeding territory sizes (excluding immature forest) of between 0.4 and 6.0 ha in Sweden, with median values of about 1.5 ha in oak and beech forest, and 2.5 ha in spruce. Wesolowski and Stawarczyk (1991) report means around 2.5 ha in ancient oak–hornbeam forest. Enoksson (1990a) found mean autumn territory sizes in

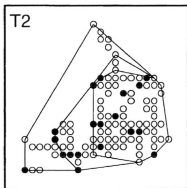
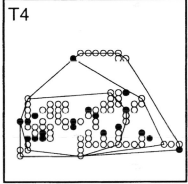
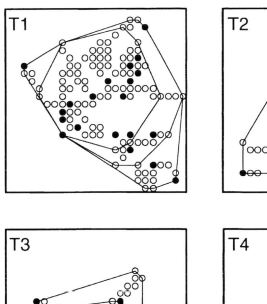
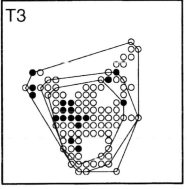

Fig. 26 *Home-ranges of four pairs (November 1988–February 1989), based on 20 observations per pair (cf. Fig. 25). Open dots indicate all grid cells (12.5 × 12.5 m) visited at least once during a 10-minute observation. Filled dots indicate the initial location for each observation. Solid lines are polygons encompassing 100%, 90% and 75% of all observations. Each panel represents a 300 × 300 m area.*

oak–birch forest of 1.5–3.4 ha. Thus, a two- to threefold variation in territory size within a population seems to be unexceptional.

Seasonal variation in territory size is not well documented. Territories probably increase somewhat in size from summer to late winter as some birds disappear and neighbours take over parts of their territories. In the Peerdsbos study this increase appeared to be limited, and territories remained very much the same throughout the year. Nevertheless, other studies suggest more important changes in size. For instance, Löhrl (1958) describes how young birds settle by encroaching on adults' territories in late summer, and in Sweden autumn territory sizes varied inversely with food availability (Enoksson and Nilsson, 1983; Enoksson, 1990a) (Chapter 7). In the course of the breeding season territory size is likely to decrease again as the pair concentrates its activity around the nest, but it is unclear

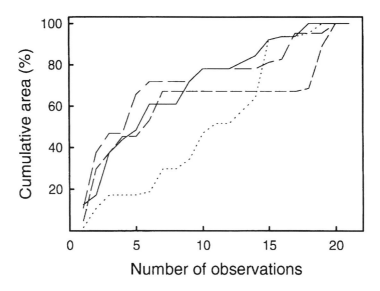

Fig. 27 *Increase in the estimated size of four winter home-ranges with the number of 10-minute observations.*

whether defence is also restricted to a smaller area. Nilsson (1976) found that territory sizes decreased by 20–30% from winter to breeding, and adults in Siberia reportedly used only 30–50% of their territory when feeding nestlings (Pravosudov, 1993a).

How the territory is used

Nuthatches are active birds which seem to use their territories rather intensively, frequently visiting all parts at relatively short intervals. There are few obvious 'outliers' in Fig. 26 that could be interpreted as casual excursions outside the normal home-range. Note that for these observations each territory was searched until the birds were found, so it seems unlikely that they would be biased towards more predictable locations near the centre of the territory. However, Lars Gabrielsen (pers. comm.) studied Nuthatches at a much lower density in mixed forest in Denmark with a lot of beech and various immature stands. With the aid of radiotracking he found that some individuals indeed travelled several hundred metres outside their territory. Moreno *et al.* (1981) suggested that male Nuthatches spent more time on the periphery of the territory than their mate, but this was based on observations of only one pair. Since I rarely observed males without their mates (1988–1989 data: two out of 89 observations on males) I cannot confirm this. Figure 28 illustrates the apparently haphazard way in which Nuthatches move through their territory in winter. In spring the activity is concentrated around the nest site. The pair in Fig. 28 used only 41% of its winter home-range in spring. From more detailed observations collected in the 1988–1989 winter I calculated an average speed of 162 metres per hour,

winter   spring

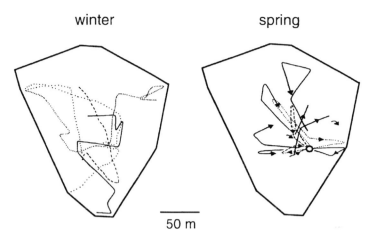

50 m

Fig. 28 *Movement patterns of a pair of Nuthatches in winter (November–December) and spring (March–April). Data for winter are based on four long (14–91 minutes) observation periods, spring observations on shorter (5–10 minutes) periods. The outer polygon is the home-range for all observations from November to April. The circle indicates the position of the nest. From Matthysen (1988).*

but this varied significantly between pairs (89–310 m per hour). Nuthatches moved equally fast when foraging on oak or on less favoured tree species but their movements were apparently slowed down when they joined a mixed-species flock travelling through their territory (149 *vs.* 213 m per hour when alone). A hypothetical explanation is that Nuthatches not in mixed-species flocks increase their speed of movement in order to locate another flock. Finally, speed of movement was slighly lower on the periphery which is the opposite of what one would expect if birds regularly 'patrol' the borders of the territory. In another study, Van de Casteele (1994) found much higher speeds (*c.* 500 m per hour), probably because he measured movements from tree to tree rather than a general trajectory.

Fighting and displays

Fighting behaviour and displays of the Nuthatch have been studied in detail by Löhrl (1958, 1988) and the following descriptions are largely based on his work. Löhrl distinguished two general kinds of fight. Border conflicts are frequent and may take up several hours per day in spring, but rarely result in changes in territorial status, and are generally limited to calling, displaying and short chases. These confrontations can last for quite a long time, up to half an hour, with both males and females participating. Real fights over ownership, in contrast, are shorter and therefore harder to observe, but also more intense. There is no evidence that fights are more intense near the centre of the territory (Löhrl, 1958). Nuthatches usually respond quickly to playback of calls or song by approaching, sometimes to a few metres, and calling or singing, and often wing-flicking as a sign of excitement. Both Schmidt (1979) and Van de Casteele (1994) found a habituation effect, meaning that birds responded less vigorously to the second or third playback trial.

Although territories are held throughout the year, intruders are not always fought to the same extent. In summer, the adult birds are quite tolerant of wandering juveniles, perhaps because they are moulting and cannot fight very well, or perhaps because food is so abundant that they can afford to tolerate intruders. Intolerance to intruders increases strongly in mid-August, and conflicts over territory boundaries are most severe in September–October and again in February–March (Löhrl, 1958). Nevertheless, both males and females reacted more intensely to live decoys in autumn than in spring (Matthysen, 1986b).

When Nuthatches hear an intruder, they often approach the site with a rapid zigzag flight, which Löhrl suggests is a ritualization derived from avoiding obstacles in flight. If intruders do not give way, or if an owner is confronted with a familiar opponent, he may resort to threat displays and/or displacement behaviour. The latter includes noisy behaviours such as pecking at twigs with dead leaves, scaling bark flakes, and throwing up leaves from the ground, within a few metres of the opponent. This behaviour probably serves as a mild threat display and may pass into normal foraging when fighting intensity decreases. There are two major threat displays which differ in intensity (Fig. 29). In the mildest form the tail is raised and spread, the bill pointed up, and the wings partly opened and

Fig. 29 *Threat displays of the Nuthatch. Based on drawings in Löhrl (1958) and Cramp and Perrins (1993).*

drooped. In the more aggressive posture, the body feathers are sleeked, the wings partly spread and the tail fanned. A high degree of excitement is often accompanied by frequent wing-beating. The chestnut flank feathers may be raised as well, mainly in the breeding season, but this behaviour is not particularly associated with the threat displays shown in Fig. 29 (see Löhrl, 1958 for more details).

Physical fights are not uncommon, both birds grappling with their feet, often falling to the ground while continuing to peck at one another. Injuries or deaths may not be infrequent, and homeless individuals – perhaps already weakened from their wanderings – may be especially vulnerable. For instance, Berndt and Dancker (1960) report on the finding of a ringed juvenile 14 km from its natal site, probably freshly killed by a territory owner, and Eriksson (1970) mentions two irruption birds being killed in Finland by already established birds. When an intruder is losing and is chased by his opponent, he often retreats into the upper crowns of the trees or escapes by flying over the tree tops. Nuthatches prospecting for vacant territories usually approach by flying over the tree tops, so that they can easily escape again when detected by the owner (Löhrl, 1958).

In the course of my own study I saw relatively few intense fights between Nuthatches, suggesting that either I missed the heaviest fighting, or that aggressiveness was much lower than in Löhrl's study. The latter possibility is consistent with the low frequency of territory take-over in my study (Chapter 7). During 5 years in the Peerdsbos I witnessed 227 aggressive encounters (or 0.16 per hour of observation), only five of which involved serious physical fighting. Aggressive encounters included 116 border conflicts, 24 cases of male–female aggression and 13 conflicts with an intruder. The majority of the remaining cases, particularly in summer, involved at least one unidentified individual.

The role of the sexes in territorial defence

Males are most active in territorial defence, but they may be joined by their females in conflicts with neighbours, and in such circumstances it is not easy to see who fights with whom. In at least some cases females attack male neighbours, for instance if a pair is attacked from different sides (Löhrl, 1958). Also, a widowed female is able to defend the territory against a neighbouring pair (Löhrl, 1958). My own observations suggest, however, that defence is to a large extent sex-specific (Matthysen, 1986b). This was shown by confronting Nuthatch pairs with male and female 'intruders' in cages at established feeding sites (Fig. 30). The owners' response varied from virtually ignoring the intruder, to vigorous displays and attempts at attack. Males as well as females responded most strongly to birds of the same sex, though male–male confrontations elicited more aggression than female–female confrontations. The seasonal difference in response level has been mentioned earlier but the explanation is not clear.

The sex-specificity of territorial behaviour is further illustrated by the existence of trios, when a single bird shares two different territories with a different mate in each (Fig. 31). Trios should not be confused with

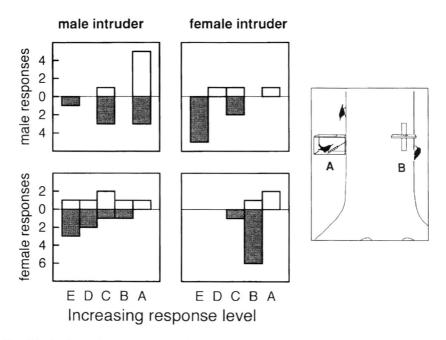

Fig. 30 *Confrontation experiments with a caged intruder (A) which is presented near a feeder (B) well-known to the local territorial pair. Response level increases from E (undisturbed foraging) to A (attack). White bars = autumn, grey bars = spring. From Matthysen (1986b).*

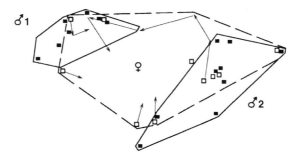

Fig. 31 *Home-ranges of a trio of Nuthatches. Filled squares and solid line polygons indicate observations and home-ranges of two males. Open squares and dashed polygon indicate the female's home-range. Arrows represent observed female movements. This particular trio lasted from approximately 5 December 1983 to 16 January 1984.*

territories inhabited by three birds, which may be called 'threesomes' (or pair plus satellite). In 3 years of observation I observed eight trios (six with a male and two with a female 'sharer') and suspected the existence of another four (Matthysen, 1986b). Trios lasted 2 weeks to more than 2 months, but all broke up before the nesting season. The 'shared' bird appeared to divide its time equally between its two partners. Trios usually originated when a bird started to use the territory of a widowed neighbour and was not expelled. They usually ended when one of the two partners disappeared (five cases) or mated with an immigrant bird (two cases). One trio broke up when the 'shared' bird returned to its original mate. Trios have also been observed in other studies, though not often (Löhrl, 1958; B. Enoksson, pers. comm.). It is not clear why trios are formed in some but not in other cases, and individual characteristics such as body size provide no clue (Matthysen, 1986b).

BEHAVIOUR OF TERRITORY OWNERS

Throughout the year, paired Nuthatches are usually found close together, often in the same or neighbouring trees. In 1988–1989, for instance, I found both members of a pair together in 93% of over 100 10-minute observations (on 10 different pairs). Limited observations on a single pair (Matthysen, 1988) suggest that the mean distance between mates decreases between November and April, but especially between November and December, which may coincide with the resurgence of singing as the first sign of the approaching breeding season. Pair members stay in contact by the constant use of low-intensity contact calls, in particular when one of them flies off (see below). (This behaviour is also of great help to the observer who is trying to find or follow the birds.) According to Löhrl (1958), individuals may even stop handling food and take it with them in pursuit of their mate. If partners do lose contact for a longer time they may call more loudly. Males may actively drive the female back into the territory if she approaches the border (Löhrl, 1958), and, a few times, I also saw a male chase the female away from an 'intruder' in a confrontation experiment, or even from a playback source. Apart from these observations, it is unknown whether males tend to follow females more often than the other way round.

Males are dominant over females outside the breeding season: females will wait for the male to leave before they approach a feeder, and are occasionally supplanted by the male when they have found a food item. During 17 hours of observation in the 1988–1989 winter I saw eight intra-pair conflicts or 0.5 per hour; this is probably an underestimate since the birds were not continuously in sight, although conflicts should have been particularly noticeable. At the time of nest-building the dominance relationship is reversed and females start dominating males, not only near the territory but also on feeders (Löhrl, 1958).

Time budget and activity patterns

At the end of winter (late January and February), Swedish Nuthatches spent about 90% of their time foraging, 6% resting, 2% in aggressive encounters, and the remaining few per cent flying, preening, bathing and (males only) singing (Enoksson, 1990b; Table 8). This corresponds well with my own estimates for November–December (1988–1989 data) with

TABLE 8: *Time budgets of Nuthatches (%) in Sweden, after Enoksson (1990b)*

| | Late winter | | Early spring | | Incubation |
	Males	Females	Males	Females	Males
Foraging	86.6	91.7	63.2	70.0	66.1
Resting	7.2	4.7	14.7	6.3	12.0
Nest-building	–	–	0.4	15.7	3.2
Singing	0.6	–	9.5	–	4.9
Preening	0.7	0.7	5.4	4.5	8.6
Aggression	3.2	1.5	2.7	1.0	0.8
Flying	1.3	1.2	1.5	1.1	2.4
Begging	–	–	–	0.2	–

92% foraging, 5% resting, 1% preening and 1% aggression. Van de Casteele (1994) found a slightly lower estimate for time spent foraging (85% for December to March). In spring, the foraging time is reduced to 60–70%, the rest being spent on resting and preening and, of course, nest-building in females and singing in males (Table 8). Rests often take the form of long pauses with one or both birds remaining motionless for several minutes without obvious reason, perched on a horizontal branch or clinging to a trunk. Such behaviour was noticed by Löhrl (1958) and has been reported for other nuthatches as well (Kilham, 1972a; Löhrl, 1988; Yaukey, 1997). In general, Swedish males spent less time foraging and more time resting and in aggressive encounters than females (Table 8; for more details see Enoksson, 1990b).

A possible reason why females forage more is that they have to compensate for a slightly lower foraging success because of their subordinate status (Enoksson, 1990b). As described in the previous chapter there is a tendency for females to forage on less preferred sites, and on top of that they have to be vigilant for the occasional theft of food by their own mates. In early spring there may be a more obvious reason as the female prepares for the considerable energetic cost of egg-laying. Curiously, the Nuthatch's time-budget seems to be unaffected by providing them with large amounts of sunflower seeds, nor is it related to temperature (Enoksson, 1990b). This may be related to their 'wise use' of stored food for times of great need. In addition, Enoksson's experiments were conducted in a relatively mild winter. It is possible that both food and temperature may have a greater effect on time-budgets in more extreme conditions.

The daily activity cycle of Nuthatches has never been studied in much detail. Löhrl (1958) provides some graphs of daily activity for specific behaviours (Fig. 32) which show a peak of activity right after sunrise and a gradual decrease throughout the day. The early morning peak in provisioning of the young also shows up on a similar graph in Polivanov (1981). The daily pattern of hoarding has already been discussed in the previous chapter. Eriksson (1970) found that irruption birds in Finland were observed particularly at bird tables shortly after sunrise and before sunset. Another reference to a possible 'pre-roosting' increase in foraging activity is Yaukey's (1997) observation that Brown-headed Nuthatches foraged

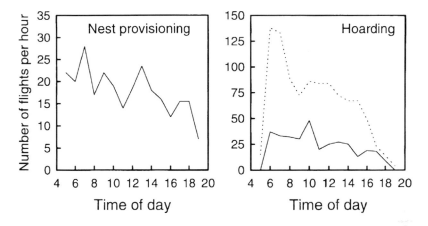

Fig. 32 *Daily pattern of nest provisioning and hoarding flights (solid line = spring, dotted line = autumn). Redrawn after Löhrl (1958).*

more on energetically rich seeds later in the day, as opposed to searching for prey on bark.

Participation in mixed flocks

Outside the breeding season Nuthatches are typically found in so-called mixed-species flocks, loose aggregations of a few to a few dozen birds which tend to move through the forest together. They follow the flock only as long as it stays within their territory. Usually various tit species make up the bulk of the flocks, plus some treecreepers, goldcrests and an occasional woodpecker. The composition of these flocks is quite similar all over Europe, if not over the entire range of the Eurasian Nuthatch (e.g. Morse, 1978; Herrera, 1979). In the 1988–1989 winter, I found that Nuthatches spent 80% of their time in mixed-species flocks, defined as two or more heterospecific birds within 25 m. Székely et al. (1989) found a somewhat lower value in Hungary (c. 60%) but used a more narrow definition of a mixed flock. In this study the Nuthatches spent less time in flocks if extra food was provided, but there was no such change in the presence of a predator (a Goshawk flown by a falconer). Since these observations were made on the same days as food was offered, the reduced flock attendance may have been mainly due to the Nuthatches' regular visits to the feeders. Berner and Grubb (1985) found a similar response to extra food by the White-breasted Nuthatch. In the Hungarian study the number of Nuthatches in mixed flocks increased markedly from September to January (Székely and Juhász, 1993).

Herrera (1979) compared the foraging behaviour of Nuthatches when they foraged alone and in mixed flocks. He concluded that the diversity of foraging sites did not differ, but that foraging success was about 60% higher in mixed flocks. Nuthatches probably benefit from the shared vigilance of other flock members, allowing them more time to forage. Carrascal and

Moreno (1992) likewise found that Nuthatches stayed longer on an artificial feeder if at least one other bird was present. Aside from the 'many eyes' effect, Nuthatches may also profit from flocking by copying the behaviour of successful flock members, or even by stealing food from subordinate species, but there are no data to substantiate this. Copying of foraging behaviour has been documented in the White-breasted Nuthatch by Waite and Grubb (1988b), in an experiment with captive birds which had to learn to locate hidden mealworms.

ROOSTING

Nuthatches almost invariably roost on their own, though in one case two birds were found sleeping together (Löhrl, 1988). They prefer to sleep in natural holes or nestboxes, but sometimes also behind loose bark or in the open. Captive birds prefer relatively small holes with wide entrances (Löhrl, 1958). Roosting in the open is probably common in summer, since many birds start roosting in nestboxes only in October or November (Radford, 1954; Busse and Olech, 1968). Some migrant Nuthatches in Finland roosted in chimneys, perhaps attracted by the warmth (Eriksson, 1970). Radford (1954, 1955) recorded the time of roosting of a female Nuthatch almost daily throughout two winters, and found that her timing corresponded closely to sunrise and sunset. In midwinter the bird left slightly before sunrise and returned slightly later as well, suggesting that she was forced to use the maximum possible daylength. Both Radford (1954) and Löhrl (1958) saw the male routinely accompanying the female to her roost site. Both sexes appear to roost in nestboxes equally frequently, and the number of birds roosting in boxes is not clearly related to habitat, altitude or weather (Winkel and Hudde, 1988). When the nest is finished, females regularly roost inside the nest cavity (Löhrl, 1988). N American White-breasted Nuthatches have a peculiar habit of taking out their faecal sacs after roosting, leaving the hole very clean (Kilham, 1971a). Curiously, Löhrl (1988) states that Eurasian Nuthatches never do this, but it has been observed repeatedly in a single bird by Radford (1954).

VOCALIZATIONS

Song

The loud but simple song of the Nuthatch is one of the most characteristic sounds in woodland in spring. The song can be heard throughout the day but is probably more frequent before noon. There is no particular song peak at dawn, and even in the egg-laying period the male is relatively silent when he waits for the female to leave the nest (Steinfatt, 1938). Song has never been recorded from females. Males do not seem to favour particular song-posts, but when the female is nest-building the male often sits singing only a few metres away from the nest site. He usually sits upright on a horizontal branch, often with the bill pointed upwards (which may be necessary to attain a high song volume; Löhrl, 1958). In Belgium, song activity starts at a low level in December and gradually increases to a peak in mid-April, which corresponds to the end of nest-building and the onset of laying (Fig. 33). Song activity then drops rapidly towards May–June, and from July to October it is heard only occasionally. Löhrl's (1958) description for south Germany is similar.

Nuthatches have five clearly distinct song types (Fig. 34). All are uniform series of loud notes with a simple structure in the same main frequency range of 2000–3000 Hz. Some of them can be easily imitated by whistling, and Löhrl (1958) used this routinely for provoking territorial males to approach and call back. All song types are more or less variable in tempo and the number of notes that are repeated. The majority of songs are relatively short and slow with three to six notes in a row, delivered at a rate of two to four per second. The slower song types can be easily characterized by the way the pitch changes within a note. Ascending Song sounds like 'whee-whee-whee', Descending like 'peeu-peeu-peeu' and Up-down has an audible up- and downward inflection ('wheeu-wheeu-wheeu'). The very rare song type Down-up (not shown in Fig. 34) has a down- and then an upward inflection. Ascending notes may have a short downward inflection at the end, but not as clear as in Up-down which appears as a symmetrical inverted 'U' on a sonagram (Fig. 34). Ascending Song is the most variable in tempo, with up to seven notes per second, and strophe lengths up to 10

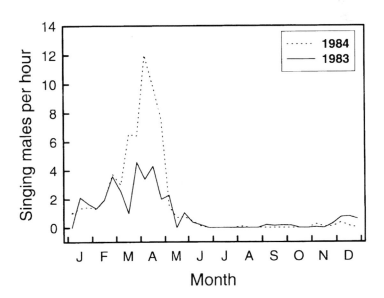

Fig. 33 *Seasonal pattern of song activity in the Peerdsbos study area, expressed as the average number of males heard singing per hour of observation in 10-day periods.*

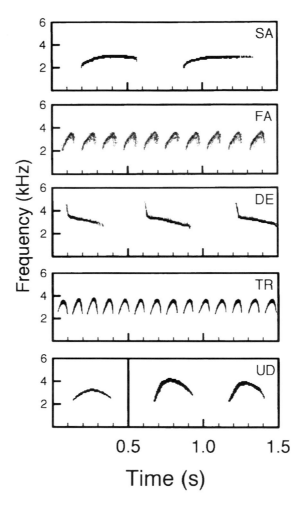

Fig. 34 *Sonagrams of the five main song types of the Nuthatch: (from top to bottom) Slow Ascending, Fast Ascending, Descending, Trill and Up-down song (two variants). Sources: recordings by author in Buggenhoutbos, March 1995, except for TR (taken over from Löhrl, 1988) and DE (recorded in Peerdsbos, April 1984).*

or more. I arbitrarily distinguish between Slow Ascending and Fast Ascending by having up to four, or more than four notes per second, respectively. Descending and Up-down Songs (and probably Down-up as well) are usually slower with about two notes per second, and rarely more than five notes in a row. The fifth common song type is a fast trill with about 10–15 notes per second and lasting up to 3 seconds. It can be distinguished on sonagrams from Fast Ascending by the shorter notes (maximally 9 *vs.* minimally 12 microseconds) and the symmetrical structure which resembles Up-down rather than Ascending (Schmidt, 1979; Fig. 34). Figure 35 shows a distinct (but uncommon) transition from Trilled to Fast Ascending song within a single strophe.

Löhrl (1967) described subsong-like 'juvenile song' (Jugendgesang) from hand-reared Nuthatches, consisting of highly variable, often quite unnuthatchlike sounds, with only minor parts that were reminiscent of true song. Löhrl heard this only occasionally from free-living birds, but captive males would 'sing' like this for several hours. Heinroth and Heinroth (1926) likewise heard an 'almost Blackbird-like song' from a captive first-

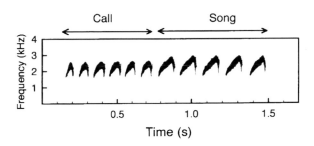

Fig. 35 *Sonagram of a transition between Trilled song and Fast Ascending song. From Schmidt, 1979.*

year male in August. In early spring, some males produce irregular variants of the normal song types, sounding as if they were practising. It seems probable that these are first-year males which have had little occasion to learn song types and/or to improve their singing ability.

Function of song

The function of song has not been properly investigated in nuthatches but there seems no close link to territorial defence, since the majority of territories are established in summer and autumn when very little song is heard. Song clearly takes over some of the territorial advertisement functions of the Excitement Call (see below), but may serve as sexual stimulation as well, with a peak in singing activity in the egg-laying period. The majority of pairs are formed before January when song is still infrequent, so the role in pair-formation must be limited; nevertheless, males that have lost their mate increase their singing rate markedly. An interesting but unanswered question is whether males that lose a partner in summer or early autumn also increase their calling activity.

The role of the different song types is also not clear. Löhrl (1958) and Schmidt (1979) stated that Trilled song was used more often in the reproductive period. However, in the Peerdsbos study (1982–1987) I found only minor seasonal changes in the use of song types, with a gradual increase in Descending Song from December (c. 10%) to May (over 20%) and a partial shift from Slow to Fast Ascending (Matthysen, 1988). In a more detailed

study, Schmidt (1979) found that song types did not follow one another in random order. If the birds were not obviously disturbed, most sequences of two or more song types followed the order Ascending – Up-down – Descending – Trilled song. If the birds were disturbed, song was usually followed by loud calling. Schmidt concluded that the Excitement Call represented the highest state of excitement, followed by the songs in the order cited above. He also demonstrated that males tend to respond to playback with the same call or song type (Fig. 36). I found the same result in playback experiments with Up-down only (Matthysen, 1997). Schmidt also found less vigorous responses to the supposedly higher intensity vocalizations (Excitement Call and Ascending song). He interpreted this as a more cautious response when the playback suggested a more highly excited intruder (Schmidt, 1979).

Geographical variation in song

In comparison to the well-documented geographical variation in morphology and plumages, there is surprisingly little information on variation in song. Löhrl (1967) suggested that Trilled song was more common in

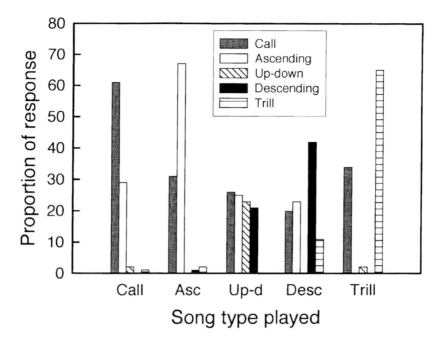

Fig. 36 *Song-matching in response to playback of different vocalizations. Data from Schmidt, 1979.*

some populations than others, but also pointed at the possible existence of annual and individual variation. The very rare song Down-up has been reported only from Morocco (Löhrl, 1982a, 1988), but it is not stated how many birds use it; I heard a very similar song just once in Belgium. Recordings from various parts of eastern Asia (Sakhalin, Amurland, China, Japan) contain versions of Trilled song, Slow Ascending, Up-down and Excitement Call that sound very similar to those of Belgian birds. According to Leonovich *et al.* (1996a), the far eastern *arctica* subspecies sings markedly different from other subspecies, but this is difficult to infer from the published sonagrams.

On a much smaller geographical scale, I discovered that Up-down song is virtually absent from a number of populations around Antwerp, while it is quite common a little further south and east (Matthysen, 1997). Near Antwerp this song type accounted for less than 1% of all songs, compared to 10–20% in other populations (Fig. 37). The exact borders of the area where the song is missing are unknown, but it covers at least 1000 km² with several hundreds of breeding pairs. Males in these populations are able to recognize Up-down as conspecific song, as shown by their responses to playback, but they are unable to reproduce it. A possible explanation is that the populations around Antwerp descend from a formerly smaller and

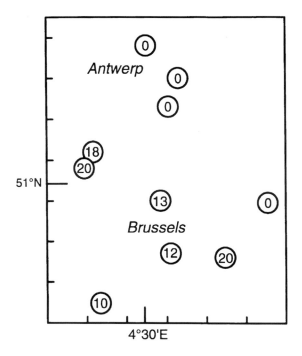

Fig. 37 *The occurrence and frequency of the song type Up-down in northern Belgium. Numbers inside circles represent the frequency of use (% of all songs heard) in 10 populations sampled in 1994 and 1995. Scale markings indicate 10 km.*

more strongly isolated population (for which there is historical evidence) where the song type was lost through a 'founder' effect. Since Nuthatches show a clear tendency to match songs, it seems probable that immigrants in an area without Up-down would gradually drop the song from their repertoire, or at least not transmit it to other individuals. In this way a relatively sharp borderline could be maintained. An intriguing question is whether there is a particular reason why Up-down has been lost and not another song type, for instance because it is a relatively complex note and therefore may be more difficult to learn.

Calls

The main call of the Nuthatch, also known as the Excitement Call (Cramp and Perrins, 1993) resembles song in that it is also a variable and loud repetition of simple identical notes. Figure 38 illustrates some of the variation in tempo and note structure. The note starts with a steeply descending part, then a more gently ascending part, and a short descending 'hook' at the end. The longer notes may therefore resemble ascending song notes, but are sonographically clearly distinct (Fig. 38). The main energy in the calls is delivered at c. 2000–2500 Hz which is slightly lower than song (2000–3500), and the presence of harmonics results in a less pure sound. The sound has been variously described as a metallic 'twit', 'dwip' or 'chwit', or as the sound of a pebble hitting ice (Walpole-Bond, 1931). The call can be produced as single notes, in short bi- or tri-syllabic units ('twit-it' or 'twit-it-it'), or in almost continuous series of variable tempo, with a maximum of about 10 notes per second. When the tempo drops, for instance at the end of a long calling bout, the notes may also become longer and drawn-out. In spring such slow calling may eventually change into Ascending Song.

Another characteristic call is the Conflict Call, a high-pitched 'sirrr' (somewhat reminiscent of the call notes of the Long-tailed Tit) (Fig. 38). This call can be uttered singly or repeated a number of times. It is given at times of great excitement, especially during or preceding a fight; Löhrl (1958) called this the 'Kampfbereitschaftslaut' or 'ready-to-fight' call. A very similar call, but somewhat softer, may be given in the presence of an aerial predator (Löhrl, 1958). The third important call is the Contact Call (Löhrl's 'Stimmfühlungslaut'). This is a high-pitched 'tsit', much like the contact calls of different tit species. It is used mainly by paired birds, and obviously serves to maintain contact (see above). Longer, more drawn-out versions of this call, often repeated a few times ('seet-seet-seet') are used in several contexts: by both partners early in the breeding season, as a

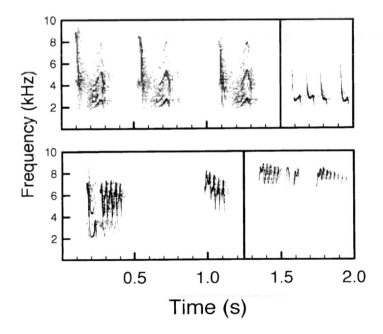

Fig. 38 *Sonagrams of two call types. Top = Excitement Call, bottom = Conflict Call (the first call preceded by a single Excitement Call note). Recordings by author in Peerdsbos, August 1983.*

begging-call by the female (during courtship-feeding) and by fledged young, and also by recently paired juvenile birds in summer. Löhrl (1958) mentions some other calls that are heard in specific contexts only. Copulation is often, but not always, accompanied by a repeated 'sië sië', not very loud and dying down with a downward inflection (although sonagrams in Glutz von Blotzheim (1993) do not clearly show this). Other various calls are associated with fighting or distress, for instance during handling, or with the feeding of nestlings, including a soft call apparently meant to stimulate begging. The begging calls of nestlings also change with age (Löhrl, 1958; Cramp and Perrins, 1993).

There are several independent reports on drumming by Nuthatches, sufficient to make it credible that this behaviour indeed occurs, albeit extremely rarely. Engel (1942) reported three observations of short drumming bouts (1–2 seconds), in two cases followed by song, of the same male. An editorial note to the same report mentions two other 'convincing' observations and one more is reported by Löhrl (1958).

Calling after release

Nuthatches have a peculiar habit in that they often start to call immediately when they are released from capture. They usually produce the Excitement Call, sometimes the Conflict Call or a combination of the two. When my attention was drawn to this behaviour it soon appeared that there was no relationship with age (as I first thought) but instead to territory ownership (Table 9). Owners called almost invariably when they were trapped and released inside their territory, rarely when they were trapped well outside the territory, and on about half the occasions when they were caught in the border zone of the territory. Interestingly, the latter category also showed many unclear responses (very short or weak calls, or after a long pause). Non-territory owners were usually silent when they were released (83%). I was able to confirm that owners did not call when I took them outside their territory, unless they were carried inside a bag or box, probably

TABLE 9: *Behaviour of Nuthatches after being released from capture (from Matthysen and Dhondt, 1988)*

Conditions of release	Number of releases	% calling	% unclear responses
BIRDS RELEASED AT THE SITE OF CAPTURE			
Owners in their own territory	445	93	2
Owners in border area	108	64	9
Owners outside their territory	31	13	0
Non-owners	90	17	9
BIRDS RELEASED ELSEWHERE			
Owners in their own territory	35	94	6
Owners in border area	36	47	11
Owners outside their territory	46	6	9
Outside territory, transported in a bag	12	67	0

because they were then unaware of where they were. Calling after release proved to be a quite reliable indicator of social status at the site of capture (Matthysen and Dhondt, 1988). The function of this behaviour, however, is still not clear. There is an obvious penalty to calling outside the territory, since this is likely to attract the local owner, but this does not explain why owners call inside the territory. It might serve to warn the mate of danger, but then Nuthatches rarely call when trapped or when handled, and solitary owners also call when released. Another possibility is that owners call in order to re-establish contact with their mate, and/or to advertise their presence after a brief period of 'absence'.

CHAPTER 6

Breeding Biology

Hij weifelt tussen het metselaars- en timmermansvak
('He hesitates between the professions of mason and carpenter')

(Thijsse, 1903)

Nuthatches belong to a large group of birds that prefer to build their nest in existing cavities in trees, the so-called secondary cavity-nesters. The obvious advantage of this habit is that it provides protection against competitors, predators and adverse weather. The major disadvantage is probably that the number of nest cavities may be limited, especially in our present-day managed forests where old decaying trees are often scarce. The habit of cavity-nesting offers to the observer both difficulties and opportunities for the study of breeding behaviour. Nests in cavities may be difficult to inspect,

86

but if a species is willing to breed in nestboxes this can be turned into a major advantage. This has made birds such as the Great Tit and the Pied Flycatcher very popular subjects for behavioural and ecological investigations. Nuthatches breed comparatively rarely in boxes, perhaps because they have taken cavity-nesting one step further by adjusting the entrance to their own size, thereby reducing both predation risk and competition. As a consequence, their breeding biology is not known in the same detail as that of other hole-nesters. The data in this chapter have therefore been compiled from a limited number of studies in natural cavities, supplemented with data from long-term nestbox programmes where a reasonable number of Nuthatch broods have accumulated over many years of study.

PRE-NESTING BEHAVIOUR

Like many other resident birds, Nuthatches start announcing the forthcoming reproductive season in midwinter by a resurgence of singing activity on mild days. This is not followed by pair-formation, however, since Nuthatch pairs remain together year-round, and new pairs can be formed at any time, particularly in early summer. As I discuss in Chapter 7, this is tightly linked with the acquisition of, and movements between, territories. Although early spring may bring about a higher level of territorial intrusions and aggression, the majority of pairs that survived the winter intact will start their reproductive activities without abrupt changes in behaviour. These activities start with the inspection of suitable nest sites on the first warm days of February or March by males as well as females. Some males, especially unmated ones, advertise the available nest sites to a female by repeatedly entering them and removing bits of material (Löhrl, 1988).

Very occasionally, a pair may not attempt to breed in a particular year. Three cases are mentioned by Nilsson (1976): one pair that stopped after walling up a nest hole, one where the male invited a female to a nest but no nest-building followed, and one pair that 'probably did not attempt to breed at all' (Nilsson, 1976). In a Siberian study two pairs also stopped nest-building although they remained on the territory (Pravosudov, 1993b). No such cases have been reported from other populations.

NEST SITES

Finding nest sites of Nuthatches is not too difficult in late March and early April, when females spend a lot of their time nest-building, often with their mate singing quite close by. Later nests, including replacements, are more difficult to find, particularly when trees come into leaf. In the egg-laying stage the birds only sporadically visit the nest site for quick inspections, while in the incubation and nestling stages the visits are more frequent but equally inconspicuous. Fortunately the birds often give themselves away by a series of contact calls before they return to the nest, often in a direct flight. This is particularly noticeable in females at the end of an incubation pause.

Inspecting a nest cavity in a large beech tree in the study area Meerdaalbos. (Photo: Frank Adriaensen).

Nuthatches never excavate their own nest, but occasionally a female may try to enlarge a small hole in decaying wood or a nestbox filled with a mixture of flour and sawdust (Löhrl, 1958, 1982b). In two unmanaged forests, about two-thirds of the nest sites were 'natural' holes, generated by cracks, broken branches or rotting (Prill, 1988; Wesolowski and Stawarczyk, 1991). Usually, however, the majority of nests are in old woodpecker holes (Löhrl, 1958; Polivanov, 1981; Nilsson, 1984; Pravosudov, 1993b; Matthysen, pers.obs.). The main source, throughout the Nuthatch's range, is undoubtedly the Great Spotted Woodpecker (also an important nest competitor and predator), but Nuthatches have been observed using cavities of at least six other woodpecker species ranging from the Lesser Spotted Woodpecker (hardly larger than a nuthatch itself) to the much larger Black Woodpecker.

The characteristics of natural nest sites have been studied in detail by van Balen *et al.* (1982), Nilsson (1984) and Wesolowski (1989). Nuthatches choose relatively high nests in comparison to other species, but mean nest height varies considerably between studies, from 3.5 m in England (Edington and Edington, 1972) and about 5 m in the Netherlands and Sweden (van Balen *et al.*, 1982; Nilsson, 1984) to more than 10 m in Germany, Belgium and Poland (Löhrl, 1987a; Matthysen, 1988; Wesolowski, 1989). In the Peerdsbos, nest heights varied between 5 and 22 m, most of them between 6 and 10 m. In Sweden, mean nest height decreased with population density, which may suggest a preference for higher nests (Nilsson, 1984). In the same study there was a higher predation level on lower nests (< 2 m), but high nests were more vulnerable to interference by Starlings. As a result, the nests at intermediate height (2–4 m) were the most successful ones. This could mean that the Nuthatch's preference for high nests has evolved in times when Starlings were much less abundant.

Nuthatches do not prefer particularly large cavities but the variation is considerable, including some very large holes that are rarely used by other small passerines (van Balen *et al.*, 1982; Nilsson, 1984). Twelve nest cavities examined in the Netherlands had diameters from 10 to 24 cm and bottom areas from 75 to 491 cm^2 (van Balen *et al.*, 1982). There is no obvious preference for small cavity entrances, but slitlike openings are avoided (Nilsson, 1984; Wesolowski, 1989).

In the Białowieza forest in Poland, a typical Nuthatch nest cavity would be situated in the trunk (79%) of a living tree (84%). Quite likely such a cavity would be in a bulge of a tree (41%) with a slight preference for an orientation to the south (Wesołowski, 1989). Löhrl (1958) explained the preference for bulges by the supposedly reduced accessibility of such sites for woodpeckers or Starlings. Nevertheless, the latter species also frequently occupies such nests (33%, Wesolowski, 1989), though not as much as the Nuthatch. The preference for south-oriented entrances probably reflects the woodpeckers' rather than the Nuthatch's choice and has not been found in other studies (e.g. Prill, 1988). Nuthatch nests are more often found in oaks than expected by chance (Matthysen, 1988; Prill, 1988) and, in England, oaks contain 40% of all nests reported to the British Trust for Ornithology (Osborne, 1982).

In the Peerdsbos study with a high density of Great Tit-type nestboxes (*c.* 10 per ha), only about one out of 10 Nuthatch pairs used these for breeding. In some German pine forests, on the other hand, two-thirds of the breeding pairs used boxes (Haupt, 1992). In Braunschweig about 2 pairs per 10 ha bred in boxes, which must be a considerable fraction of the breeding population (Winkel, 1989). It is unclear to what extent these differences between areas reflect preferences for different nestbox types, since most studies do not specify nestbox sizes or heights. Some experiments on preferences for nestbox types are summarized in Fig. 39. When given the choice, Nuthatches clearly preferred wider boxes, smaller entrances, and boxes higher in the tree, but they avoided very deep boxes. Boxes with entrances smaller than 30 mm are used only occasionally (Hertzer, 1972; Matthysen, pers. obs.).

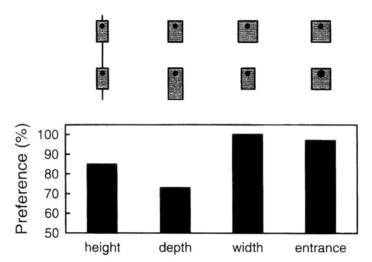

Fig. 39 *Preference of Nuthatches for different nestbox types, based on data in Löhrl (1966, 1977, 1987a). Bars express the preference for boxes that are higher above the ground (above/below 10 m), less deep below the entrance (14 vs. 19 cm), wider (20 vs. 11.5 or 14 cm) and have a smaller entrance (32 vs. 60 mm).*

Pairs often re-use the same nest site in successive years. In the Peerdsbos study this was 69% of pairs that bred in two successive years, and 76% of pairs consisting of one old bird and one first-year bird. However, even pairs consisting of two first-year birds usually nested in a site that had been used by a different pair the year before (75%) (Matthysen, 1988). These may have been guided in their choice by the remains of old nest material, but perhaps good sites were simply limited. Over a 5-year period the mean number of broods per cavity was 2.0 (Matthysen, unpubl.). One case has been reported of a cavity that was still used by Nuthatches 26 years after the first observed breeding (Glutz von Blotzheim, 1993). Pravosudov (1993b) found that the tendency to re-use old nests increased with age in males, but not females, for unknown reasons.

A large variety of 'unusual' nest sites has been recorded in this species. These include nestboxes for owls and Kestrels; letterboxes; holes in walls, rock cliffs and even once in a lighting pole 40 cm above ground; a stone kiln; nests in the ground between the roots of a tree or in a steep road verge; and old nests of Sand Martin and Magpie. The most remarkable nest was one built entirely of mud in the side of a haystack, weighing 5 kg (Witherby *et al.*, 1940).

NEST-BUILDING

The female starts her preparations for nest-building by cleaning the cavity, throwing out bits of old nest material and decayed wood. In this phase she may still switch between cavities. Proper nest-building usually starts in the second half of March, in southern Germany as well as Belgium. Early nest-builders are usually older females (Löhrl, 1958). From the beginning, the female alternates between walling up the nest entrance and bringing in nest material, but the latter gradually takes up more time as nest-building advances. The male takes little part, at most provisioning his mate with bits of material, but never entering (Venables, 1938; Löhrl, 1958; Matthysen, pers. obs.). However, after the beginning of egg-laying, his share in providing nest material increases and may surpass that of the female; later on the male may even participate in the plastering work when repair is needed (Steinfatt, 1938).

The plastering is done mainly with mud, but when this is in short supply (for instance in dry weather) dung of deer, fowl or domestic animals may be used as well. Occasionally the mud may be mixed with little pieces of wood, twigs or dry leaves (Venables, 1938; Löhrl, 1958). Bussmann (1946) suspected one female had mixed in a resinous substance collected from flower buds of a pear tree. This is not implausible since some related species also mix resin with mud (see Table 1 in Chapter 1). The total amount of mud may weigh up to 1 or 2 kg if large openings have to be closed, such as an entrance to a Tawny Owl nestbox 13 cm in diameter (Delmée *et al.*, 1979). On the other hand, some small holes may have no visible mud from the outside, though it may be present inside (Neumann, 1961).

The mud is collected by the female in small pellets, carried to the nest and put in place with short thrusts of her bill. She then repeatedly pokes into it until it becomes a thin mud layer. The bill marks often remain visible for a long time. The mud is not smeared out as some authors have suggested, neither is saliva used (Glutz von Blotzheim, 1993). The material hardens in a few days to a substance that, at least for human hands, is difficult to remove without tools. The work usually starts on the inside upper rim of the entrance tunnel, invisible to the outside, and continues until the female can barely pass through. The floor of the finished entrance is often the bottom of the original opening, probably because any mud applied to the bottom is continually eroded as the female enters and leaves. Occasionally a nest will have two entrances. Often the Nuthatches will plaster some crevices inside the cavity as well, or even outside the cavity, for instance between the back of a nestbox and the tree trunk. There are also two intriguing observations on cavities that were entirely sealed with mud but otherwise nearly empty (Utley, 1944; Weller, 1949).

The foundation of the nest consists of pieces of rotten wood or bark, often several centimetres long, filling up the bottom of large cavities. On top of this comes a layer of lining material, typically bark flakes of Scots pine or other trees. In several cases, the birds are known to have travelled several hundred metres to obtain this material (Bahr, 1907; Henze, 1940; Löhrl, 1958). Other, but less favoured, materials are dry leaves, lichens, conifer needles and flower buds of beech (Venables, 1938; Henze, 1940; Löhrl, 1958; Pravosudov, 1993b). One nest in a large cavity contained no fewer than 11 440 bark fragments and 850 fragments of wood (Olsson, 1957), and another in a Tawny Owl nestbox had a layer of material 30 cm

A nestbox with a half-open front, narrowed with mud.

deep (Delmée *et al.*, 1979). The nest material is placed in more or less concentric rows with a nest cup in the middle, but this arrangement is easily disturbed, especially when the nestlings start moving around. Typically, the top of the nest almost reaches the level of the entrance tunnel, such that the eggs or nestlings are directly visible through the entrance. This greatly facilitates nest inspection as well as removal of nestlings for ringing.

It has often been suggested that Nuthatches prefer pine bark because of its anti-parasitic properties. In Sweden, Nuthatch nests contained fewer fleas than Great Tit nests in similar nestboxes and habitat, and more fleas were found in a nest built of leaves instead of pine bark (B. Enoksson, pers. comm.). The loose nest material may also provide fewer opportunities for ectoparasites to hide. Many other nuthatches use similar woody materials such as pine seed wings (e.g. Chapter 14). Clark and Mason (1985) suggested that whereas most cavity-nesters use fresh plant material for their volatile anti-parasitic compounds, N American Red- and White-breasted Nuthatches use resin or crushed insects for the same purpose. Pine bark may have the same function as well.

Nest-building typically takes a few weeks, but may last more than a month if birds start early. Before and during egg-laying, females in Sweden spent 10–20% of their time nest-building (Enoksson, 1990b). Observations by Löhrl (1958) suggest that plastering activity is greatest in the early afternoon (12:00–15:00 h) while nest material is collected mainly in the late morning. The plastering is often discontinued during dry weather but resumed immediately after rain (Löhrl, 1958), and may then be finished within a few days. Additional plastering occurs throughout the nesting period, even up to the day of fledging (Glutz von Blotzheim, 1993).

Competition for nest sites

Apart from keeping out predators (see later in this chapter), a major function of plastering is undoubtedly to prevent larger nest competitors from taking over the cavity. In forests where agricultural land is nearby, the Starling is usually the most common competitor. Their preferred nest cavities are quite similar in dimensions and height above the ground to those used by Nuthatches, more than any other species (van Balen *et al.*, 1982; Nilsson, 1984; Wesolowski, 1989). Löhrl (1956) described the interactions between the two competing species in lively detail. When both claim the same cavity, Starlings may remove the nest material and the freshly applied mud each morning while the Nuthatches continue to build throughout the day. The dispute is decided when either the Nuthatches manage to make the entrance too small for the Starlings to enter, or when the Starlings start to bring nest material of their own. However, Starlings may still take over cavities when the entrance is completely narrowed (14% of completed nests; Nilsson, 1984). Another nest usurper that is also an important predator is the Great Spotted Woodpecker (Löhrl, 1958; Matthysen and Adriaensen, unpubl.). This species may force its way through a plastered entrance or even dig a new entrance hole. On the other hand, Ludwig (1978) described a case where a Nuthatch pair was successful in driving a woodpecker out of its own nest. There are a few reports of competition with other species as well. In Germany's Black Forest, Nuthatches are the main nest competitors for Pygmy Owls (König *et al.*, 1995). Schandy (1981) saw a pair of Green Woodpeckers destroying a Nuthatch nest and removing the plastering, but then moving on to another cavity. Tree Sparrows and even House Sparrows may take over a nest even after completion of the wall (Löhrl, 1958, 1988). Schönfeld and Brauer (1972) reported two Pied Flycatcher nests built on top of a Nuthatch clutch, but gave no information about what had happened to the Nuthatches themselves. Several species of dormouse, and even bumblebees, are potential nest competitors as well (Löhrl, 1958).

Some birds pay heavily for interfering with the Nuthatch's nest choice. Wydler (1973) found a dead Tree Sparrow in a nestbox of which the opening was completely sealed with mud, and there are two independent observations of adult female Tengmalm's Owls being trapped with their nestlings inside a walled-up cavity (König, 1968; Diemer, 1985). On the other hand, there are several records of peaceful cohabitation with other birds in the same nest tree, including two reports of Nuthatches, Great Spotted Woodpeckers and Starlings within a single tree (Löhrl, 1958; Gogel, 1973) and one case where woodpeckers and Nuthatches were breeding 'back to back' on different sides of a tree (Prill, 1988). One male Nuthatch has been observed provisioning a Starling nest in the same tree (Vieweg, 1989).

There are several reports of Nuthatch broods containing one or more young of other species, possibly because they built on top of their nests without removing the eggs. For instance, one pair observed by Brauer and Kiesewetter (1983) took over a Great Tit nest with six eggs, added five eggs of its own, and succeeded in raising most of the young, although the Great Tits fledged somewhat earlier. Similar cases have been reported of nests containing Great, Blue, Coal and Willow Tit nestlings (Arn, 1955; Chappuis, 1970; Hellebrekers, 1970; Löhrl, 1988; Matthysen, pers. obs.).

Nest defence

Nuthatches can be quite aggressive at nest sites and physically attack enemies as large as Great Spotted Woodpeckers (Löhrl, 1958, 1988), or even a human observer (Bussman, 1943). When humans approach, the birds may start pecking at the trunk above the nest causing pieces of bark or moss to fall down (Bussman, 1943; Löhrl, 1958; Matthysen, pers. obs.). This may be merely displacement behaviour, but might also help to distract real enemies. Specific aggressive displays seem to be rare except for a few anecdotal reports. One bird, in reaction to a stuffed woodpecker, adopted a head-down posture with wings and tail spread, and side-to-side Pendulum movement of the head (Löhrl, 1988; see also Fig. D in Cramp and Perrins, 1993). A similar display is found much more commonly in the White-breasted Nuthatch (Chapter 14). There are two reports of possible injury feigning with birds running on the ground with wings spread and drooping (Bentham, 1946; Took, 1946). In contrast to many tit species, females disturbed on the nest have no specific display.

COURTSHIP AND COPULATION

Nuthatch copulations are not seen very often. They usually occur within 50 m of the nest and may be initiated by females or males (Steinfatt, 1938; Löhrl, 1958; Cramp and Perrins, 1993). According to Löhrl (in Glutz von Blotzheim, 1993), the female responds to the male's solicitations only a few days before egg-laying starts. The female solicits copulations by crouching, shivering with her wings and pivoting the head from side to side (the so-called 'Pendulum' display). The male then adopts a striking upright posture, the head stretched obliquely upwards, and making the same Pendulum movement but swaying more heavily, while also shivering his wings. He typically sits in front of the female, sometimes behind her, but always facing away. During this display his chestnut flanks are clearly visible, and both sexes utter the copulation call (see previous chapter). In a slightly different account, both sexes sit upright (like 'small penguins') and circle around each other (Enoksson, cited in Cramp and Perrins, 1993). The display may last for up to half a minute, until the male turns, the bill still pointed upward, and mounts the female (Löhrl, 1958). Yates (1983) describes two observations on a single pair where a Pendulum-like display was not followed by copulation but by 'body-brushing', i.e. the male circling

Pendulum display.

a few times around the female in close body contact. In other instances normal mating was observed. Mating is usually repeated two or three times (Löhrl, 1958) with additional displays between each copulation. Cuypers (1944) described a sequence of five copulations in a row with pauses of about 30 seconds. During courtship the male sometimes pursues the female before copulating, and this may be accompanied by loud cries, zigzag flight and also a peculiar gliding flight (Durango and Durango, 1942; De Vries, 1977). The gliding flight may also be observed later in the breeding season (Durango and Durango, 1942). Little is known about copulations outside the pair bond. Lars Gabrielsen (pers. comm.) observed an extra-pair copulation in Denmark, and I once saw a male intruder displaying to a female near her nest site. Some males which Löhrl (1958) observed near a neighbouring nest and interpreted as 'provoking' their neighbour, may have been attempting to gain access to another female.

The Pendulum display seems to be widespread in the nuthatch family. Löhrl (1988) observed it in all species where he studied copulation behaviour, except for the Eastern Rock Nuthatch. He suggested that it might have evolved as a ritualization of the bird's zigzag foraging motions when swinging the head from side to side to inspect different sides of a branch. The Eurasian Nuthatch has the most highly ritualized display, which may be associated with its higher level of aggressiveness (Löhrl, 1988). Pendulum movements have also been observed in young pairs in late summer (Löhrl, 1958).

EGG-LAYING

The female lays her eggs in the early morning before leaving the nest, normally one a day. Laying interruptions of one or two days are rarely recorded (Schmidt and Hamann, 1983). Courtship-feeding usually starts only a few days before egg-laying, and only if the female begs (Glutz von Blotzheim, 1993). Begging increases towards the end of the laying period, and some of the females observed in this period may appear in unusually poor condition, to the point of having difficulty with flying (Löhrl, 1958). The clutch is covered with nest material from the first egg onwards, and the female continues to do so throughout laying and incubation. This habit is unusual in comparison with other songbirds; tits, for instance, cover the eggs during egg-laying but not incubation. Haftorn and Slagsvold (1995) suggested that, for Nuthatches, the cost of rearranging the nest material is much lower than in other birds because it is more loosely packed. The most likely function is to make the eggs invisible to predators. Until the clutch is completed and incubation starts, the female sits on the edge of the nest cup while roosting (Löhrl, 1958). The eggs are white with reddish speckles, the latter usually concentrated at the broad end, and are similar to the eggs of tits. Mean egg weights vary in parallel with body size, from about 1.8 g in *asiatica* to 2.0 g in *caesia* and 2.2 g in *europaea* (Cramp and Perrins, 1993; see also Makatsch, 1976). Egg length varies from 16.5 to 23.1 mm and egg width from 12.7 to 16.5 mm (Cramp and Perrins, 1993).

First-egg dates

In central and western Europe, Nuthatches typically start laying during the second half of April, with little variation in mean first-egg dates from southern France (Mont Ventoux) to Sweden (Table 10). In Germany and Belgium they lay slightly earlier than Great and Blue Tits (Table 11), a difference much smaller than the 5–10 days reported from Germany by Kiziroglu (1984). Slightly earlier laying dates have been reported, without details, from south-west Germany (first half of April; Löhrl, 1967) and southern France (end of March to mid-April; Géroudet, 1963). In Iran, laying is probably concentrated in March, since fledged young have been recorded as early as April in the south and mid-May in the north, several weeks before most other populations (Sarudny and Härms, 1923). Except for Sweden, northern populations appear to breed later, with laying dates from 15 April to 9 May in the Moscow region (Ptushenko and Inozemtsev, 1968; Cramp and Perrins, 1993), from 20 April to 10 May in Norway (Haftorn, 1971), 24 to 28 April in Ussuriland, south-eastern Siberia (Polivanov, 1981), 20 April to 10 May in Japan (Haneda and Rokugawa, 1972) and around 20 May in far eastern Siberia, a full month later than in central Europe (Pravosudov, 1993b). In western Siberia eggs are also laid in May, probably late May in the mountains (Johansen, 1944). In the Moroccan mountains the breeding season is late as well, with laying dates from the end of April to the end of May (Heim de

TABLE 10: *Breeding parameters of Nuthatches in different study areas*

Country	Source	Date of 1st egg	Clutch size	Breeding success (%)	Fledglings per brood	Fledglings per clutch	Number of nests
IN NESTBOXES							
Belgium	Matthysen, unpubl.	17 Apr	7.2	86	5.8	5.0	22
France	J. Blondel, pers. comm.	18 Apr	6.0	76	5.0	3.8	29
Germany	Henze, 1940	26 Apr	7.7	–	6.0	–	27
Germany	Löhrl, 1967	–	7.1	~93	~5.9	5.5	80
Germany	*Löhrl, 1987c	–	6.5–7.8	96	5.2	5.0	327
Germany	†Zang, 1988	23 Apr	6.8	88	5.4	4.8	95
Germany	Schmidt et al., 1992	19 Apr	6.5	77	5.1	3.9	400
Germany	Haupt, 1992	16 Apr	7.6	88	6.2	5.5	137
Poland	Sikora, 1975b	–	7.7	77	6.5	5.1	66
Slovakia	A. Krištín, pers. comm.	15 Apr	7.5	90	6.3	5.7	42
UK	†BTO nest cards	–	5.9	91	–	–	–
UK	Cramp and Perrins, 1993	–	7.6	73	6.6	5.5	25
(MAINLY) IN NATURAL CAVITIES							
Belgium	Matthysen, 1989b	c. 20 Apr	–	#91	–	–	78
Belgium	Matthysen and Adriaensen, unpubl.	10 Apr	–	#75	6.5	4.9	159
Germany	Löhrl, 1967	–	–	83	–	–	–
Poland	Wesołowski and Stawarczyk, 1991	c. 15 Apr	7.1	#65	^4.4	^3.1	174
Siberia	Pravosudov, 1993b	20 May	6.5	76	5.1	3.7	37
Sweden	Nilsson, 1987	29 Apr	–	72	5.1	3.7	119
Sweden	†Enoksson, 1993	20 Apr	–	80	5.3	4.3	177
UK	BTO nest cards	–	–	79	–	–	–

NOTES:

~ estimate based on Löhrl (1957) cited by Cramp and Perrins (1993).

* see Fig. 42 for variation with habitat and nestbox type.

† based on hatching dates, assuming 16 days incubation and a mean clutch size of seven eggs.

proportion of successful pairs.

^ based on 17 nests in one year only.

1. Note that sample sizes are usually smaller for brood size than those given in the last column.

2. Habitat is mainly deciduous forest except for Henze (mixed) and Haupt and Pravosudov (conifers).

3. Breeding success = proportion of clutches from which at least one young fledged, except where otherwise indicated.

TABLE 11: *Mean first-egg dates of Nuthatches and several tit species in two nestbox studies: Frankfurt, Germany (Schmidt, 1984; Schmidt et al., 1992) and Peerdsbos, Belgium (Matthysen, unpubl.)*

	Frankfurt	Peerdsbos
Nuthatch	19 April	17 April
Blue Tit	24 April	19 April
Coal Tit	21 April	–
Great Tit	20 April	23 April
Marsh Tit	17 April	–

Balsac and Mayaud, 1962). It should be remembered that some of the apparent variation between populations – especially from short-term studies – may be due to annual variation. Near Antwerp, for instance, mean laying dates were about a week earlier in 1991–1994 than in 1983–1987, perhaps because the latter years were characterized by colder winters.

Within populations, first-egg dates may differ by up to 3 weeks (Matthysen, 1989b). In exceptional cases a cold spell may arrest egg-laying and result in a bimodal pattern over time (Wesołowski and Stawarczyk, 1991). Some exceptionally early broods have been reported where the first eggs must have been laid in early February and young birds fledged in late March (Boiteau, 1991; Steinparz, 1954 cited by Löhrl, 1958). Several cases of broods fledging in the second half of April were reported by Löhrl (1958) and Géroudet (1991).

Several studies have revealed differences in laying date between forest types, but no general picture has emerged. In the largest analysis to date (400 broods observed over 16 years; Schmidt *et al.*, 1992) the earliest nests were found in oak-dominated forest, followed by conifers, oak–beech, mixed (coniferous and deciduous) and beech-dominated forest. On a smaller scale, Nilsson (1976) and Matthysen (1989b) found earlier laying dates in territories dominated by conifers than those with oak and/or beech, though Nilsson provided no statistical comparison. The most important conifer species were spruce and Scots pine, respectively. Yet another study, in the German Harz mountains, found that clutches in spruce forest were about a week later than in beech-dominated forest at the same altitude (Zang, 1988). Within the same beech forests, laying was delayed by about 1 day per 100 m altitude. Pravosudov (1993b) found no significant difference in laying date between upland and riparian forests in Siberia, both of which were dominated (but to a different degree) by larch. Finally, laying dates did not differ between oak forest and parks in Belgium (Matthysen and Adriaensen, unpubl.).

Annual variation in laying is clearly related to spring weather, in particular to mean March/April temperatures (Matthysen, 1989b; Schmidt *et al.*, 1992; Enoksson, 1993). In Sweden, for example, laying advanced three days per degree Celsius in April (Enoksson, 1993). The difference between years may amount to 2 weeks (e.g. Nilsson, 1976; Wesołowski and Stawarczyk, 1991; Enoksson, 1993; Pravosudov, 1993b). Schmidt *et al.* (1992) found that laying dates did not vary in parallel between years in different habitats, but were unable to identify the causes.

Younger females tend to lay later than older females (3–4 days in Enoksson, 1993), a pattern that is found in several studies (Matthysen, 1989b; Schmidt *et al.*, 1992; Pravosudov, 1993b). Except for Schmidt *et al.*, none of these studies found an effect of male age. However, the pattern may be more complex than a simple advancement of laying as females grow older. When laying dates of the same individuals were compared in different years, no age difference was found (Matthysen, 1989b; Schmidt *et al.*, 1992). This could be explained if early-laying birds also have a higher chance of survival and are therefore overrepresented in older age classes. One reason may be that early breeders live in better territories. In the Peerdsbos study there were indeed consistent differences in laying dates between territories even when the owners changed, but the relationship with territory quality was not obvious (Matthysen, 1989b; see also Chapter 7). Schmidt *et al.* (1992) found that age-related differences in laying were most pronounced in the most heterogeneous habitat types, which again suggests a relationship between the age effect and variation in territory quality. There is also some individual variation in laying date that cannot be explained by age and/or territory quality. In the Peerdsbos study, individual pairs tended to start laying at the same time in different years, and this effect diminished if either of the partners was replaced, even though the territory remained the same, suggesting that both males and females influence the timing of egg-laying (Matthysen, 1989b).

Clutch size

A complete clutch usually contains five to nine eggs with a maximum of 12 (Zang, 1980), and population means lie between six and eight (Table 10; more data on clutch sizes can be found in Glutz von Blotzheim, 1993). There is no clear geographical pattern (Table 10), possibly because of confounding effects of habitat, year and cavity type. For instance, Löhrl (1987b) found that clutches in enlarged boxes (diameter 20 cm) contained 0.6–0.9 eggs more than standard ones (11.5 cm), while clutches in deeper boxes (14 instead of 11 cm below the entrance) also contained 0.8 eggs more. However, Pravosudov (1995) reported no effect of the size of natural cavities on clutch size or on fledging success. Other factors that contribute to variation in clutch size are discussed in the section on breeding success (see below).

INCUBATION

Females usually incubate for 13–18 days with mean values of 15–16 (Henze, 1940; Glutz von Blotzheim, 1993). It is not clear to what extent some extreme observations such as those by Ptushenko and Inozemtsev (1968; 13–14 days in Russia) or Haneda and Rokugawa (1972; 18 days in Japan) represent geographical differences or merely anecdotal variation. Incubation is usually started after the last egg is laid, but occasionally a few days before or after the last egg, especially in spells of adverse weather when the delay may be up to 8 days (Löhrl, 1958, 1967). Löhrl (1958) was able to infer the start of incubation from slight changes in the behaviour of the female, who already spends long periods in the nest before incubating: she starts to use the typical juvenile begging call rather than the contact call, preens more frequently, and often uses a gliding flight when moving between trees.

An incubating female spends two-thirds to three-quarters of her time on the nest, alternating 20–30 minute incubation bouts with pauses of about 10 minutes (Steinfatt, 1938; Löhrl, 1958; V. Pravosudov, pers. comm.). In Siberia the periods on and off the nest alternated more rapidly as incubation progressed (V. Pravosudov, pers. comm.). When she returns to the nest, the female lowers herself on the clutch and turns about in half-circles until the eggs are free from nest material. She continues to change position in this manner every few minutes, sometimes reordering the nest material as well. She sleeps for very brief periods of 1 or 2 minutes, especially after long incubation pauses, and also preens regularly (Löhrl, 1958). Males feed their mate, without entering the nest, between 5 and 20 times per day throughout the incubation period (V. Pravosudov, pers. comm.). The female often follows her partner and begs for food as well as foraging on her own.

THE NESTLING STAGE

Generally all eggs hatch within 24 hours, occasionally on different days. After the young have hatched, the female shatters the remains of the eggshells and swallows them. In the first few days she still spends one-third of her time brooding the young, sometimes even returning to the nest without food, but brooding decreases rapidly to only a few minutes per hour at the end of the nestling stage. Throughout the nestling period she spends more time in the nest than the male (Löhrl, 1958; V. Pravosudov, pers. comm.). Up to day 15 after hatching, the male regularly passes food to the female when she is present, rather than entering. The female may roost with the young for up to 16 days after hatching (Steinfatt, 1938). The young stay in the nest for about 20–26 days, on average 24.4 days in central Europe (Glutz von Blotzheim, 1993). Extremes of 19 and 29 days are mentioned by Henze (1940) and Haftorn (1971). In Siberia the nestling time seems to be appreciably shorter with a mean of 20 days (range 18–22; Pravosudov, 1993b).

Nestlings appear to be fed with whatever food the parents are able to find (see Chapter 4), and large prey items are not necessarily broken up to make them easier to swallow. In the White-breasted Nuthatch, parents have been seen feeding the young with large numbers of dragonflies,

A c. 10-day-old nestling with colour-rings.

without removing the wings (Ingold, 1977a). The unusual amount of saliva in the nestlings' bills may be an adaptation to swallowing large prey (Löhrl, 1958). There is no obvious relationship between nestling age and prey size (Kaczmarek *et al.*, 1981; Krištín, 1992a). The rate of provisioning, to which both parents contribute, increases with nestling age from about 10 feeds per hour in the earliest days to about 20 in the last week. The number of feeds per hour in three different studies is comparable (Fig. 40). In Siberia the males fed the young on average 10.1 times per hour, the females 6.7 (V. Pravosudov, pers. comm.). The provisioning rate decreases gradually throughout the day (see Fig. 32 in the previous chapter). Usually one prey item is delivered per feed, but sometimes parents bring 2 or 3 items at the same time (Steinfatt, 1938). Sikora (1975a) estimated that a single brood received about 38 500 insects before fledging. Typical prey sizes are in the order of 50–100 mg (Pravosudov *et al.*, 1996).

Up to the fourth day the parents consume the faecal sacs of the young. Later they take out the larger sacs (on approximately one out of six visits; Löhrl, 1958) and soiled nest material as well (Enoksson, 1993). They continue to do so until the very last days before fledging (Löhrl, 1958). Faecal sacs are often smeared on branches rather than dropped at random (De Vries, 1977), a habit that is observed in other nuthatches as well (e.g. Krueper's, Corsican; Löhrl, 1988).

Nestling development

Young Nuthatches develop more slowly than other passerines of comparable size (Bussmann, 1943; Löhrl, 1958; Winkel, 1970). The first traces of feathers appear only on the third day (Table 12; see Glutz von

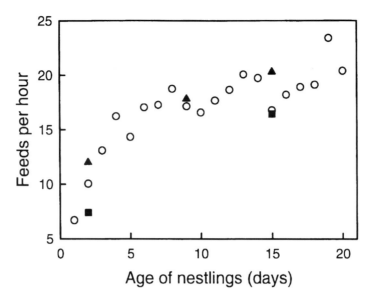

Fig. 40 *Provisioning rates (of both parents combined) of nestlings in Siberia (open symbols) and Germany (filled symbols). Data from V. Pravosudov (pers. comm.), Steinfatt (1938, squares) and Löhrl (1958, triangles). Siberian data points are means over eight broods, German data represent full-day observations of a single brood each.*

TABLE 12: *Plumage development with nestling age, according to Winkel (1970)*

Age	Developmental stage
Day 3	First feathers appear as dark spots
Day 6–7	Primaries extend over the margin of the wing
Day 9	First feathers open
Day 11	Appearance of primaries and primary coverts
Day 11–12	Appearance of secondaries
Day 12	Appearance of secondary coverts
Day 13–14	Appearance of tail-feathers
Day 17–18	Greater secondary coverts reach secondaries
Day 19–20	Greater primary coverts reach primaries
Day 20	Wing completely feathered

Blotzheim, 1985, for a description of newly hatched young). On the sixth day the young are able to lift their cloaca in the direction of the parent when they produce a faecal sac. The eyes open on day 9, and around day 10 the calls become louder and acquire a rolling sound. The first feathers emerge on day 9, the first flight feathers only on day 11. At 14 days of age the primary feathers have opened from their sheaths and the feathers on the back are partially opened. At this age preening behaviour starts and the weight starts to approach its asymptote (Matthysen, unpubl.) which is, on average, 24.1 g in Belgium (Matthysen & Adriaensen, unpubl.) and around 20 g in Siberia (V. Pravosudov, pers. comm.). Around day 20 the typical nest-begging call is replaced by another that is more similar to the Contact Call (see Chapter 5). At day 20 the wings appear fully feathered. Some pictures of young Nuthatches of different ages are shown in Heinroth and Heinroth (1926).

BREEDING SUCCESS

Basic statistics on breeding success in different studies are given in Table 10. The proportion of successful broods, i.e. those that produce at least one fledgling, is generally high, in several cases over 90%. The mean number of fledglings per successful brood is rather similar in most studies, usually between five and six (Table 10), which means that, on average, even successful broods lose one or two eggs. As a consequence, a high success rate is not necessarily accompanied by large broods (correlation coefficient $r = 0.27$, $N = 15$ studies, $p > 0.1$). The number of young produced per pair varies almost twofold between studies, from 2.9 to 5.5. This is, on average, 66% of the eggs laid.

Figure 41 summarizes the fate of 400 nests analysed by Schmidt *et al.* (1992), and indicates the main sources of variation in losses that have been identified. The majority of other studies on Nuthatch breeding success have either considered only a few sources of variation (Löhrl, 1987c), or

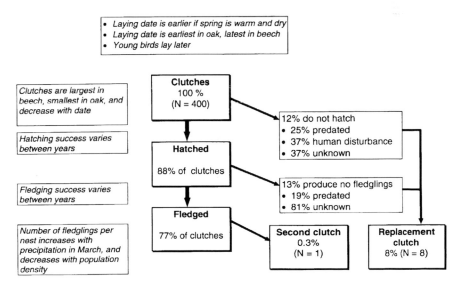

Fig. 41 *Summary of the fate of 400 broods in a nestbox study near Frankfurt. Data from Schmidt* et al. *(1992). Boxes to the right indicate the main sources of variation in breeding parameters.*

were based on relatively small samples (see Table 10). In the following sections I explore the causes of variation in breeding success (habitat, laying date, weather, population density) in more detail. I have chosen to group the available information by effect, rather than breeding stage, in order to highlight the factors that are important to the Nuthatch's overall performance. Effects of forest fragmentation on breeding success (which are limited anyway) are discussed separately in Chapter 10.

Causes of brood failure

The rate of total brood failure varies greatly between studies (4–35%), in nestboxes as well as natural cavities. Failures tend to be more common in studies in cavities than in boxes (means = 22% and 15%, *t*-test on log-transformed proportions: $p = 0.06$). In the 'fragments study' in Belgium, nests in natural cavities also failed more often than those in boxes (37% *vs.* 13%). However, since many early failures were followed by a successful renesting attempt, the success rates of pairs that built their first nest in a cavity or a nestbox differed much less (26% *vs.* 20%; Matthysen, unpubl.). In a study in Poland, broods failed mainly between hatching and fledging (30%, compared to 10% before hatching; Wesołowski and Stawarczyk, 1991). In Germany, however, losses were equally frequent in both stages (Fig. 41). In Sweden, 40% of the failures occurred even before incubation, and equal proportions before and after hatching (30% each; Enoksson, 1993).

The causes of brood failure are varied. A major cause is undoubtedly nest take-over by competing species. In two studies in Sweden, interference by Starlings was the main identified cause of failure (27%, Nilsson, 1987; 9%, Enoksson, 1993). In the former study this amounted to 14% of all nest attempts (Nilsson, 1984). In the fragments study area, 16% of failed nests were taken over by Starlings and 13% by Great Spotted Woodpeckers, and these take-overs occurred more frequently in small forest fragments (see Chapter 10). Two nests have been reported to fail because of take-over by Pied Flycatchers (Nilsson, 1987; Haupt, 1992). Another important cause, but one that is often difficult to prove, is predation of eggs or young. Minimum estimates vary from 18% (Nilsson, 1987; Wesołowski and Stawarczyk, 1991) over 26% (Enoksson, 1993) to 48% (Haupt, 1992). The maximum estimates, however – if all 'unknown' causes were attributed to predation – would vary from 42% to 71% in the same studies. Great Spotted Woodpeckers are probably the main predators, but there are some records of mammal predation as well, including certified records of marten, weasel and fat dormouse (Löhrl, 1967; Andresen, 1989; A. Krištín, pers. comm.). In a nestbox study in pine forest, martens were the sole predators (Haupt, 1992). Nevertheless, Nuthatches suffer less nest predation than other secondary cavity-nesters (Nilsson, 1984) probably because of the low accessibility of the nest hole. Other causes of failure that have been mentioned in the literature are listed in Table 13. One observation (not in the table) suggests infestation with a parasitic fly (probably *Protocalliphora*; editor's notes to Radford, 1957).

TABLE 13: *Causes of breeding failure other than nest take-over or predation*

Causes of failure	Number of cases
Probable starvation	13
Female died or disappeared	4
Desertion for unknown reasons	3
Flooding	2
Hatching failure	1
Treefall	1
Invasion by ants	1
Human disturbance	1
Premature fledging	1

NOTES:
Compiled from Nilsson (1987), Matthysen (1989b), Wesołowski and Stawarczyk (1991), Haupt (1992), Enoksson (1993) and Pravosudov (1993b).

Differences between habitats

Breeding success may depend on a variety of environmental factors related, for example, to food abundance or predation risk, which one might expect to be correlated with the type of forest. However, even if one considers only studies where different forest types are compared with a similar methodology and over the same study period, no general patterns emerge. Löhrl (1987c) found larger clutches in riverine and oak habitats, and smaller ones in mixed broadleaved and spruce–fir habitats (Fig. 42). Schmidt *et al.* (1992) found the largest clutches in beech forest, intermediate ones in conifers and the smallest in oak and oak–beech forest. Zang (1988) found no difference between beech and conifer forests at the same altitude, although clutch size did decrease with altitude by about 0.4 eggs per 100 m. In Siberia, clutches were smaller by about one egg in riparian than in upland coniferous forests (Pravosudov, 1993b).

No studies have reported differences in failure rate between forest types, and even if we look at the number of fledglings per nest or per pair, no general pattern emerges. Nilsson (1976) found the most fledglings per pair (3.3) in oak forest, followed by spruce (2.9) and beech (1.4). However, the degree of variation between years was highest in spruce and lowest in oak forest. In Löhrl's (1987c) study, more young fledged per clutch in mature pine forest, followed by riverine forest, spruce–fir, oak, and mixed broadleaved forest. This variation did not parallel the differences in clutch size (Fig. 42). Zang (1988) found more young per successful nest in beech than in spruce, decreasing by about 0.5 fledglings per 100 m altitude in deciduous forest. Schmidt *et al.* (1992) and Pravosudov (1993b) reported no differences in brood size between habitats. Finally, Matthysen and Adriaensen (unpubl.) found larger broods in oak forests than in parks, but, since brood failures tended to be less common in parks (though this was not statistically different), not more fledglings per pair. They also found no relationship between brood size and the age of oak stands.

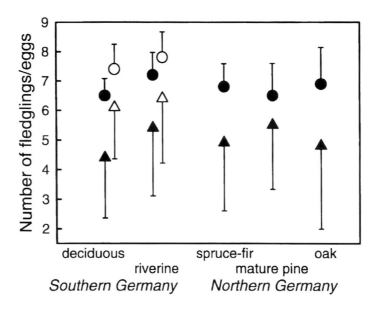

Fig. 42 *Clutch sizes (circles) and number of fledged young per nest (triangles) (means and standard deviation) for five habitats in southern and northern Germany. Data from Löhrl (1987c). Open symbols are broods in enlarged boxes (see text).*

In summary, there is no evidence that breeding success is higher in the most preferred habitat types (broadleaved forest, and oak in particular). This does not necessarily mean that there is no variation in habitat quality, since this may be compensated by higher breeding densities, and density itself has a negative effect on breeding success within study areas (see below). When habitat variation is considered at the scale of individual territories, two studies found evidence that breeding success is higher in more preferred territories (Nilsson, 1987; Matthysen, 1990b). I shall return to the problem of habitat and territory quality in Chapter 7.

Seasonal variation

Four studies showed a seasonal decline in clutch size with laying date by about one egg per month, although very early clutches may be smaller as well (Zang, 1988; Schmidt *et al.*, 1992; Glutz von Blotzheim, 1993; Pravosudov, 1993b). In none of these cases, however, did the number of fledglings decrease with date (not tested by Glutz). Matthysen and Adriaensen (unpubl.) found that brood size decreased by about 0.08 fledglings per day, but had no data on clutch size. Schmidt *et al.* (1992) found a higher breeding success in early years (early mean laying dates) but no difference in clutch size. Pravosudov (1993b) found the opposite, i.e. larger clutches in late breeding seasons (5 years of data). The conclusion is that the effect is not always clear, but that if anything, early nests tend to produce more eggs and/or young. A clear advantage of early nests, however, is that these appear to be more likely to produce recruits, more than might be expected from a slightly larger brood size (Fig. 43). This effect can probably be explained by competition for territories, which is explored further in the next chapter.

Weather

The effect of weather on breeding success is not well documented, apart from suggestions by Löhrl (1958, 1966) that failures are more common in cold and rainy weather. Schmidt *et al.* (1992) analysed the effect of temperature and rainfall in spring on various components of breeding success using an 11-year data set. Although their results are not entirely conclusive, they suggest a higher breeding success if March and April are warm but rainy, whereas weather conditions in May – when the majority of pairs have nestlings – seem to be less important. The positive effect of rainfall in early spring is probably mediated through effects on leafing and insect populations, but this remains to be studied. In contrast, two studies found no variation in breeding success between years at all (Enoksson, 1993; Pravosudov, 1993b).

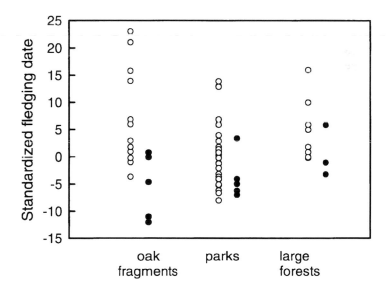

Fig. 43 *Standardized fledging dates (i.e. relative to annual mean) of nests that produced at least one recruit (filled dots) and nests without recruits (open dots). Recruits are juveniles that survived long enough to establish a territory. Data were collected from 1990 to 1993 in fragments of oak forest (see Chapter 10), in parks and in two large forests (Meerdaalbos and Peerdsbos). Logistic regression analysis showed a significant effect of date independent of habitat type.*

Breeding density

Two studies have shown a decrease in breeding success with the size of the breeding population. Nilsson (1987) found that more broods failed at higher densities, and offered two possible explanations: at higher population densities competition with Starlings for cavities increases (see above), and/or more birds are forced to breed in lower quality territories (see also Chapters 7 and 9). The number of fledglings per successful pair also decreased with density, but not in a statistically significant way. Schmidt *et al.* (1992) also found better reproduction in low density years, mainly because more young were raised per successful nest (Fig. 44). They also attributed this to density-dependent variation in mean territory quality. In both studies, a twofold increase in the number of pairs reduced the number of fledglings by approximately one-third. A 10-year study by Haupt (1992) in pine forest showed no effect of density, however (a correlation of $r = -0.18$ was calculated using the original data). A possible explanation is that his study area is more homogeneous and territory quality less variable.

Other sources of variation

Neither Schmidt *et al.* (1992) nor Pravosudov (1993b) found a relationship between parental age (male or female) and clutch size or breeding success. In Sweden, however, older females raised on average 0.7 fledglings more than first-year females (in successful broods) (Enoksson, 1993). Löhrl (1987b) demonstrated an effect of nestbox size, since pairs using enlarged boxes raised on average 1.7 and 0.9 more fledglings (in two different study areas) which

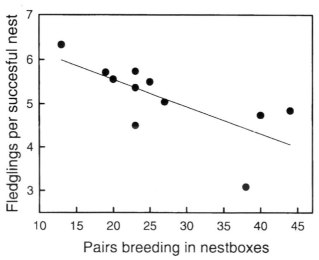

Fig. 44 *Breeding success in relation to population density. From Schmidt* et al. *(1992)* ($R^2 = 0.49$). *Note that the legend in the original paper erroneously stated 'fledglings per clutch' instead of 'per successful nest'.*

is much more than the difference in clutch size (0.9 and 0.6 eggs) (Fig. 42). Hence, more eggs survived to fledging in enlarged boxes, perhaps because these were less sensitive to overheating. Finally, Nilsson (1976) suggested a positive effect of food availability: in two years with an abundant spruce cone crop in spring, six out of seven pairs raised fledglings, compared to one out of five in a year without cones (including two that probably did not even lay eggs). Otherwise there are no studies on the effects of food availability.

SEX RATIOS

In comparison to many other passerines, nestling Nuthatches are easy to sex, which can be done by flank colour when about two weeks old (Chapter 2). In a data set from the Harz and Braunschweig areas, Zang (1980) found a slight surplus of male nestlings (53%). In Frankfurt there was virtually no excess (51%; Matthysen and Schmidt, unpubl.), but in two other studies with smaller samples the proportion of males was much higher, in both cases 63% (Matthysen, 1988, $N = 67$ nestlings; Pravosudov, 1993a, $N = 41$). The most plausible, but unverified, explanation is that female nestlings suffer a higher mortality because of competition with male siblings (Lessells *et al.*, 1996). This is supported by data on sex ratios in relation to brood size (Fig. 45) which suggest that there is no excess of males in very small broods, where mortality has been high and perhaps less selective, or in large broods where little or no mortality has occurred. In support of this idea, a later analysis of the Braunschweig data found no deviation from a 1:1 sex ratio in broods where all young survived (Winkel, 1996).

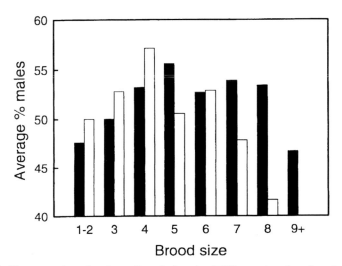

Fig. 45 *The proportion of male nestlings in broods of different sizes, based on data from Zang (1988; filled bars) and unpublished data from Frankfurt (E. Matthysen and K.-H. Schmidt; open bars). Minimum sample sizes per size class are 42 and seven for the two studies, respectively. A single brood of nine in the Frankfurt study is included here in class 8.*

THE POST-FLEDGING PERIOD

In the last few days before fledging, young Nuthatches often appear in the nest opening. When they leave they typically climb around on the nest tree for a while, although they are usually well able to fly. There seems to be no characteristic time of day for fledging. Radford (1957) and De Meersman (1981) described the fledging of three broods. In all three cases the first young hesitated for some time before emerging, but then the others followed rapidly. One brood observed by Löhrl (1958) fledged within 14 minutes. De Meersman (1981) describes how the adult birds seem to encourage the fledglings by calling and fluttering around in front of the nest (cf. Ingold, 1977a, for similar observations on White-breasted Nuthatches).

There are no detailed observations on the behaviour of pairs with dependent young. In the first few days the young spend most of their time sitting high in the canopy waiting to be fed, and, although the family can be located by the begging calls and the parents' activity, the young remain difficult to find. Later the family roams over a relatively large area, sometimes beyond the usual territory borders. Broods probably do not split up between parents (D. Currie, pers. comm.). In contrast to other nuthatch species, newly fledged young do not associate closely in captivity and may even show threat behaviours in close contact (Löhrl, 1988). The risk of predation is probably severe at this stage of the bird's life, and this may explain why, in a study by Currie and Matthysen (unpubl.), 42% of the fledglings disappeared by day 8 after fledging, presumably before much dispersal could have occurred. In this study two attacks, by a Jay and a Sparrowhawk, were witnessed. Pairs with recently fledged young can often be found alarm-calling and chasing Jays or other corvids. Even Little Owls sometimes take young Nuthatches, as shown by a ring found in an owl's nestbox in early July (Cohen, 1960).

Hand-reared fledglings started to search for their own food about 6 days after fledging, and as early as day 4 they started to hoard some of the food they received (Löhrl, 1967). After 8–10 days the young birds are largely independent. Currie and Matthysen (unpubl.) saw no parental feeding after day 11, although begging was seen at least occasionally until day 18. At about 10 days after fledging, young birds in captivity enter a phase of restlessness which was noticed by both Heinroth and Heinroth (1926) and Löhrl (1958). The birds become shy and restless and start flying about the cage for hours at a time. Incidentally, this was not observed in the closely related Chestnut-bellied Nuthatch, and was seen to a lesser degree in Kashmir Nuthatches (Löhrl, 1988). This phase lasts 1–2 weeks and corresponds to the normal timing of dispersal, which is discussed further in Chapter 8.

SECOND AND REPLACEMENT BROODS

After a successful nest is raised, very few Nuthatches produce a second clutch. As far as I know, only nine such cases have been reported, usually within the same nest (Henze, 1940; Löhrl, 1967; von Knorre *et al.*, 1986; Haupt, 1992; Schmidt *et al.*, 1992; A. Krištín, pers. comm.). In four nestbox studies the combined frequency was less than 1% (4 out of 486 successful first broods) (Henze, 1940; Haupt, 1992; Schmidt *et al.*, 1992; A. Krištín, pers. comm.). This is a minimal estimate, however, since some pairs may have moved to a natural cavity and not been detected in the study. Some very late broods may also be explained by second clutches, e.g. eggs laid in late June (Melchior *et al.*, 1987), incubation in mid-July (Becker, cited by Glutz von Blotzheim, 1993) and 18-day-old nestlings in early October (Bohác, 1965). In two cases the second clutch was initiated about a week after the first brood fledged (Haupt, 1992; Schmidt *et al.*, 1992). Boiteau (1991) reported a probable second brood which followed an exceptionally early brood (see earlier in this chapter). Replacement clutches also seem to be infrequent: in Schmidt's study only eight of 95 pairs started a new clutch after the first one was lost, but again others may have been missed. Laying dates of these replacement clutches varied from 22 April to 1 June. Estimates in natural cavities are also low but probably more reliable: one out of eight (Matthysen, 1989b) and three out of 13 failed nests were replaced (Nilsson, 1984). Zang (1988) and Möckel (1992) recorded no second or replacement broods at all. The study by Haupt (1992) in pine forest is exceptional in this respect with 13 out of 17 clutches being replaced. Whether this signals a more prolonged peak in food availability in coniferous forest is an intriguing question for further study.

Replacement clutches are typically smaller than first clutches, by 1.2 and 3.0 eggs, respectively, in studies by Schmidt *et al.* (1992) and Haupt (1992). In Schmidt's study these nests were nearly as successful as first clutches, but Haupt reported a considerable difference (5.4 fledglings from first clutches, 2.1 from replacement clutches), though this difference was not tested statistically. Such a difference is in agreement with the earlier mentioned seasonal decline in clutch or brood size with date.

CHAPTER 7

Finding a Territory

We are not very sociable creatures, we Nuthatches; two is company and three is always a crowd

Allen (1929) on the White-breasted Nuthatch

After the last broods of the season have been raised, the rest of the summer is often a somewhat neglected period in ornithological field studies. While field researchers take a well-deserved break or start analysing their data, the subjects of study enter their annual moult and may be quite hard to find and observe, especially in the dense summer forest. The period right after fledging is of crucial importance to many young birds, however, and this is particularly true for young Nuthatches. The task faced by a newly fledged bird is clear: to acquire a territory and a partner as quickly as possible. This task is not simple, because the adult birds remain on their territories throughout the summer and are not inclined to share it with their offspring for more than a few weeks after fledging.

In this chapter I describe the possible behavioural routes to territory ownership that are used by young Nuthatches. In contrast to the previous chapters, I draw almost exclusively on my own observations, and defer comparisons with other studies to the last section of the chapter. Unless otherwise mentioned, all results are from the 1982–1987 study in the Peerdsbos area (mainly after Matthysen, 1987, 1989c, 1990b).

TERRITORY ESTABLISHMENT

The most straightforward, if not most common, route to territory ownership for young Nuthatches is to take up a territory of their own shortly after becoming independent in June or July. The Peerdsbos population nearly doubled each summer by the addition of new territorial pairs (from 10–15 to 20–25). The majority of these were juveniles, with the exception of three adult males paired to young females (4% of 84 settlers). Some young birds did not settle in a vacant site right away but acquired a territory by replacing an owner that disappeared during the summer; the previous history of these late settlers, however, was not always known. A number of birds were also observed to settle in the area in a small home-range without defending it. These non-territorial residents, and their eventual fate, are discussed later in this chapter.

Timing of establishment in summer

When I started the Peerdsbos study, I had no reason to doubt that Löhrl's (1958) description of summer behaviour would also apply to this population. According to Löhrl, young birds would wander around for several weeks and only start to take up territories in August (see also below in this chapter). However, I soon found out that some of them appeared to have settled in a particular home-range or even a territory as early as July or even June. This is best illustrated by the detailed observations made during 1985 and 1986. In 1985 the first juvenile settlers appeared on 8 June, only 9 days after the earliest brood had fledged. One settler was a young male paired to a widowed female, the other a young female that was seen 2 days later with a young male. I found three more new pairs on 11, 14 and 15 June, before the last brood in the area had fledged. In 1986 I had even better proof of early establishment from observations on three birds hatched in the same nest (Fig. 46). One female (F118) was found 200 m outside the natal territory as early as day 8, within her later territory. On day 10 she was last seen with her parents (begging for food) on the border of the territory, and later that day she was seen with a young male in the territory where she remained throughout the summer. Her sister (F119) likewise moved into a vacant area next to the territory between 6 and 8 days after fledging. A male sibling from the same brood established a territory about 1 km from the nest between 10 and 15 days after fledging. Note that it is exceptional for young

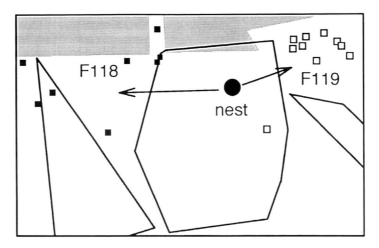

Fig. 46 *Observations on two fledglings from the same nest that settled very rapidly after fledging on 4 June 1986. Squares mark observations from 10 June to 2 July. Solid lines = breeding territories, shaded area = non-forest.*

birds to settle this close to the parents' territory (see Chapter 8). With a few exceptions including those described here, the majority of settlers immigrated from outside the study area.

I estimated from the 1985–1986 observations that about half of the new territories were established within 2 weeks of the earliest broods fledging, and almost all within 1 month. Each year one or two juvenile pairs did not establish a territory before late July or mid-August, and these usually settled in low-quality territories. In Chapter 10, however, I discuss observations in small patches of forest where the pattern of establishment showed some differences from that portrayed here.

The fact that most available vacancies may be taken up by the time the last broods have barely left the nest implies a strong selection on early fledging, and hence on early breeding by the parents. In addition, the best territories are taken up first (see below). This is in agreement with the observation that early broods are more likely to produce recruits. In turn, this provides a further explanation for the near absence of second and replacement clutches in most populations (Chapter 6).

Behaviour accompanying establishment

Except for the timing, very little is known about the way in which territories are established. New owners were always observed in pairs from the first or at most from the second observation, and even in the latter case I could not exclude the possibility that a partner had been overlooked on the first occasion. In the fragments study, however, several territories were established by single individuals (Chapter 10). In four cases a male settled first and was joined by a female 10–20 days later; and in three cases (two females, one unknown) no second bird arrived at all (Matthysen and Currie, 1996). This suggests that birds of either sex may establish a territory before they have a mate. I have no evidence that pairs are formed by 'landless' birds before they establish a territory. This would be hard to document, however, since the majority of juveniles are ringed only after they have settled. I have little or no evidence that juveniles would aggressively take over territories (or parts of them) from adult owners, in contrast to Löhrl (1958; see below).

All observations indicate that territorial and pair behaviour is fully developed from a very early age. Juvenile pairs behave in a similar way to adults from the very first day they are observed, foraging close together while maintaining vocal contact, and actively defending the territory. Several juveniles formed a pair with widowed adult males or females, some of them probably less than a few weeks after fledging. The contact calls of young pairs are initially reminiscent of the juveniles' begging calls, being more drawn-out and somewhat louder than the typical Contact Calls.

The establishment of new territories is often accompanied by intense fighting. Of the five physical fights I witnessed during a 5-year study, four involved juvenile birds in June and early July. Juvenile owners (and some adults as well) are highly responsive to playback in this period (and can be caught in this way). Once territories were established, the young birds appeared well able to defend their property, since territories changed little in location or size.

Territory shifts

Few of the young birds that established a territory in summer and survived bred within the same territory. Only seven out of 24 surviving juveniles (29%) did so, but none remained with their original mate. In comparison, 95% of all surviving adults nested in last year's territory. There were three kinds of circumstances in which young birds moved to another territory: (1) moving as a pair (five cases); (2) a recently widowed bird went to join a solitary neighbour (18 cases); (3) a real 'divorce', that is, a bird that left its partner to join a solitary bird in another territory (nine cases). Birds always shifted to territories that had a vacancy for their own sex. They usually moved to the nearest territory, except for two cases where one territory lay in between (Fig. 47). Males and females shifted equally often. Figure 48 further illustrates the history of a number of territories in the course of a year, with a few examples of territory shifts and of trios (see Chapter 5). It also shows that the majority of birds settled in the area either in summer or in early spring.

Territory shifts have not been studied in detail by other authors, but the data presented by Enoksson (1987) suggest that 53% of the juveniles moved to another territory between early autumn and spring. In the fragments study, however, very few juveniles shifted to another territory (29%) (see Chapter 10).

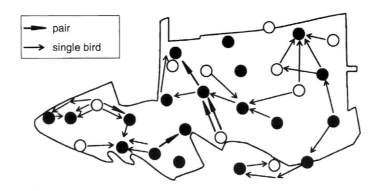

Fig. 47 *Movements of pairs or individual birds between territories of different quality in the Peerdsbos study. Quality is indexed by occupation rate (white = 0 or 1 breeding seasons out of 5, grey = 2 to 4, black = 5).*

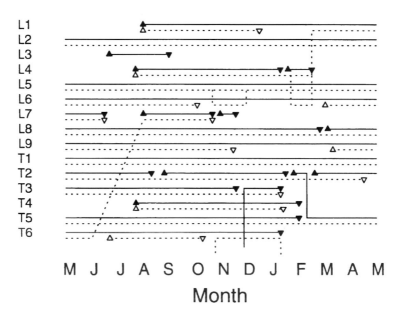

Fig. 48 *One-year histories (1984–1985) of a set of territories in the Peerdsbos study area. Full and dashed lines indicate the presence of males and females, respectively. Triangles mark the arrival and disappearance of individual birds. Birds moving between territories are indicated by vertical lines. Lines branching into two lines represent birds participating in a 'trio' (see Chapter 5). A female that moved from territory T6 to L7 (oblique line) was not observed during the intervening 3 months.*

Non-territorial residents

Many juvenile Nuthatches do not establish a territory in their first summer. A total of 58 first-year birds were seen just once or a few times, mainly from June to October, and were classified as 'non-residents' or transients (see also Chapter 8 on dispersal). However, each year a small number of settlers temporarily occupied small home-ranges which they did not defend (29 birds in five summers). I defined *non-territorial residents* as birds that were observed over a period of at least 10 days, at sites no more than 500 m apart, and that were socially subordinate to local territory owners (Table 14). Their home-ranges were usually smaller than 2 ha, but this might be an underestimate because I usually had only a few observations per individual. All birds of known age were in their first year. Their settling dates were mostly unknown, but six that were observed for the first time in September or October were probably not present before that time.

The behaviour of non-territorial birds proved to be difficult to study. They were less conspicuous than owners, rarely called, and were often detected only by captures or observations at feeding sites (66% of all observations). The majority of these birds lived solitarily in a small area that overlapped with one or more territories (several examples are shown in Matthysen and Dhondt, 1983). In some cases, a male and female occupied the same small area and were regularly seen together but not with other birds ('paired' in Table 14). A third category ('satellites') were always observed within a particular territory, often in company with the owners

TABLE 14: *Behavioural status of first-year Nuthatches when they settled in the Peerdsbos study area (after Matthysen, 1989c)*

	Males	Females	Total (%)
SUMMER TERRITORY OWNERS	45	50	95 (55)
Owner of newly established territory	39	42	81 (46)
Paired to widowed adult	6	8	14 (8)
SUMMER NON-TERRITORIAL RESIDENTS	18	11	29 (17)
Solitary	8	4	12 (7)
Paired	4	7	11 (6)
Satellite	2	0	2 (1)
Putative satellites	4	0	4 (2)
POST-SUMMER IMMIGRANTS	26	24	50 (29)
Paired to widowed owner	16	24	40 (23)
Possible take-over	4	0	4 (2)
Individual territory	4	0	4 (2)
Non-territorial resident	2	0	2 (1)

NOTES:
1. Numbers of males and females do not necessarily match, since a number of females paired with already settled and/or adult males.
2. Individuals that changed their behaviour (e.g. from solitary to satellite) are included only once.

and without overt aggression (Fig. 49). In five summers I observed only two such cases with certainty, plus four doubtful cases that were observed only a few times (all males) (Table 14). Some non-territorial residents changed their home-range or even their behaviour during their stay, an example of which is given in Fig. 50. A few birds lived in the same home-range for

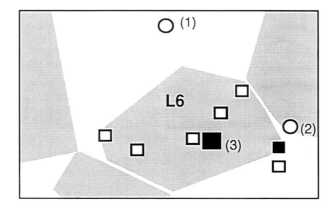

Fig. 49 *Observations on a male satellite in and around territory L6 from 9 September to 10 December 1986. White squares indicate observations with the owners of L6, white circles are observations alone (1) or with a neighbour (2), black squares indicate capture sites with one and four (3) observations, respectively. Shaded areas represent territories.*

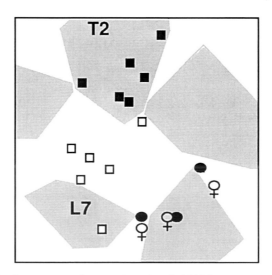

Fig. 50 *Successive home-ranges of non-territorial male M136 in autumn and winter 1986. This male was a satellite in territory T2 from July to August (black squares), paired to a non-territorial female in September (black circles; observations of partner are also mapped) and lived solitarily from November to January (empty squares). In January he replaced the owner of territory L7. Shaded areas represent territories.*

several months without defending a territory, five of them until January. However, most of the others either moved into a vacant territory in the course of the summer, or quickly disappeared.

Post-summer settling

After the initial settling peak in June–July, some immigration continued through early autumn, reaching a minimum in November and December and followed by a resurgence in spring (see Chapter 8 for more details). The majority of these immigrants settled by pairing with a territory owner (Table 14). In 13 cases no aggression was involved, since the previous owner of the same sex had disappeared some time before. In 31 cases the newcomer was present on the first visit after the last observation of the previous owner, therefore aggressive take-over could not be excluded. In four cases there was indirect evidence for take-over, since the previous owner was known to be alive after the immigrant settled (all males). One male retreated to part of its territory while the immigrant took over the other part and the female. Two males later reappeared elsewhere, and a fourth male was last seen on the day the immigrant was first observed. Four immigrant males settled in unoccupied areas and their behaviour suggested establishment of a new territory. They occupied small home-ranges for a few weeks or months, were involved in conflicts with neighbouring males and often called or even sang but remained unmated. Two other males were best categorized as non-territorial residents. Of these six males two managed to settle in a pair territory by replacing the owner. In summary, whereas the majority of immigrants in summer established a new pair territory, post-summer immigrants usually joined the owner of an established territory. Also, becoming a non-territorial resident was a much more common strategy among summer immigrants than in spring. A likely explanation is that settling as a non-territorial resident while waiting for a vacancy to appear becomes a less rewarding strategy as the breeding season approaches, because of the increasing chance that no vacancy will be found.

TERRITORY QUALITY

After studying the same population in consecutive years, I noticed identical patterns of occupation of the same areas in different years by different individuals. For instance, in early April of both 1984 and 1987, a pair of first-year birds shifted from territory T4 to the neighbouring territory T3, while in the summers of both 1982 and 1983 a first-year male settled as a non-territorial resident between territories L8 and L9, and was joined by a female in September or October. A more general observation was that particular areas were usually settled by juvenile pairs in summer, but were invariably vacant again by spring. It appeared as if some territories were good enough to support a territorial pair in late summer, but not during the breeding season. Or, more generally speaking, that some patterns of territory occupation and behaviour were influenced by differences in quality between territories.

The challenge then was to prove that territories indeed differed in quality, and if so to find out which characteristics determined this quality. The most straightforward solution would be to search for relationships between vegetation characteristics or food availability and occupation rates, survival or breeding success. Unfortunately, the Peerdsbos study area is not very suitable for quantifying the vegetation on a territory. As described in Chapter 5, the area is a patchwork of small stands separated by lanes, and although the boundaries of territories were more or less known, I did not know how frequently different parts were used. I therefore examined differences in quality from the birds' point of view, by studying correlations between different parameters of attractiveness. I used four such parameters: the order in which vacant territories were occupied by juveniles in summer; the order in which territories with a solitary male were joined by a female immigrant in spring; shifts between territories; and the number of years that a territory was occupied (Matthysen, 1990b). This approach was possible because territories remained more or less constant in position over the years, and could be treated as permanent entities that were either vacant or occupied at any given time (see Chapter 5).

Settling order in summer and in spring correlated with occupation frequency in previous years (Fig. 51) and birds tended to shift towards territories with a high occupation rate (Fig. 47). This allowed me to use occupation frequency as an indicator of territory quality, and this parameter appeared to be related to some other observations. For instance, non-territorial residents were found more often in or near high-quality territories, and territory owners were more likely to intrude into territories of better quality than their own (Matthysen, 1990b). This might indicate that owners of low-quality territories prospected the surrounding high-quality territories for vacancies.

The advantage of living in a high-quality territory is best documented by data on body condition. First-year birds in high-quality territories (occupied for four or five breeding seasons) lost on average 0.24 g body weight from summer to early winter, while birds in low-quality territories lost 0.86 g. Survival rates were also higher in 'good' territories, as shown by positive correlations with occupation rate, settling order and the direction of territory shifts (Matthysen, 1990b). For instance, all juvenile males that settled before 15 June survived, but none of the later settlers did (both $N = 8$). The same difference, but less pronounced, occurred in females (57% and 22%; Matthysen, 1989c). Pairs in low-quality territories also had a lower chance of raising a brood and tended to lay somewhat later. Neither of the latter relationships was entirely convincing, however, because of possible age-related effects (Matthysen, 1988, 1989d). Since birds shift to better territories and also appear to survive better in good territories, old birds tend to accumulate in the better territories. Nilsson (1987) found more convincing evidence in Sweden for a relationship between territory quality and reproductive success (see below).

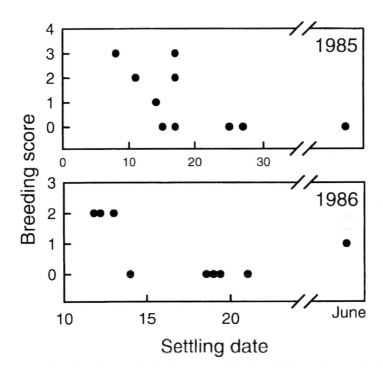

Fig. 51 *Settling dates of juveniles in early summer in relation to 'breeding score', i.e. the number of years the territory has been occupied in previous years. Both rank correlations are statistically significant (see Matthysen and Currie, 1996).*

The logical next step was to study how territory quality was related to variation in habitat. Although a quantitative analysis was not possible, it became clear that the best territories all contained either a large patch of oak or an area of garden habitat (with large broadleaved trees), while the lowest quality territories were often dominated by conifers (pine or spruce). Territories with beech or with mixtures of different broadleaved trees were of intermediate quality (Matthysen, 1990b). However, the quality of territories with beech varied between years, depending on the availability of beech mast (see Chapter 9). Territory quality was not related to territory size, which was to be expected in a very heterogeneous habitat. A possible exception were some very small 'marginal' territories, which were occupied in only one or two summers, and appeared to be squeezed between more permanent territories. They may have represented an intermediate stage between non-territorial pairs and true territories. They were not associated with particular habitat types, however.

Other studies on territory quality

A few studies provide additional evidence, though in less detail, on the importance of variation in territory quality. The best data are provided by Nilsson's (1987) study in Sweden. Here, pairs in frequently occupied territories (5–8 years out of 8) raised on average 1.3 fledglings more than pairs in less frequently occupied territories (corrected for variation between years). Some indirect evidence for relationships between territory quality and breeding success or laying date is discussed in Chapter 6. Nilsson also found longer-winged males in territories dominated by oaks, and, on a somewhat larger scale, Zang (1988) found larger males in deciduous than in coniferous forest. However, this may simply represent the association between age and territory quality. Löhrl (1958) wrote that high-quality territories were more often the subject of aggressive take-over, but gave no details.

DIFFERENT 'STRATEGIES' TO TERRITORY OWNERSHIP?

The breeding birds in the Peerdsbos study could be categorized into three groups according to whether they had settled as territory owners in the summer (58%), as non-territorial residents in the summer (4%), or as post-summer immigrants (38%). An intriguing question is whether these groups represent different behavioural strategies with equal pay-offs, or whether some of them should be considered 'second-best' options for birds that failed to obtain a territory in the first place (Matthysen, 1989c). Post-summer immigrants did not differ in body size or weight from the other categories (Matthysen, 1988). It is difficult to compare their overall success with summer settlers because their previous history and mortality rate are unknown (see also Chapter 8). Many male immigrants disappeared again within a month of arriving, which is much more than expected from the mortality rate of resident first-year birds at the same time (Matthysen, 1988). This suggests either that they survived less well, or that some of them dispersed further.

The success of territorial versus non-territorial birds can be compared using data on local survival, condition and quality of their breeding territory. Somewhat surprisingly, the comparison shows few obvious advantages for the territorial mode of settling. During August, more non-territorial residents disappeared than territorial juveniles, but this may reflect higher dispersal as well as mortality (see Chapter 8). The remaining birds survived equally well, did not lose more body weight from summer to winter, and ended up breeding in territories of the same quality as birds that established a territory in summer (Matthysen, 1989c). This may be explained in part by their tendency to occupy the better parts of the study area.

Despite their good prospects (at least after August), the small number of non-territorial residents per summer (1–8) and the strong correlation between their number and the preceding breeding density (Matthysen, 1989c) suggest that this behaviour is still a 'second-best' option when territorial vacancies become scarce. It therefore seems plausible that territory ownership has an important advantage in early summer, when comparison with non-owners is extremely difficult. The crucial factor determining whether a bird can obtain a territory is probably fledging date, since non-territorial residents do not seem to be at a disadvantage in terms of body size or condition (Matthysen, 1989c).

Territorial and non-territorial residents were in significantly better condition than transients (defined as being present for less than 2 weeks and/or roaming over a large area; Table 15). Interestingly, in females this difference increases from June to August, which suggests that females in particular may suffer from not having a territory. This may be explained by their socially subordinate position to males.

TABLE 15: *Body size and condition of resident (including non-territorial) and transient juvenile Nuthatches*

Variable	Sex	Residents	Transients	Difference
Wing length (mm)	M	86.4	86.1	0.3
	F	83.5	84.1	−0.6
Tarsus length (mm)	M	18.05	17.86	0.19
	F	17.83	17.76	0.07
Body weight (June) (g)	M	22.0 (11)	21.4 (7)	0.6
	F	20.3 (7)	20.5 (6)	−0.2
Body weight (August) (g)	M	23.5	22.9 (8)	0.6
	F	22.3 (45)	21.4 (11)	0.9*
Body weight (total) (g)	M	23.1	22.2	0.9***
	F	21.9	21.2	0.7*
Condition (g/mm)	M	1.28	1.24	0.04*
	F	1.23	1.19	0.04*

NOTES:
1. M – males; F = females.
2. Condition is weight divided by tarsus length.
3. Sample sizes are 57–75 individuals for residents and 15–18 for transients, except where indicated in parentheses.
4. Body weight averages all weight records except for roosting controls.
5. *t*-test: * $p < 0.05$, *** $p < 0.001$.

OTHER ACCOUNTS OF TERRITORY ESTABLISHMENT

Germany

The description of territory establishment by Löhrl (1958) diverges profoundly from my own observations in a number of ways. Löhrl's juveniles did not attempt to establish territories before the end of July, but during that time wandered around or associated temporarily with adult pairs in 'pseudofamilies'. At the end of July, juvenile pairs settled at the borders of existing territories, and in August these new pairs gradually expanded their areas without obvious conflicts. In early September, however, males, in particular, initiated conflicts to expand their boundaries at the expense of the adults' territories. As a result, former breeding territories were subdivided into smaller parts, and territory size varied greatly between seasons (Fig. 52). This period of fighting usually ended with the onset of cold or rainy

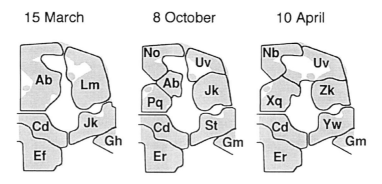

Fig. 52 *Territory boundaries in the course of 1 year in a park in Ludwigsburg. After Löhrl (1958). Upper and lower case characters refer to individual males and females, respectively. Suitable habitat is shaded.*

weather in October, and territory boundaries then changed little over the next few months.

Löhrl does not mention non-territorial residents as a specific category. The 'pseudofamilies' in early summer might be interpreted as adult pairs with one or more satellites, but apparently these associations were rarely continued after July. Löhrl does mention some juvenile females living inside a territory for some weeks or perhaps months, without being particularly associated with the owners. Males were apparently much less tolerated, and could maintain themselves for some time only in so-called neutral areas.

In late February there was a revival of fights over boundaries without major changes in ownership. In March and early April, however, some owners were challenged by single males or pairs, and often defeated. The challengers could be local non-territorial birds, immigrants, or birds that gave up their own territory, but no further details are given. Löhrl observed five successful take-overs by single males and four by pairs (two in November, six in March, one in April). The defeated males did not challenge other owners, but retreated to neutral areas and waited for a vacancy to arise. In this study, aggressive take-over was the most common type of settling behaviour (13 cases), followed by 'squeezing-in' of pairs between territories in summer (12 cases), pairing with a widowed bird (seven cases) and settling in an empty territory (one male, later joined by a female) (see also Table 16). As a consequence of the high incidence of aggressive take-overs, few pairs remained together until their second breeding season (one out of eight pairs).

It seems unlikely that the discrepancies between Löhrl's and my own observations are due to differences in observation methods or interpretation. Löhrl apparently saw no new territorial pairs before August whereas this was the commonest type of settling in the Peerdsbos. On the other hand I have no observations that match the 'squeezing-in'. Aggressive take-over was very common in Löhrl's study but limited in the Peerdsbos study. There is some agreement between the studies in the occurrence of settling by joining widowed birds (though this was more common in the Peerdsbos) and in the occurrence of non-territorial residents. Unfortunately Löhrl provided few details of the latter. I discuss possible explanations for these differences later in this chapter.

Sweden

Bodil Enoksson studied Nuthatch territoriality in Sweden as part of a series of food addition experiments (Enoksson and Nilsson, 1983; Enoksson, 1990a). Perhaps her most important finding with respect to settling behaviour is that territory size was related to food supply. In years with a rich food supply, either natural (beech nuts) or artificial (sunflower seeds), formerly large territories became subdivided (Fig. 53). It would have been interesting to know exactly how and at what time this subdivision took place, but unfortunately no early summer observations are available. Some more detailed maps in Enoksson (1987) also give evidence for the reverse: as individuals disappeared throughout the winter months, territories grew larger and filled up most of the space that became available. This pattern of flexible, compressible territories seems to be more compatible with 'squeezing in' as settling behaviour (cf. Löhrl's study), than with the more rigid and permanent territories in the Peerdsbos study. In Chapter 9, I return to this difference in the discussion on population regulation by territoriality.

In years with a rich food supply, up to half of the territories in Sweden contained one, or rarely two, satellites (or 'extra-birds' as Enoksson named them; Fig. 53). On average they made up 19% of the autumn population. In low-food years they were found in 11% of the territories, which is comparable to the Peerdsbos in autumn (c. two non-territorial birds per year on 20 territories in early October). There is little information on their behaviour; at least in some cases they were not the pair's offspring (B. Enoksson, pers. comm.).

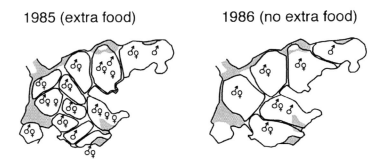

1985 (extra food)   **1986 (no extra food)**

Fig. 53 *Autumn territories in a study area in Sweden in years with and without food supplementation. After Enoksson (1990a). Note the satellite birds in several pair territories. Shaded area = unsuitable forest.*

Siberia

The study by Pravosudov (1993a) provides yet another different account on settling by first-year birds. In this population young birds acquired a territory only by replacing a territory owner that disappeared. In the meantime they all settled as non-territorial birds within adult territories, some of them in pairs. Apparently, as many as 10 could be found within a single territory. The home-ranges of these non-territorial birds covered only part of the adults' territory but overlapped extensively with one another (Fig. 54). They did not associate with one another or with the owners. These non-territorial birds are therefore comparable

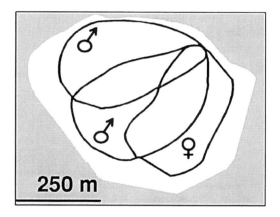

Fig. 54 *Home-ranges of three non-territorial juveniles in Siberia inside a pair territory (unshaded area). After Pravosudov (1993a). Note the large size of the territory (cf. scale bar).*

with the solitary and paired non-territorial residents I found in the Peerdsbos, but not with satellites. The main difference is their number and the longer residence times. Although some probably moved out of the area in autumn or spring (and, to some extent, were replaced by new birds), several survived as non-territorial birds until the breeding season and even beyond. For instance, five out of 11 non-territorial birds that settled in a particular territory in early autumn continued to live there until the next autumn. In another case, a non-territorial male stayed in a territory for two consecutive winters. At least one juvenile pair lived within a territory of an adult pair throughout the breeding season but did not breed. Pravosudov also found that early settling birds had a higher chance of becoming owners than late ones, which suggests that they were socially dominant over other non-territorial birds. However, there was no relationship with body size, nor was there a difference between resident and transient juveniles, in contrast to the Peerdsbos study.

How to explain differences between studies?

The four studies on settling behaviour of juvenile Nuthatches present a substantially different picture, which is summarized in Table 16. The only strategy that occurred commonly in all populations was for juveniles to form a pair-bond with a widowed territory owner. In Siberia this was the main possibility; in the other studies the alternatives were to settle in vacant areas, or to fight with local owners to obtain part or all of their territory. Various non-territorial strategies were possible in all studies (Table 16). One of the major factors explaining this variation seems to be the availability of territorial vacancies.

TABLE 16: *Main modes of establishment of young birds in four different studies*

	Belgium *Matthysen* *(1989c)*	*Germany* *Löhrl* *(1958)*	*Sweden* *Enoksson* *(1990a)*	*Siberia* *Pravosudov* *(1993a)*
ESTABLISHING A TERRITORY				
Pair settles in vacant territory	+++	–	++	+
'Squeezing in'	–	+++	+++?	–
Joining widowed owner	++	++	++	+++
Aggressive take-over	+	+++	+?	–
NON-TERRITORIAL STRATEGIES				
Solitary/paired	++	+?	–?	++
Satellite	+	+?	+++	–?

NOTES:
+++ = very frequent, ++ = frequent, + = rare, – = not observed.

In comparison with the German and Siberian studies, the Belgian population clearly represents an unsaturated breeding habitat, where there is ample space for young birds to set up territories of their own. A possible explanation is the low survival rate (*c.* 50%) in Belgium compared to Siberia (67%; no data are available for Germany; see Chapter 9). If adults live longer, their territories become available at a slower rate and juveniles have less chance of obtaining a territory early in life. Of course, the degree of saturation of a particular study area may also be influenced by habitat selection at a larger scale; a high-quality area surrounded by inferior habitat will attract more birds and support a higher population density throughout the year, and may present a different behavioural pattern from a less optimal area.

Spatial and temporal variation in territory quality may also play a role, however. In the Peerdsbos study area, low-quality territories provide a temporary refuge for juvenile birds and become vacant again each year, since they never support a breeding pair. In a sense, juvenile pairs in these territories have not yet achieved full territory ownership (i.e. of a territory that allows reproduction) but are waiting for a better vacancy to appear. In a more homogeneous area without low-quality sites, it seems more likely that vacant territories are divided among neighbours and that, each summer, juveniles have to fight to get their share. This situation may apply in Germany and to some extent in Sweden. The Swedish study further points out the importance of annual variation in food supply. When food is abundant territories are smaller but many juveniles still have to settle as subordinates. Whether the latter is caused by reduced territorial defence by adults, increased aggression by juveniles, or both, remains unknown. If food is less abundant, the territories expand and the number of non-territorial birds is reduced as well. One aspect that remains unclear is why the Siberian juveniles do not attempt to take over parts of the adults' territories.

PAIR-BOND AND TERRITORY

The different ways in which juvenile Nuthatches may achieve pair-territorial status again illustrate that pair-bond and territory are independent to a considerable degree. It is true that new territories are typically established by pairs, and there are a few cases (particularly in Löhrl's study) where birds move to a different territory together. On the other hand, individual birds readily switch to a different territory without their mate, and the same kind of opportunistic 'divorce' occurs in non-territorial pairs as well. Individuals can also be evicted from their territory while their mate accepts the newcomer. These observations are in agreement with the sex-specific aggression seen in confrontation experiments (Chapter 5). Thus, while

there is good evidence that Nuthatches actively choose between territories of different quality, there is no evidence that they have preferences for particular mates. Without good measures of individual quality and no details of how juvenile pairs are formed, the existence of mate choice remains an unsolved problem in Nuthatch natural history.

The high frequency of territory and mate changes in first-year birds sharply contrasts with the faithfulness of adult birds. During 5 years I observed only one territory shift by an adult individual (after losing its partner) compared to over 30 in juveniles, and no divorces occurred among the 16 pairs where both members survived from one breeding season to the next. It could be argued that adults already live in good territories and can rarely benefit by moving. Adult pairs in Siberia were also extremely faithful, however (Pravosudov, 1993a), even though they could not shift to better territories during their first year of life. Of course, the Siberian birds would face competition from many non-territorial individuals if they attempted to move to another territory. Thus, it is still unclear whether there is a real advantage to site and mate fidelity for older birds with more experience.

In the light of the above, the 'permanent' pair-bond of Nuthatches should probably be interpreted as an opportunistic union between two individuals, both with a long-term interest in defending the same territory. Once they are living in the same territory there may be extra benefits in having a close social bond: short-term benefits of foraging together and sharing vigilance, or long-term benefits from experience in breeding and defending a territory together (see Matthysen, 1993 for an extensive discussion on non-breeding pair-bonds in birds). These benefits can perhaps be studied properly only by experiment, removing single birds from territories and studying the reaction of the remaining partner.

CHAPTER 8

Dispersal and Migration

Right outside the lazy gate of winter's summer home
wond'rin' where the nuthatch winters
wings a mile long just carried the bird away

From the song *Eyes of the World* on the Grateful Dead's 1973 album *Wake of the Flood*

The ability of birds to travel long distances is well illustrated by the annual migration of many bird species. Nuthatches take little part in these movements, with the exception of a few Asiatic subspecies, and even then in some years only. However, like many other animals, Nuthatches may under-

130

take movements of a more modest kind but of equal importance, when they move from their natal site to a place to reproduce. These more or less permanent movements between one home-range and the next are known as *dispersal*. Dispersal movements are important because they provide a link between the dynamics of different populations and result in exchanges of genetic material (*gene flow*) between populations. Dispersal is also a crucial phase in an individual's life history. A dispersing bird has to find a place to live while risking the dangers associated with travelling through unknown or even hostile habitats, and when it eventually settles it has to integrate itself into a social group or establish a territory. The fate of dispersers is an important, but often poorly understood, element in the demography of populations. In this respect, the present chapter is a necessary introduction to population dynamics and the effects of forest fragmentation (Chapters 9 and 10).

SOURCES OF INFORMATION ON DISPERSAL

Despite its obvious importance, dispersal is a difficult process to study, because of the large spatial scale on which it operates and because of the elusiveness of the dispersal act itself. If, for instance, a Nuthatch moves over a distance of three territories in an unknown direction, an observer would have to search approximately 30 territories to be certain of finding it. If the bird moves as far as 10 territories the search would need to cover a few hundred territories. Very few population studies cover this kind of area in sufficient detail to locate all individuals after dispersal. If finding dispersers is difficult enough, observing them 'en route' is virtually impossible without the aid of miniature transmitters, and even this has so far been achieved for only a handful of vertebrate species except for the largest mammals and birds (e.g. Small *et al.*, 1993).

Given these constraints, the present chapter on Nuthatch dispersal has necessarily been assembled from two quite distinct sources of information. First, detailed local studies, such as Löhrl's or my own, can provide information on the arrival pattern of dispersers into a small study area: when, how many, and how they behave. However, no information is obtained on the origin of such immigrants. Considering that this kind of study is often done in areas with no more than 20 breeding pairs and surrounded by other suitable habitat, any disperser settling in the study area might just as well have been hatched only a few territories away. Emigration is even harder to document in this kind of study, and often indistinguishable from losses due to mortality.

The second source of information is recapture data obtained from large-scale ringing studies, where birds are ringed and recaptured or observed at different locations. Standardized nestbox programmes are particularly useful in this respect if they include a large number of study sites and if distances between sites are of the same order of magnitude as typical dispersal distances. Such studies may provide a good estimate of distances moved, but offer little or no detail about when the movement took place, the bird's social status before and after moving, or the route it might have followed. Since Nuthatches are infrequent users of nestboxes, only a few large-scale studies provide enough data to draw any conclusions (Matthysen and Schmidt, 1987; Winkel, 1989). Recapture data from international ringing schemes have the advantage that they cover the full range of dispersal distances, but otherwise provide even less detail. For instance, when we analysed recoveries in the Belgian and Dutch Ringing Schemes (Matthysen and Adriaensen, 1989b) we had to disregard all recoveries within 20 km of the ringing location, because of the large uncertainty about the actual site coordinates within the data bank.

NATAL DISPERSAL DISTANCES

Natal dispersal is usually defined as the distance between an individual's site of birth and the place where it first reproduces (Greenwood and Harvey, 1982). I use an extended definition here, as the distance between the natal site and the first site where a bird settles, that is, where it is found in late summer or later. In most cases the difference is probably insignificant since movements after the first summer are limited.

The first dispersal data on Nuthatches were published by Berndt and Sternberg (1968) using results from the Braunschweig nestbox programme. Winkel (1989) published a more detailed analysis using more recoveries from the same area, which showed the same general picture. The median dispersal distance between hatching and first breeding was 700 m in males and 825 m in females (not significantly different). Twenty-five per cent of all individuals were recovered within 400 m (males) and 465 m (females), and only 25% moved farther than 1250 m (males) and 1650 m (females). Two birds had a dispersal distance of zero (one male and one female) since they bred in the same nestbox where they hatched. The longest recorded movement within the study area was 27 km, but a considerable number were found outside the area with a maximum of 290 km. However, as the chance of recovery is clearly much lower outside the area with nestboxes, it is not possible to give an accurate estimate of the distribution over the whole range of distances. In a different part of Germany, a comparable but smaller data set gave very similar results (Fig. 55). Here maximum distances (within the study area) were 6.1 km in males and 11.3 km in females with median values of 1200 and 800 m, again without a significant difference. Most of these individuals were trapped at feeding sites or when roosting in nestboxes, rather than breeding. Dispersal distances appeared not to be related to the date of fledging (Matthysen and Schmidt, 1987).

The fact that males and females disperse equally far contrasts with many other bird species where females move farther. Greenwood (1980) attributed this general pattern to the need for males, but not females, to establish a territory. Since both sexes of the Nuthatch are territorial, their dispersal pattern does not contradict this hypothesis (*cf.* the Magpie; Eden, 1987). Nevertheless, Matthysen and Schmidt (unpubl.) found that female, but not

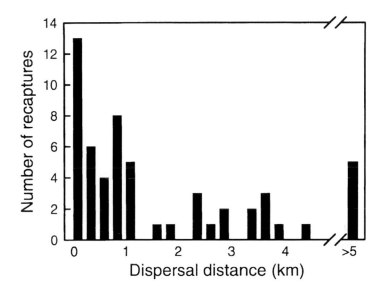

Fig. 55 *Frequency distribution (per 250 m class) of natal dispersal distances in the Frankfurt study (same data as in Matthysen and Schmidt, 1987).*

male, dispersal distance was correlated with the number of breeding pairs in the population, suggesting at least some difference between the sexes. A speculative explanation is that transient females suffer more than males from their subordinate position (*cf.* previous chapter) and will more rapidly leave areas without vacancies.

In agreement with their generally limited movements, the homing capabilities of Nuthatches appear to be relatively small, though only one experiment appears to have been published. In Latvia none of the birds released at more than 20 km of the ringing site returned (Vilks and Vilka, 1961), and of those released at 1–4 km, only about half returned (Vilks, 1966).

In many cases it is impossible to estimate mean dispersal distances since a large but unknown proportion of dispersers moves out of the study area. In the fragments study, for instance, I estimated that as many as 80% of all dispersers left the study area of *c.* 200 km². I tentatively estimated the median distance to lie somewhere between 3 and 10 km. This estimate is several times higher than those of Matthysen and Schmidt (1987) and Winkel (1989), probably because of the more open landscape in which the data were collected (see Chapter 10).

When we move to the scale of general ringing schemes, it is again impossible to estimate mean dispersal distances since short movements are less likely to be detected or reported. The only detailed analysis on the Nuthatch was by Matthysen and Adriaensen (1989b) using movements by Belgian and Dutch birds. Only 12 out of 122 birds moved more than 20 km,

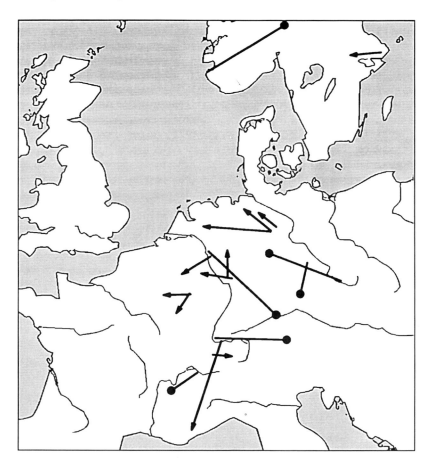

Fig. 56 *Ringing recoveries over more than 100 km. Arrows indicate recapture sites of birds ringed as nestlings, dots are recaptures of birds ringed as full grown individuals (from Zink 1981).*

with a maximum of 102 km. Again there was no correlation between fledging date and recovery distance. There are various additional sources for large dispersal distances, as exemplified by Fig. 56 showing 17 recoveries over 100 km. The record holder is a Nuthatch that moved 460 km south-south-west from the upper Rhine to south-eastern France. In Britain the longest recorded movement is 87 km south-south-west (Mead and Clark, 1991) and only six movements exceeded 10 km (Cramp and Perrins, 1993). Movements over 100 km are also reported from Sweden, Norway, Luxemburg and Switzerland (Plattner and Sutter, 1947; Haftorn, 1971; Melchior *et al.*, 1987; Stolt, cited by Glutz von Blotzheim, 1993). Long-distance movements are also implied by observations of white-breasted birds well into the brown-breasted range. These include observations (perhaps not all of them equally reliable) in northern Switzerland (Hartert, 1910–1922), Thüringen (Kleinschmidt, 1928), the Netherlands (van den Brink, 1951), Niedersachsen (north-west Germany; two individuals in close company, Scherner, 1983) and the Isle of Man (Nixon and Nixon, 1985). These birds must have travelled between 200 and 700 km from the nearest white-breasted populations in south Norway, Denmark or eastern Europe.

THE TIMING OF NATAL DISPERSAL

Most authors agree that young Nuthatches leave their parents within a few weeks of fledging. In the previous chapter I mentioned two fledglings that settled close to the natal territory about 8 days after leaving the nest. There are two cases of nestlings recovered more than a kilometre from the nest only 10 days after fledging (Stechow, 1937; Matthysen, unpubl.). The onset of dispersal at 8–10 days corresponds with a phase of restlessness, as described in Chapter 6. When monitoring the presence of colour-ringed fledglings on their natal territories, Currie and Matthysen (unpubl.) found a marked acceleration in the rate of disappearance around day 10 which probably also marks the onset of dispersal. However, there was considerable variation in timing, with some young staying up to 28 days after fledging (Fig. 57). At least some of this variation reflects within-brood differences, since two fledglings were known to have left the territory while other siblings remained behind (see Stechow, 1937, for a similar case). Some birds

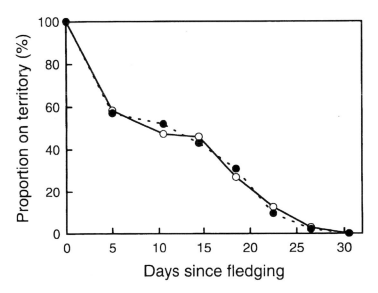

Fig. 57 *Decrease in the proportion of fledglings on the natal territory in a large forest (filled circles, means of six families) and forest fragments (open circles, eight families). Currie and Matthysen, unpubl.*

do travel over considerable distances in the first few days of dispersal: four juveniles in Braunschweig travelled 5–15 km before the end of June (Winkel, 1989), and a nestling in Antwerp had travelled 8 km when found dead 10 days after fledging (Matthysen, unpubl.).

Several lines of evidence suggest that, as summer proceeds, dispersal distances continue to increase, and probably the number of dispersers as well, leading to a second dispersal peak in late August and September. For instance, in the Belgian and Dutch ringing records, the proportion of juveniles recovered far from the natal site (i.e. in another municipality) increased from 30% in June and July to 60% in late summer (Matthysen and Adriaensen, 1989b). In the Peerdsbos I found no immigrants ringed outside the forest in June or July, but five of them between late August and October. Also, Berndt and Dancker (1960) found that mean recovery distances of birds ringed in Braunschweig increased from about 15 km in August and early September to nearly 100 km for birds found after mid-September. In areas with few breeding birds the number of observations typically increases in August–September (e.g. Fig. 58). In Latvia, Vilks (1966) was able to trap 28 individual Nuthatches at a single feeding site between mid-August and mid-November, simply by removing all newly caught birds immediately. After the peak in August–September dispersal seems to decrease again rapidly, as illustrated by the number of new birds settling per month in the Peerdsbos (Fig. 59) as well as the number of transients per month (Fig. 60). Note that the latter numbers are corrected for observation time, since transient birds easily escape detection.

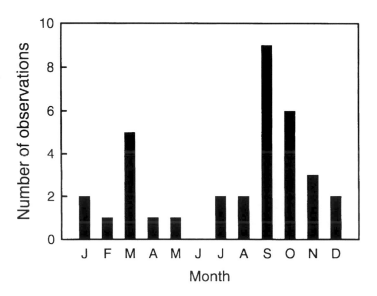

Fig. 58 *The number of Nuthatches observed per month in the city of Amsterdam from 1910 to 1979. Data from Schoevaart (1981).*

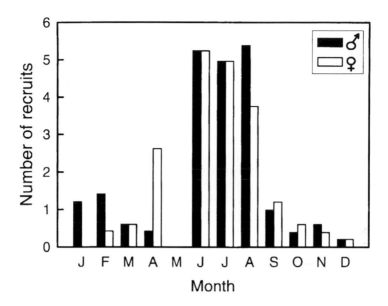

Fig. 59 *The number of male and female Nuthatches settling per month in the Peerdsbos study area per month (means for 1982–1987). After Matthysen (1988).*

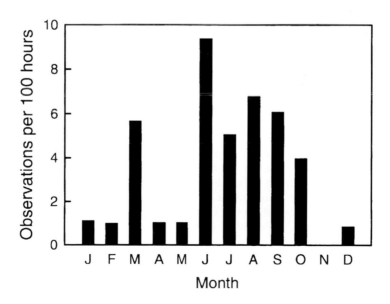

Fig. 60 *The number of transients observed per month in the Peerdsbos study. After Matthysen (1988).*

138 *Dispersal and Migration*

The amount of movement in late summer and autumn seems to vary markedly between years. Delmée (1948, 1949) reported an unusual number of observations in north-west Belgium in the autumn and winter of 1947–1948, which was followed by breeding attempts well outside the birds' normal range. Similar peak years seem to occur regularly in western and central Europe (e.g. Zink, 1981) (see below for a discussion of 'irruptions'). The autumn of 1996 was the most recent such year in Belgium, with Nuthatches appearing in several small forest patches or parks where they are not normally observed. One bird stayed for several weeks at the University of Antwerp campus, the first such observation in 15 years.

In summary, the available data suggest a rapid initial dispersal phase in June and July during which juveniles leave the natal territory, and a second phase involving longer distances in August–September. This implies that some birds settle temporarily in early summer and resume their wanderings later, and is in accordance with the disappearance of many non-territorial residents in August. The rapid initial dispersal is probably a response to the limited availability of high-quality territories (see previous chapter). It is less clear why there should be a second dispersal phase in late summer. Perhaps there is a tendency to postpone further movements until the end of moulting. On the other hand, birds that have waited for a vacancy to appear may 'decide' to move elsewhere before winter approaches.

DISPERSAL OR IRRUPTION?

Some characteristics of the autumn dispersal movements have led authors to consider them as invasions or irruptions. The main reasons are the increase in movements in high-density years (e.g. Fig. 61) and the dominant west or south-west direction of movement. The latter tendency is very noticeable in longer movements, both in the Braunschweig study and in the analysis of Belgian and Dutch ringing records (Fig. 62). In some autumns Nuthatches are even seen on well-known migratory routes, behaving in the same way as other migrants. In the Schwäbische Alb in Switzerland, for instance, up to 40 birds passed south-west from September to early October each year, their number closely correlated with south German population densities (Gatter, 1974). Some birds flew more than 1 km without resting, flying at heights between 20 and 40 m, but most stayed close to cover. The peak occurred in the morning between 09:00 and 10:00 h. In south-eastern France on the Col de la Golèze, movements also peaked in the second half of September but somewhat later in the day, between 10:00 and 12:00 h (Frelin, 1975). Migration-like passage of Nuthatches has also been observed occasionally in south Norway, south Sweden (Falsterbo) and the Rybachy islands in the Baltic Sea (Eriksson, 1970; Griffin *et al.*, 1984). Lack and Lack (1953) saw a Nuthatch that 'behaved like a migrant' crossing a bay 1 km wide north of the Pyrenees in October, and Moore (1969, cited by Cramp and Perrins, 1993) saw one flying in off the sea in Devon (south-west England). Vagrant Nuthatches have also been recorded

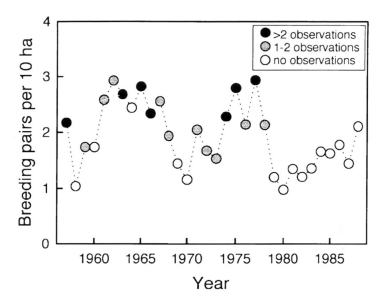

Fig. 61 *The occurrence of long-distance dispersal (> 5 km) in relation to population density (pairs in nestboxes) in the Braunschweig area. Data from Winkel (1989).*

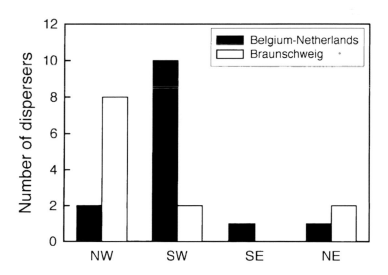

Fig. 62 *The number of long-distance dispersers by compass quadrants in Belgium and the Netherlands (Matthysen and Adriaensen, 1989b) and Braunschweig (Winkel, 1989). Minimum distances are 20 and 30 km, respectively.*

offshore on Helgoland, Hilbre Island in northern England, the Isle of Man and (unconfirmed) on Fair Isle in Scotland (Harrap and Quinn, 1996).

What is remarkable, however, is that the dominant orientation to the west can also be found in dispersal data at smaller scales (less than 100 km) (Table 17). There is no similar directional bias in movements in later life, although it must be admitted that the number of observations is rather small, and the west/east ratio is not significantly different from that of earlier movements (Table 17). In any case, it seems that the tendency to move

TABLE 17: *The proportion of dispersal movements to the west (vs. east) in dispersal studies at different scales*

Area	Distance range (km)	West (%)		Source
		Natal	Later	
NW Europe	100–460	92 (12)*	50 (6)	Zink, 1981
Germany	30–290	83 (12)*	–	Winkel, 1989
Benelux	20–102	86 (14)*	33 (3)	Matthysen and Adriaensen, 1989b
Germany	0–11	69 (42)*	44 (9)	Matthysen and Schmidt, unpubl.
Belgium (fragments)	0–9	73 (15)	60 (5)	Matthysen, unpubl.
Belgium (Peerdsbos)	0–8	43 (7)	60 (5)	Matthysen, 1988
All studies	0–460	75 (102)*	50 (28)	

NOTES:
1. 'Natal' refers to dispersal by birds caught before their first autumn, 'Later' to birds caught in their first autumn or later.
2. Binomial test: * $p < 0.05$, significantly different from the expected 50%.
3. Sample sizes are shown in parentheses.
4. The total proportions for 'Natal' and 'Later' dispersal do not differ statistically from one another (Fisher's exact test, $p > 0.1$).

west or south-west is not necessarily associated with long-distance movements, and should not necessarily bring dispersers into more favourable areas, which is usually assumed to be the purpose of irruption movements. Furthermore, Matthysen and Adriaensen (1989b) showed that long-distance movements in Belgium and the Netherlands did not correspond closely with the irruption years mentioned for western and central Europe by Zink (1981). This again suggests that 'irruptions' may in fact be fairly local phenomena, triggered by high population densities and perhaps, but not necessarily, accompanied by a stronger than usual tendency to move west to south-west.

A good explanation for the non-random orientation of movements is still lacking. One might argue that moving south-west offers an advantage to some long-distance dispersers, which can spend the winter in a slightly more benign climate, but for the majority of birds this advantage must be very small indeed. Also, there is no real evidence that dispersers may return east or north-east after the winter, though a few records of eastward movements of birds captured in autumn are suggestive (Vilks, 1966; Zink, 1981; Matthysen and Adriaensen, 1989b). Another possibility is that dispersing Nuthatches have no real preference but tend to be 'dragged along' with other migrating birds, for instance when hesitating to cross large open spaces (Matthysen and Adriaensen, 1989b). However, the majority of Nuthatches passing in autumn appear to travel singly (e.g. Gatter, 1974).

MOVEMENTS AFTER SUMMER

After the end of the summer/early autumn dispersal period, immigrants continued to appear in small numbers in the Peerdsbos study area throughout winter and early spring, the majority of unknown origin (Figs 59, 60). There was no evidence that other birds emigrated in the same period, but this would have been difficult to detect anyway. Five immigrants had been ringed at distances between 1 and 8 km outside the study area. Two were first-year birds when they immigrated, two were of unknown age, and one was at least 4 years old. Four 'immigrants' had been present in the study area the previous summer as either residents or transients; it is not unlikely that they spent the intervening period just outside the borders of the study area and then moved a short distance into the study area again.

Whereas the arrival of male immigrants did not show a distinct seasonal pattern, there was a clear immigration peak in females in early spring, reaching its height in April (Fig. 59) with a similar pattern shown by the transient birds but with a peak in March (Fig. 60). As a consequence, in some years there was an excess of solitary males in early spring, which were joined by immigrant females later on (Matthysen, 1988). There is no clear explanation for this difference in timing between the sexes. Spring movements have been observed in other populations as well, but with no comparable sexual difference. Löhrl (1958) gives no quantitative information but it is clear that at least some birds, including pairs, entered his study area in this period. In Sweden, immigration started later with few or no immigrants before March (Enoksson, 1987). In the latter study there was evidence that some birds left the area in autumn to spend winter in a nearby area, and returned in spring. There is other anecdotal evidence that Nuthatches may spend autumn or winter in areas where they do not breed, for instance in Slovakian spruce forest (Turček, 1956), in holm oak forest in central Spain (Tellería and Santos, 1995) and in high-altitude conifer forest in the Harz mountains (Zang, 1988). In Britain, observations of Nuthatches in rural gardens were more common in winters with a poor beech mast crop (Thompson, 1988, cited by Marchant *et al.*, 1990). Ptushenko and Inozemtsev (1968) likewise comment that part of the Nuthatch population (in the Moscow region) moves to human habitation in winter. Finally, Melchior *et al.* (1987) have reported groups of 10–50 individuals in Luxemburg in particular winters, which might also represent aggregations of dispersers in areas unsuitable for breeding. In none of these cases (except Enoksson's) is there solid evidence that the Nuthatches actually left these areas to breed elsewhere, the obvious alternative being that few of these birds ever survive to breed in marginal habitat conditions.

Some studies, including those by Pravosudov (1993a) and Matthysen and Schmidt (1987), found very limited movements between winter and spring. Even in the fragments study, which was designed to find movements over short to moderate distances, very few birds moved to another territory between summer and spring (see Chapter 10).

BREEDING DISPERSAL

Breeding dispersal is defined as a movement between one breeding site and the next (Greenwood and Harvey, 1982). In the Peerdsbos study there were various sources of evidence for such movements by Nuthatches. First, there were at least five certain cases of immigration by adults: one ringed female in October (see previous section), and one unringed female and three unringed males in June–July. Second, two females left the area in spring and returned the following summer. Third, one adult female was seen for the last time about 300 m outside her former territory before she disappeared in August, suggesting that she had started to wander away. Other studies have also provided anecdotal evidence for breeding dispersal. In the Frankfurt study, a female moved over 1300 m between her second autumn and her fifth spring (Schmidt *et al.*, 1992). The Belgian ringing records include a female ringed as an adult in June and recovered in October almost 100 km to the west (Matthysen and Adriaensen, 1989b). In the fragments study, a second-year female moved 500 m to a different forest patch after the breeding season (Matthysen *et al.*, 1995). Finally, in a study in Latvia, nine out of 65 ringed adults were recovered between 500 m and 3 km from the ringing site, which suggests some short-distance dispersal (Vilks, 1966).

Two major points emerge from this list of anecdotal records. The first is that most movements seem to occur between June and October, suggesting a distinct post-breeding dispersal phase in summer and early autumn. Second, breeding dispersal appears to be more common in females (8 out of 11 cases) which is in accordance with the higher rate of disappearance of adult females than of males in summer (see next chapter). It is also remarkable that of the 13 adult females that disappeared in the Peerdsbos in summer (over 4 years), only two had bred more than once in the area, and seven were spring immigrants of the same year. Thus, summer movements might be typical of second-year birds. Whether breeding dispersal is also influenced by low breeding success, as in other bird species (e.g. Greenwood *et al.*, 1979) remains unknown. Immigration by some adults in June–July suggests that they may not have raised a brood in that year. In Siberia, Pravosudov (1993a) found that birds with low reproductive success had a lower local survival rate, which could indicate a higher tendency to disperse. I will briefly return to the issue of breeding dispersal in the next chapter, when discussing variation in local survival between sexes and between studies.

IRRUPTIONS OF *ASIATICA* IN EUROPE

In comparison to the rather modest autumn movements of the West- and central European Nuthatches, the Siberian *asiatica* subspecies is a true 'irruption' bird. Although the nearest breeding populations are in the Ural mountains, these birds appear almost yearly in Finland in small numbers, and can be quite numerous in some years. The earliest known irruptions date from the late 18th century (Eriksson, 1970). Notable years of irruptions (as far as they were reported) were 1900, 1944, 1951, 1962, 1963, 1976, 1983, 1987 and 1995 (Svärdson, 1955; Eriksson, 1970; Hildén, 1977; Hildén and Saurola, 1985; Glutz von Blotzheim, 1993; *EuroBirdNet Finland*, November 1995). In such years, *asiatica* birds may be found all over Finland and often in northern Sweden (occasionally Norway), and the Baltic States as well (Glutz von Blotzheim, 1993). The largest irruptions reported are those from 1976–1977, when several thousands were reported from Finland and over 900 from Sweden (Hildén, 1977; Blomgren *et al.*, 1979), and 1995–1996 when nearly 5000 birds were reported in Finland by the end of November (*EuroBirdNet Finland*, November 1995).

The origin of birds in these irruptions is not very well known except that they all belong to *asiatica*. The larger *europaea* breeds to the south and east of Finland and is observed only occasionally, and not particularly in years of irruptions. Invading birds probably follow the taiga belt in a west-north-west direction until they reach Fennoscandia. Eriksson (1970) suggested that

asiatica *(top)* and europaea *(bottom) subspecies.*

their origin should be sought relatively far east, since they have distinctly smaller bills than specimens from western Siberia and the Altai mountains. However, given the considerable variation in bill length with season and probably diet (see Chapter 4) this might not be a very strong argument.

The 1962–1963 and 1963–1964 irruptions in Finland were exceptionally well documented, thanks to repeated appeals to the public in newspapers and ornithological journals. Eriksson (1970) analysed over 300 amateur observations from these years, complemented by 21 days of observation on the behaviour of birds during irruptions. Additional data on the same irruption from Sweden were published by Wahlstedt (1965). The irruptions started in late August but in most places no birds were seen before October or November. This does not necessarily indicate the timing of their arrival, since many were probably attracted to bird tables when temperatures dropped sharply in October. By December, the majority of birds appeared to have settled. They were observed mainly at bird tables near houses surrounded by well-managed open woodland or wooded parks, but did not appear to be very selective in their habitat choice (Eriksson, 1970). Irrupting birds were particularly numerous in the northern parts of Finland

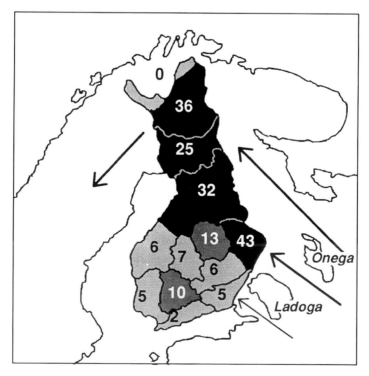

Fig. 63 *Probable invasion routes of* S. e. asiatica *into Fennoscandia. Numbers and degree of shading indicate the number of observations per 100 000 inhabitants in each administrative district. Note the presence of Lakes Onega and Ladoga to the east. Based on Eriksson (1970).*

(Fig. 63), a pattern that was seen in later irruptions as well (Hildén and Saurola, 1985) and may be due to the barrier effect of the lakes Onega and Ladoga which may deflect movements to the north-west. Very few, if any, attempted to cross the Gulf of Bothnia, since no observations were reported from the Åland Islands, and observations in Sweden were restricted to areas north to north-west of the Gulf (Eriksson, 1970). Even in 1976–1977 when Nuthatches were found much further south in Sweden (down to 62°N) relatively few were reported from coastal areas (Risberg, 1977; Blomgren *et al.*, 1979). In 1983–1984, however, at least one Nuthatch did cross the Gulf (Hildén and Saurola, 1985).

By January of these irruption years, many Nuthatches had disappeared from their usual observation sites (Eriksson, 1970). Several were found dead with no visible injuries so they probably died of exhaustion or starvation, and the number of observations was reduced to about one-third by late January. This was perhaps caused by a very cold spell (−20°C) in mid-December, although later cold spells in January and February did not seem to have the same effect (Eriksson, 1970). In April the numbers decreased again, and only a few remained by the end of the month. This should not necessarily indicate return movements, since it could also be due to a shift away from human habitation in search of breeding areas.

Invading Nuthatches did not appear to be very social. The largest groups recorded during the 1962–1963 irruption contained five and six birds, both of them observed in September. Invaders behaved aggressively towards conspecifics and other species such as tits. They had relatively large home-ranges, on average about 5 ha, sometimes more than 10 ha. They were mainly observed feeding on anthropogenic food, at bird tables, compost heaps and slaughterhouse offal. Some even entered houses and learned to take food from the hand. Only about 10% were seen in pairs, and this particularly in high-density areas. Some pairs were reported as early as October, but most of them in December and January. Males sang from January onwards, and in April some showed interest in nestboxes (Eriksson, 1970). After the 1976–1977 irruption at least three pairs attempted to breed, one of them with success (Blomgren *et al.*, 1979).

In the autumn and winter following the 1962–1963 irruption, Nuthatches were again observed but in much smaller numbers, and this time with a stronger concentration in the southern and central parts of Finland. Eriksson (1970) suggested that these were birds that remained from the previous irruption, and had moved southwards after the breeding season. Similarly, in the winter and spring following the 1976–1977 irruption in Sweden several individuals were seen at bird tables in Sweden, and two pairs bred in 1978 (Blomgren *et al.*, 1979).

IRRUPTIONS IN OTHER POPULATIONS

Given the large range of the *asiatica* subspecies throughout most of Siberia, it would be interesting to know whether irruptions also occur within the Asian part of the distribution, but little information is available.

There is one report of 'several dozens of thousands' of Nuthatches flying southwards from Mimoye in central Siberia (Harrap and Quinn, 1996). Johansen (1944) observed an increase in movements in western Siberia in September and October, and thought it possible that north–south movements were involved. Irruptions may also be expected from the *arctica* subspecies in north-eastern Siberia, but again little is known except that it is believed to make limited movements south and east (to *c.* 60°) in winter (Harrap and Quinn, 1996).

Further to the south-east, however, the subspecies *amurensis* makes notable but highly irregular irruptions. Its population seems to be largely sedentary with some annual migration southward to Korea, but mass irruptions have been recorded at least twice this century. In 1944 large numbers of Nuthatches moved north-east along the shores of the Sea of Japan in Ussuriland, mainly from 26 September to 2 October, often together with Marsh Tits, and sometimes crossing open areas (Belopolskiy, 1950, cited by Polivanov, 1981). In October 1980, massive south-west migration was recorded along the same coast, with an estimated number of 30 000–40 000 passing within a few days (Banin *et al.*, 1984). Some small groups moved along creeks and rivers in woodland, but migration was particularly noticeable along the rocky shoreline, with birds passing in 'waves' of 40–50 individuals. The number of migrants gradually decreased during the day. While on the move they did not associate with migrating Coal and Marsh Tits, but during short foraging stops they did join mixed-species flocks. Of 21 captured birds, 10 had average to high fat loads (Banin *et al.*, 1984). Polivanov (1981) never observed such mass movements during 16 years in the same region, which underlines their occasional nature. It seems plausible that irruptions in this subspecies (and perhaps *asiatica* as well) are triggered by widespread seed crop failure, in particular of the Korean pine (Banin *et al.*, 1984). Apparently some individuals also straggle northwards during autumn migration, since every year a few *amurensis* are caught in the Magadan area further north, within the breeding range of *asiatica* (V. Pravosudov, pers. comm.). A puzzling question that remains unanswered, however, is why a north-east direction is cited for the 1944 irruption.

To conclude this section, there is an interesting observation mentioned by Brehm (1920) from the area of Sambor, now south-eastern Ukraine, within the range of the white-breasted *europaea* subspecies. After an explosion of bark beetles in 1834, Nuthatches appeared in very large numbers (see citation introducing next chapter) and foraged largely on the adult beetles, which they consumed 'by the millions'. They stayed for up to 4 months and gradually left the area by October. Given the early onset in July, this remarkable phenomenon appears unlikely to have been an irruption of *asiatica*, but rather a local concentration of dispersing juveniles from a wide area. No other observations of this type have been reported.

Young Nuthatch chased by Jay.

CHAPTER 9

Population Dynamics

Da erschienen im Juli desselben Jahres Kleiber in unglaublicher Menge, so daß alle Wälder und Gärten von ihnen wimmelten, ja sie flogen vielfach in die Zimmer

('In July of the same year, Nuthatches appeared in incredible numbers, so that all woods and gardens were swarming with them; they even flew often into rooms')

Brehm (1920)

Natural populations of animals change in numbers between seasons and years because of the processes of birth, mortality, immigration and emigration. A population can be stable in the long run only if reproduction or mortality are regulated by population size, that is, if fewer individuals are born and/or more of them die when population density increases. Immigration and emigration can also have regulating effects, but on a local scale only. For instance, populations in low-quality habitat may not be viable by themselves ('population sinks'), but may persist because they are continually supported by net immigration from more productive areas ('sources'). In Chapter 10 I discuss the possibility that small forest remnants represent such a population sink.

147

The effects of population density on Nuthatch breeding success are discussed in Chapter 6. In two studies, the number of fledglings per pair decreased by about 30% for a twofold increase in population density (Nilsson, 1987; Schmidt *et al.*, 1992). Nevertheless, as in many other bird species, many more juveniles fledge than would keep the population at equilibrium. If about 50% of the adults die each year, only one juvenile per breeding pair in a stable population should survive. Consequently, mortality in the first year is severe, and it is easy to imagine that population changes are more sensitive to variation in survival rates, both of young birds and adults, than in reproductive rates. The first part of this chapter deals with causes of mortality and survival rates. The second part addresses the question of population regulation, drawing heavily on Nilsson's (1987) study in south Sweden, complemented with information on social behaviour as a regulating mechanism. The third part of the chapter discusses trends in population sizes in various parts of the Nuthatch's range.

CAUSES OF MORTALITY

As for many other bird species, the causes of death for Nuthatches are known only from some anecdotal observations. Probably the most important predator in most of Europe is the Sparrowhawk, even though Nuthatches make up only a small part of its diet. Uttendörfer (1952) found 257 remains of Nuthatches in 50 000 Sparrowhawk catches and 49 victims in 6000 Tawny Owl catches, in both cases less than 1% of all prey items. In southern Germany the Pygmy Owl is known as a predator as well (8 of 900 catches or 0.8%; König *et al.*, 1995). Other predators include Little Owl, Long-eared Owl, marten and feral cat (Cohen, 1960; Delmée *et al.*, 1972; Zukal, 1992; Yanagawa and Shibuya, 1996). Cats are thought to be particularly dangerous to females when these are searching for nest material on the ground (Glutz von Blotzheim, 1993). Predation by Sparrowhawks and corvids on newly fledged Nuthatches has already been mentioned in Chapter 6. Nuthatches typically 'freeze' for a few minutes when a predator appears. Wilde (1973) describes how a Nuthatch being pursued by a Sparrowhawk suddenly dived and landed on a trunk just above the ground, head downwards, with the body at an angle of about 60° from the trunk; he interpreted this as a camouflage posture.

Remarkably, Nuthatches were found to be common victims of collision with windows or buildings in Japan, and this was the major cause of death reported for birds found dead (15 out of 19; Yanagawa and Shibuya, 1996). The contribution of disease and parasitism to mortality has not been studied. There are some reports of bill deformities (Nowak, 1965; Dhondt, 1967) and one of a leg deformity caused by a mite in the White-breasted Nuthatch (Hardy, 1965). One female was found dead with her toes stuck together with resin (Nothdurft, 1978). The maximum recorded age is 9.5 years (Glutz von Blotzheim, 1993).

SURVIVAL BETWEEN BREEDING SEASONS

Since Nuthatches are not very likely to leave their territory once they have completed the first breeding season (see previous chapter), adult survival can be reasonably estimated from 'local survival', that is, the proportion of birds remaining on the same territory (or at least within the study area) between breeding seasons. Table 18 shows a compilation of such local survival rates. On average slightly more than half of the adult birds survive to the next breeding season, which is comparable to other non-migrating temperate-zone passerines (Dobson, 1987). Nevertheless, there is considerable variation between studies, from 42% in Sweden to 67% in Siberia (Table 18). This variation is more pronounced in females than in males (Fig. 64): if survival is high both sexes survive equally, but if survival is reduced this is more pronounced in females. One possible explanation is that females suffer a higher mortality in harsh conditions because they are socially subordinate and have more restricted access to food. Nilsson (1987; no details given) indeed found that females survived less well than males in cold winters without beech mast. However, it is not clear whether this hypothesis explains the differences between studies, since survival is actually highest (and equal in the two sexes) in Siberia where winters are most severe – though the availability of food there is not known. An alternative hypothesis is that some of the variation in Fig. 64 reflects differences in breeding dispersal, which is more common among females (Chapter 8). Thus, populations with low local survival may in reality have a higher incidence of breeding dispersal, and this would be particularly pronounced in females. Two arguments give further support to this hypothesis. First, from the Peerdsbos study we know that sex-related differences in local survival are most pronounced in summer (see below) which coincides with the timing of breeding dispersal. Second, in the fragments study, breeding dispersal is thought to be very limited (see Chapter 10) and in this study both sexes did survive equally well. The Siberian data would also seem to fit the hypothesis since, in this population, where competition for territories is very intense (see Chapter 7), few individuals should risk leaving their territory even after breeding failure. Unfortunately there appears to be a

TABLE 18: *Annual local survival rates (in %) of adult Nuthatches, measured between breeding seasons*

	Males	Females	Mean	Source
Belgium (Peerdsbos) (Bp)	55	47	51	Matthysen, 1988
Belgium (fragments) (Bf)	59	60	59	Matthysen, unpubl.
Sweden (Sn)	51	44	47	Nilsson, 1982
Sweden (Se)	46	39	42	Enoksson, 1988b
Poland (P)	54	45	49	Wesołowski and Stawarczyk, 1991
Siberia (SI)	65	70	67	Pravosudov, 1993a

NOTE:
Letter codes in parentheses refer to Fig. 64.

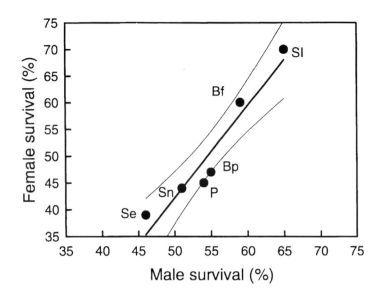

Fig. 64 *The relationship between male and female survival rates in six studies (data and codes from Table 18). Note that the regression line ($R^2 = 0.92$) is significantly different from the diagonal where male and female survival are equal, as shown by the confidence interval (curved lines).*

contradiction here with the observation that, in this study, more individuals disappeared after breeding failure, which was thought to reflect dispersal (Chapter 8).

An alternative method of explaining differences in survival between the sexes is to compare the age structure of breeding birds that were ringed as nestlings. In the Braunschweig study area, male breeders were indeed on average older than females (2.5 *vs.* 1.9 years), supporting the idea of lower female survival (Winkel, 1989). Another testable prediction is that lower survival of females, in combination with the slight excess of males at the fledgling stage (see Chapter 6), should lead to a surplus of males in the population. This has been confirmed by observational data from different populations. In the Peerdsbos study I observed 23 solitary resident males but only four solitary females (at any time of the year; Matthysen, 1988), in the Netherlands Alex Schotman (pers. comm.) observed 38 single males *vs.* four single females in the breeding season, and in south Sweden Hans Källander (in press) found five single males but no females. Of course, single females may more often remain unnoticed than males, but still these differences are remarkable.

SURVIVAL IN THE FIRST YEAR

Survival in the first year of life is more difficult to study because of the higher mobility of first-year birds. Rough estimates can be made, however, based on the proportion of first-year birds in the population in the course of the year. Appendix V provides a tentative life-table for the Peerdsbos population. Using the assumption that 60% of all fledglings survive to independence (Currie and Matthysen,unpubl.), I estimate that about 57% of them will survive from independence to the end of summer, and 55% of these from the end of summer to the next spring. As the Appendix shows, the survival rate changes relatively little after the end of the first summer. In total, less than 20% of all fledglings will survive to breed the next year, which makes sense if, in a stable population, only one fledgling per pair (from an average brood of five or six) survives to replace an adult that dies.

SEASONAL AND ANNUAL VARIATION IN SURVIVAL

Seasonal patterns of survival have been studied in Belgium (Matthysen, 1988) and two areas in Sweden (Nilsson, 1982; Enoksson, 1988b). In the Peerdsbos study (Fig. 65) the majority of birds disappeared in winter with little difference between the sexes, except that more adult females disappeared in summer (26% between 1 June and 1 September, 6% in males). The most likely explanation seems to be that some females emigrate shortly

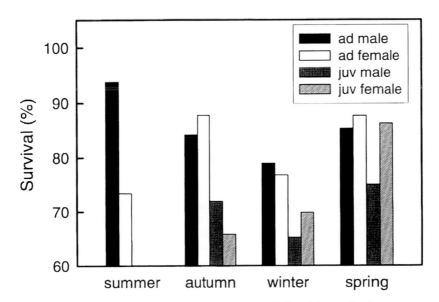

Fig. 65 *Local survival per season, sex and age class in the Peerdsbos study. Seasons start on 1 June, 1 September, etc. From Matthysen (1988).*

after the breeding season (see also the previous section and Chapter 8). Local survival of juveniles was lower than that of adults, particularly in autumn but also in winter. In Nilsson's (1982) study there were more disappearances in early winter (November and December) than in the Peerdsbos, which may reflect a true difference in mortality, owing to the harsher Swedish climate (Table 19). On the other hand, fewer birds in Sweden disappeared in summer (April–September), but this is more likely to reflect a difference in breeding dispersal. Enoksson's (1988b) data are not directly comparable with the other studies, but also show a high rate of disappearance from December to March.

TABLE 19: *Seasonal survival (%) of adult birds in Belgium (Peerdsbos; Matthysen, 1988) and Sweden (Nilsson, 1982)*

Period	Sweden	Belgium
1 June–1 September	95 (60)	84 (97)
1 September–1 November	91 (57)	93 (84)
1 November–1 January	83 (52)	90 (78)
1 January–1 April	77 (43)	76 (83)
1 April–1 June	100 (30)	89 (65)

NOTE:
Sample sizes in parentheses.

The most likely causes of variation between years are winter harshness and food availability. The best available data are those of Nilsson (1987), who calculated winter survival by comparing the numbers of birds present in autumn with those in spring. He found that these overwinter losses were closely related to the mean winter temperature from December to March (Fig. 66), but not to the presence of beech mast. Enoksson (1988b) found a similar result based on survival rates of marked individuals: in three cold winters (mean temperature between –5 and –6°C) only 57% of adults and 31% of juveniles survived, whereas in two warm winters (temperatures –1 to +2°C) 83% and 57%, respectively, survived. Again, the amount of food (in this case, a provisioning experiment with sunflower seeds) did not influence survival.

The data from the Peerdsbos study are not very conclusive, perhaps because they are based on only 5 years (Matthysen, 1989d). Juveniles tended to survive better in mild winters and when beech mast was available, but no trend was apparent in adult survival (Matthysen, 1989d). However, survival during autumn (from 1 September to 1 December) was markedly higher in years with beech mast, particularly in first-year birds: 48% and 65% survived in two autumns without beech nuts, *vs.* 77% and 80% in two mast years (Matthysen, 1989d). The difference in survival was found only in territories that contained beech trees, which supports the idea of a causal relationship between beech mast and survival, which need not necessarily be the case (see Tinbergen *et al.*, 1985, for an example in Great Tits). This

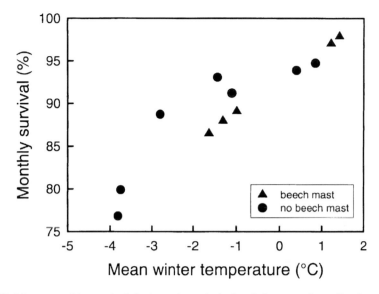

Fig. 66 *Mean monthly survival during winter (calculated from numbers alive in autumn and spring) in relation to mean winter temperature in Sweden (December–March) and the presence of beech mast. After Nilsson (1987).*

was no longer true in winter, hence the tendency for better winter survival in mast years remains unconfirmed (Matthysen, 1989d). Thus, three different studies failed to find a correlation between food abundance and survival in winter, but at least in one case there is such a relationship during autumn. The relationship between food abundance and autumn population size, and the possible roles of mortality and emigration, are taken up later in this chapter.

RECRUITMENT AND POPULATION REGULATION

Figure 67 shows how the Peerdsbos population increased in size each summer, with, on average, the number of birds present in late summer about twice that of the breeding population. This period of summer recruitment was followed by a gradual loss over the winter. Some of the variation between summers is due to the number of non-territorial residents which varied between one and eight per year. The density of territorial pairs, on the other hand, changed relatively little between summers (from 5.4 to 6.3 pairs per 10 ha) and was more constant than the spring population (Fig. 67). This suggests that regulating factors may influence the population in the summer period. The very existence of non-territorial birds – the number of which was related to the size of the adult population, hence to competition for territorial vacancies – suggests that there is an upper limit to the size of the territorial population (Matthysen, 1989c).

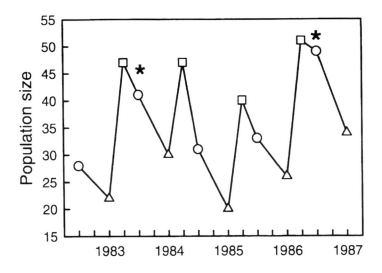

Fig. 67 *Seasonal variation in population size during 5 years in the Peerdsbos study. Squares = 1 September, circles = 1 December, triangles = 1 June. Asterisks indicate autumns with beech mast. After Matthysen (1989d).*

The idea of regulation in summer through territorial behaviour has been demonstrated even more clearly by Nilsson (1987). In this study, the increase in population size from spring to autumn was not closely related to the number of young produced, but rather showed an inverse relationship to the breeding density: the more adults present, the fewer juveniles were added. However, recruitment was also higher in years when food (beech mast) was abundant. Nilsson further showed that variation in winter mortality was not density-dependent, and concluded that recruitment during summer was the only factor regulating the size of the population. As a consequence, the breeding population was more variable than the autumn population, and at least some of this variation could be attributed to the severity of the winter.

RELATIONSHIP BETWEEN FOOD AND AUTUMN POPULATION SIZE

Several studies have shown that Nuthatch population sizes in autumn or even in spring are correlated with food supply. Nilsson (1987) found a correlation between numbers in autumn and the size of the beech crop, and Enoksson (1990a) with the crop of hazelnuts. In the Peerdsbos I also found a larger population size in late autumn in beech mast years (Matthysen, 1989d). Earlier, Källander and Karlsson (1981) demonstrated a correlation between spring population sizes in Germany and beech crops in Denmark. While they had no German beech crop data available, they assumed that crops vary in parallel over wide areas. Data from the Common Bird Census suggest similar patterns in Britain (Marchant *et al.*, 1990). Within Sweden, the correlation with beech mast was better in census plots with more beech trees (Källander and Nilsson, cited by Nilsson, 1987), again suggesting a causal relationship. Further evidence for a direct relationship is the absence of a correlation with the number of acorns (Nilsson, 1987), since these are not normally used by Nuthatches.

Enoksson and Nilsson (1983) and Enoksson (1990a) performed a series of food addition experiments to confirm the causal relationship between food abundance and population size. In the first experiment, food was added between November and February (Enoksson and Nilsson, 1983). The food consisted of sunflower seeds, which were continuously available on a number of feeders throughout the study area. The extra food did not affect winter survival, but, in the following spring, territories were significantly smaller than in a year without food addition. In the second experiment food was added from June to November. This resulted in a larger autumn population (Fig. 68) with smaller territories (see also Chapter 7), and also more non-territorial 'satellites'. Enoksson (1990a) concluded that Nuthatches adjust the size of their territories in autumn to the expected food supply in winter. If there is less food, territories become larger, which presumably prevents a number of individuals from taking up territories. This effectively results in a limitation of population size in relation to food.

My data from the Peerdsbos suggest a somewhat different mechanism for the adjustment of population size to food. I found that summer population

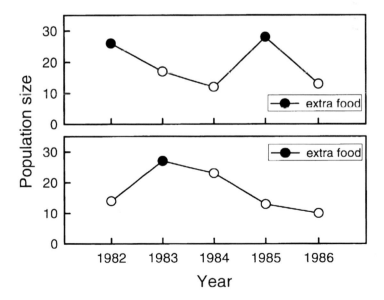

Fig. 68 *Autumn population sizes in two study areas in Sweden in years with and without extra food. Data from Enoksson (1990a).*

size – counted much earlier in the season than in Enoksson's case – was not larger, nor were territories smaller, in beech mast years. In such years, however, more juveniles remained on territories with beech trees throughout the autumn and winter (Matthysen, 1989d). Hence the food supply appeared to influence not the number of juveniles settling, but the proportion remaining through autumn. Since Enoksson's study provides no information on how territories were established, the possibility cannot be excluded that, in Sweden also, more birds disappeared in early autumn in non-mast years, before the census was taken. The major difference would then be whether the remaining birds enlarged their territories into the vacated areas. One possible explanation for the lack of change in territory size is that, in the Peerdsbos study, beech trees were highly clumped and concentrated in a few territories. Thus, birds in beech territories had the opportunity for expansion but still disappeared, while birds in non-beech territories had little reason to expand into areas with low food availability. On a subjective note I would add that it would seem unlikely for territory sizes in summer to be adjusted to food availability, for two reasons. First, young birds settle very rapidly in early summer – at least in Belgium – perhaps without much opportunity to evaluate the expected food level and hence the optimal area to defend. Second, even well-established juvenile territory owners did not visit bird tables in Belgium before early August, long after the territories were settled. The way to resolve this discussion would, of course, be to repeat the food addition experiment in Belgium and in Sweden, and to observe the initial settling behaviour in summer as well as the changes in numbers and territory boundaries thereafter.

CHANGES IN POPULATION SIZE

From the previous discussion it follows that fluctuations in breeding population sizes are mainly determined by (1) the available food in autumn and (2) the severity of winter which influences survival. This has indeed been demonstrated in several studies. Nilsson (1987) found a relationship between the breeding population size and mean winter temperature as well as the beech mast crop. Källander and Karlsson (1981), using population data from Braunschweig (Berndt and Winkel, 1979), found that numbers correlated well with both winter weather and food (beech crop, in this case measured in Denmark). These factors explained 25% and 18% of the variation in spring population size, respectively. Svensson (1981) also found a correlation between breeding census data in south Sweden and winter temperatures, in particular temperatures in December and February. In this study the Nuthatch population fluctuated in parallel with those of Great and Blue Tits, but much less with Coal Tit, and not at all with Marsh Tit populations. Källander (in press) found a correlation with winter temperature in a single study plot in south Sweden, but since winters became gradually milder over the study period (1977–1991) it is difficult to distinguish between weather effects and a long-term population trend. Finally, Zang (1988) also found a good correspondence between mean winter tempera-

ture and the change in Nuthatch breeding population size in the Harz mountains (5 years of data). One study does not conform to the overall pattern, that in Białowieza (Poland) by Wesołowski and Stawarczyk (1991) and Wesolowski (1994). Here no relationships were found between population size or population change and winter cold, seed crop or caterpillar abundance in the previous year. There is no clear explanation for this, except that this population has been steadily increasing during the study period (see also below in this chapter). In such circumstances other factors such as production of young or immigration may have a larger influence on population changes than mere winter survival rates.

Although most Nuthatch populations show larger overwinter losses in cold winters, they do not seem to suffer particularly badly from extremely cold winters. The exceptional 1870–1871 winter resulted in a population decline in Thüringen, according to Brehm (1920). The 1916–1917 winter had noticeable effects on Nuthatches in some English counties, but not in others (Jourdain and Witherby, 1918). The 1962–1963 winter, however, was followed by only moderate decreases (−8 to −16%) in three German studies (Fig. 69), unlike the dramatic losses recorded at the same time in other resident bird species. In the same winter, no particular effect was noticed on Nuthatches in England (Dobinson and Richards, 1964). This does not exclude the possibility of severe local declines, as exemplified by a small nestbox plot in central Germany where the number of broods in boxes fell from seven to zero (Schönfeld and Brauer, 1972). The 1978–1979 winter did show a general pattern of decrease (Fig. 69; the mean change in 10 studies = −20%) but with a lot of variation, including a 43% increase in the Netherlands. Finally, the 1984–1985 winter showed no general trend (the mean change in 9 studies in Fig. 69 = +4%).

LONG-TERM POPULATION TRENDS

In a recent account of Nuthatch populations in central Europe, Glutz von Blotzheim (1993) concluded that there have been no long-term trends in population sizes. As Fig. 69 shows, the picture is a little more compli-

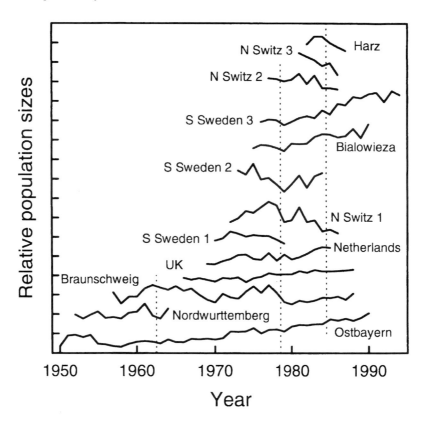

Fig. 69 *Changes in population size in different areas in Europe since 1950. One unit on the vertical axis corresponds to a 100% increase or decrease. Vertical dotted lines indicate three severe winters (1962–1963, 1978–1979, 1984–1985). Data sources: Ostbayern, Germany: Bäumler (1993); Nordwurttemberg, Germany: Löhrl (1967); Braunschweig, Germany: Winkel (1989); UK: Marchant et al. (1990); Netherlands: Jonkers (1990); southern Sweden: 1. Svensson (1981), 2. Nilsson (1987), 3. S. Svensson (pers. comm.); Białowieza, Poland: Wesołowski and Stawarczyk (1991); northern Switzerland: 1. Witvliet (1987), 2. Widmer (1987), 3. a compilation of Lienhardt (1987), Buff (1987) and Kohler (1987); Harz, Germany: Zang (1988). Data are based on nestbox surveys (Switzerland and Germany except Harz), large-scale monitoring (UK, Netherlands, S Sweden except 2) and local censuses (S Sweden 2, Białowieza, Harz).*

cated. During the past few decades populations have been steadily increasing on a countrywide scale in the UK, the Netherlands and Sweden (since c. 1980; see also Källander, in press), and on a more local scale in Ostbayern (Germany) and Białowieza (Poland). Local decreases in the early 1980s have been suggested in northern Switzerland and the German Harz mountains (Fig. 69). In other areas, no consistent trend can be seen. In a study area in Slovakia (not on Fig. 69) the number of breeding pairs was nearly constant from 1988 to 1995 (10–11 pairs per 10 ha, A. Krištín, pers.

comm.). We can conclude that most of the available data show increases or at least stability in Nuthatch numbers, and that decreasing trends have been documented only locally (see also Bauer and Berthold, 1996 for additional information). As I show below, a plethora of reasons has been put forward to explain these trends, but none of them has been thoroughly documented.

In some parts of Europe the increase in numbers has been accompanied by a moderate expansion in range. This is clearly the case in Great Britain where the Nuthatches reach the north-western edge of their distribution. Nuthatches were more common in northern England (north to Durham) in the 19th century than at the beginning of the 20th century when they had retreated into Yorkshire. After 1920 the species expanded again and by 1940 they had occupied their former range, as well as extending their range in Wales (Witherby *et al.*, 1940). Since then the slow northward spread has continued, especially in Cumbria and Northumberland (Parslow, 1973; Murray, 1991; Gibbons *et al.*, 1993). As confirmation, a regression analysis on the Common Bird Census data (Gooch *et al.*, 1991) showed that numbers increased significantly only in the northern parts of England. In Scotland the number of Nuthatch sightings has increased from about two per decade prior to 1960 to more than one per year in the last few decades, culminating in two successful broods in 1989 (Murray, 1991). Even within the London area there may have been an increase since the 1950s (Montier, 1977). The reasons for this increase remain a matter of speculation. One explanation that has been put forward for the increase in the UK is the spread of Dutch Elm disease since 1969, leading to an increase in insect populations on some trees (Marchant *et al.*, 1990). Another possibility is the Nuthatch's increased use of gardens and bird tables (Parslow, 1973). Despite their low population densities, Nuthatches rank as the 17th most frequent species on bird tables, though of course never in large numbers (Thompson *et al.*, 1993). Other artificial food sources may also have increased over time; in the Netherlands, for instance, Nuthatches often forage on maize heaps around farms adjacent to woodlots (A. Schotman, pers. comm.). Their disappearance and later partial recovery in the London area, and their general scarcity in the neighbourhood of large industrial towns, may also be connected with the contamination of trees by soot which reached its peak several decades ago (Parslow, 1973).

Apart from the data for Ostbayern shown in Fig. 69, three other German studies have found increases in the last decades. Busche (1993) attributed the increase in western Schleswig-Holstein (1960–1990) to changes in forestry with the preservation of older trees. The upward trend recorded in Denmark (DOFF, cited by Marchant *et al.*, 1990) may be connected to the increase in Schleswig-Holstein. Möckel (1992) thought that the altered forest structure caused by forest damage (more light gaps), could explain the recent population increase in spruce forests in the western Erzgebirge (1974–1988). Kooiker (1994) found a significant increase in Nuthatch numbers in the urban area of Osnabrück between 1986 and 1993. In other parts of Germany there has been no clear trend or at most a slight increase (Fig. 69). For instance, around Bonn the number of occupied 2.2 km^2 grid cells increased from 83 in 1975 to 96 in 1985 (Erhard and Wink, 1987). Although the data for Braunschweig in Fig. 69 show no clear trend, a marked increase has occurred in at least some areas since the mid-1980s (Bauer and Berthold, 1996). In Switzerland there has been a weakly decreasing trend between 1969 and 1987 but not in all areas (Blattner and Speiser, 1990). A few local studies in northern Switzerland found marked decreases up to 1986, as shown in Fig. 69.

The best documented, but equally unexplained, increase in Nuthatch numbers has occurred in the Białowieza forest, where numbers have increased twofold from the late 1970s to the late 1980s (Fig. 69). The increase seems unrelated to mild winters, changes in habitat quality or predation pressure (Wesołowski and Stawarczyk, 1991). In fact, the data suggest that the increase has not been continuous, but occurred mainly during the early 1980s and now fluctuates around a new, higher equilibrium level (T. Wesołowski, pers. comm.). Of course, local studies cannot exclude the possibility that changes elsewhere have resulted in an increasing 'spillover' into the study area. It is interesting to see that the south Swedish population has started to increase at roughly the same time as the Białowieza population. The Swedish increase, for which there is no satisfactory explanation either, has been accompanied by a gradual range extension to the north. Nevertheless, most of the change in numbers can be attributed to sites where the species was already common (S. Svensson, pers. comm.).

In Belgium and the Netherlands there is good evidence for both increases in numbers and small-scale range expansions. If one compares the tentative distribution map for Belgium in Delmée (1948) with counts from the 1970s (Devillers *et al.*, 1988), it appears that in the northern part of the country, the species has expanded from a few strongholds to a much wider distribution (Fig. 70). This impression is supported by several regional surveys. For instance, in the forested hilltops of north-west Belgium known as Vlaamse Ardennen, Nuthatches first bred in the 1960s

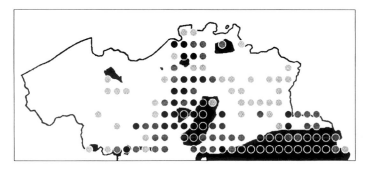

Fig. 70 *Past and present breeding distribution of Nuthatches in northern Belgium. Black areas (background) represent the approximate breeding range in the 1940s after Delmée (1948). Dots (foreground) indicate the number of pairs in 8 × 10 km² rectangles censused in the 1970s. Data from Devillers et al. (1988). Light grey = < 5 pairs, dark grey = 6–25 pairs, black = > 25 pairs.*

and have since gradually increased to over 100 pairs, most markedly in the late 1980s (Menschaert, 1991). In the north-eastern province of Limburg there has been a similar increase from the first breeding record in 1949 to an estimated 150 pairs in 1982 and over 800 pairs by 1992 (Gabriels, 1985; Gabriels *et al.*, 1994). Even in the Peerdsbos forest where I did most of my own studies, Nuthatches had been quite rare in 1958 with less than 1 pair per 10 ha (Van Styvendaele, 1963). The same pattern is seen in some Dutch provinces. In the southern province of Noord-Brabant numbers increased from 20–30 in 1965 to 200–225 in 1990 (Post and Ongenae, 1990). In the north-eastern province of Drenthe numbers increased from 2–3 in 1970 to 8–10 around 1980 and 60–100 pairs in the 1980s (van Dijk, cited by Glutz von Blotzheim, 1993). These dramatic increases are usually explained by the general ageing of forests and the increase in deciduous forest at the expense of conifers. Nevertheless, the rate of change is rather impressive for a bird with a moderate reproductive output and high degree of site fidelity.

NUTHATCHES AND PEOPLE

Nuthatch populations are as much influenced by human activities as are many other plant or animal species, at least in most of Europe where hardly any patch of truly natural forest remains. The large-scale deforestations that followed the spread of human civilization during the last few millenia have undoubtedly severely reduced the availability and quality of forest habitat. The increasing patchiness of the landscape, notably the increasing isolation of forest remnants, appears to affect the remaining populations even further, which I discuss in the next chapter. However, it is less clear to what extent Nuthatches may benefit or suffer from more recent changes. As the previous paragraphs have shown, populations are stable or increasing in most parts of Europe, for which various explanations have been suggested, including changes in forest management (older stands, shifts from conifers to broadleaved forest) and access to anthropogenic food. The large-scale degradation of coniferous forests in central and eastern Europe, probably caused by air pollution, may to some degree benefit Nuthatches by creating more light gaps (Möckel, 1992). However, severely damaged forests may become unsuitable, as suggested by census data from fir–beech forests in the northern Czech Republic and Slovakia (Flousek *et al.*, 1993). Otherwise little is known about the possible effects of environmental pollution on Nuthatches (but see above). An isolated note by Henze (1985) suggests that deformed eggs may be caused by increasing air pollution. Maréchal (1992) cites a few observations that may suggest birds develop respiratory problems in areas with high air pollution, but this remains no more than anecdotal information.

Even though Nuthatches have recently become popular as a model species in research on the effects of forest fragmentation, the species itself is obviously thriving and has received no particular attention with respect to nature conservation. There have been at least a few attempts to release

TABLE 20: *Population estimates for different European countries*

Country	Breeding pairs × 1000
Austria	300–500
Belgium	23
Denmark	10–50
Germany	750–1500
Netherlands	10–17
Norway	10–100
Sweden	100–500
UK	50
Slovakia	700–1000
Czech Republic	600–1200
Spain	550–1200
Bulgaria	500–1000
Romania	400–600
France	100–1000
Belarus	300
Croatia	200–300
Turkey	10–100
Russia	500–5000

NOTE:
Based on Glutz von Blotzheim (1993) and Hagemeijer and Blair (1997).

Nuthatches in sites where they did not occur, however. An adult and a juvenile pair were released in Kensington Gardens in London in April 1956, and apparently bred there the same spring (Meaden, 1970). Another introduction experiment, on the small island of Hven in the Öresund between south Sweden and Denmark, was less successful. Ten birds were released in the winter of 1973–1974, but they all disappeared (Alerstam and Winge, 1975). The authors suggest that the lack of stored food, and lack of familiarity with the new habitat, may have been the cause.

To conclude this chapter, population estimates for a number of European countries are given in Table 20. The largest populations are found in Russia, Germany, the Czech Republic, Slovakia, Spain and Bulgaria. The total population for Europe is estimated between 6 and 8 million breeding pairs (Hagemeijer and Blair, 1997). The Asian population is difficult to estimate but must be at least of the same order of magnitude.

Starling at Nuthatch nest.

CHAPTER 10

Nuthatches in Forest Fragments

One poorly understood but critical question is that of under which conditions a species that has evolved in a more continuous habitat will survive, or evolve to survive, in the more fragmented environment

Heywood and Watson (1995)

If one theme can be said to have dominated field studies on Nuthatches in the last 10 years, it is the study of the effects of habitat fragmentation. There are ongoing or recently completed projects in Belgium, Denmark, the Netherlands and Sweden, and this includes empirical work as well as modelling studies. Habitat fragmentation implies the breakdown of large stretches of habitat into a number of smaller patches (or fragments) that are isolated from one another by less suitable habitat (often agricultural or residential areas). Ecological research has focused on the effects of the size and degree of isolation of the remaining patches on the presence and abundance of the species living in this habitat. Nuthatches are often seen as representatives of an entire group of species, belonging to mature forest, that have moderate mobility and are thus at risk from the effects of forest fragmentation. Indeed they appear to be among the most sensitive species to the effects of patch size and isolation (van Dorp and Opdam, 1987). For instance, in a small-scale study near Antwerp, the Nuthatch was the only resident forest bird that was clearly less abundant in the smallest most isolated woodlots (Nour, 1997). In this chapter I review the available information on the effects of forest fragmentation on the ecology and behaviour of Nuthatches, and discuss some of the modelling work that has been based on the available field data.

NUTHATCH ABUNDANCE IN FOREST FRAGMENTS

Ever since Wilson and Willis (1975) first drew attention to the parallels between real islands and so-called 'habitat islands', an increasing number of studies have explored the relationships between the size and isolation of habitat patches and the diversity and abundance of animal and plant species in them. Forest birds have been particularly well studied in this respect, at least in forests of the northern temperate zone (e.g. Opdam, 1991). Several studies, in particular in the Netherlands, have shown that the presence and/or abundance of Nuthatches is reduced as forests become smaller and/or more isolated from one another (Table 21). The tendency of Nuthatches to be absent from open landscapes with scattered but suitable forest fragments has also been noted in England (Gibbons *et al.*, 1993). Their more frequent occurrence in larger patches is in itself not a very informative result, since this kind of relationship may be nothing but a sampling effect (Haila, 1983). However, some studies have also found a reduced presence in more isolated forests, and three studies found that patch characteristics affected not only presence but also the abundance, i.e. breeding density (Table 21). In my own study area in northern Belgium, which I describe in more detail later in this chapter, the population density of Nuthatches in a set of forest fragments dominated by oaks was about half that in three large, but otherwise very similar, forests (Table 22). Also, the population density in these fragments declined away from the nearest large forests, even though there was no relationship with fragment size or isolation on a smaller scale (Matthysen, unpubl.).

In the Netherlands there also seems to be a regional effect of forest fragmentation, since densities were about three times higher in the eastern and southern parts of the country, where a higher proportion of the landscape is wooded, than in the north and west, and this did not depend on fragment size and isolation (Opdam *et al.*, 1993). Interestingly, the effects of fragmentation are not limited to forest 'islands' surrounded by open landscape. Indeed, Enoksson *et al.* (1995)

TABLE 21: *Effects of landscape variables on the presence/absence and population density of Nuthatches in forest fragments*

Study	Country	Min area (ha)	Presence/absence	Density
Moore and Hooper, 1975	UK	1.25	+ size	
van Dorp and Opdam, 1987	Netherlands		+ size, isol, corr	
Opdam *et al.*, 1984	Netherlands		− isol	
Opdam *et al.*, 1985	Netherlands		+ size, isol	
			− corr	
Ford, 1987	UK	2.4	+ size	
Van Noorden *et al.*, 1988	Netherlands			+ corr
				− size, isol
Verboom *et al.*, 1991a	Netherlands		+ size, isol	+ size, isol
Verboom and Schotman, 1994	Netherlands			+ corr
~Tellería and Santos, 1995	Spain	100	+ size	
Komdeur and Gabrielsen, unpubl.	Denmark		+ size, isol − corr	
Matthysen, unpubl.	Belgium	1.5		− size, isol

NOTES:
~ in holm oak forest in winter, probably including many altitudinal migrants (see also Chapter 8).
1. + effect, − no effect.
2. Size = forest fragment size, isol = distance to nearby forest(s), corr = presence or density of corridors.
3. There is partial overlap between the data sets used by Opdam *et al.* (1985), van Dorp and Opdam (1987) and van Noorden *et al.* (1988).

TABLE 22: *Breeding population densities (pairs per 10 ha), breeding success and survival rates in different forest types (Matthysen and Adriaensen, unpubl.)*

	Density	% successful pairs	Fledglings per brood	Fledglings per pair	% adult survival	% juvenile survival
Oak-forest fragments	1.8	69 (52)	7.2 (20)	5.0	60 (57)	54 (13)
Parks	2.7	84 (25)	6.0 (35)	5.0	61 (42)	44 (9)
Large forests	3.4	76 (58)	6.8 (18)	5.1	32 (19)	14 (7)

NOTES:
1. Survival is calculated between breeding seasons for adults, and from late summer to breeding for juveniles.
2. Sample sizes are given in parentheses.
3. Survival data are compiled for oak-forest fragments and parks.

found comparable effects in small deciduous stands of birch and aspen surrounded by coniferous forest. These deciduous patches more often contained Nuthatches if they were located within a few hundred metres of other similar stands.

Although Nuthatches are often considered 'forest interior' birds – in contrast to so-called edge species – there is little evidence that they avoid the forest edges. Pairs living in woodlots or parks smaller than 2 ha effectively find themselves in a pure 'edge' habitat. I found several of them breeding in narrow belts of trees less than 50 m wide and surrounded by built-up areas, sometimes no more than a double row of large beeches or lime trees. Some territorial pairs in summer lived in small parks with less than 0.5 ha of canopy cover (see also Ito and Fujimaki, 1990). Even in larger fragments, nests may be found in trees at the very edge of the forest, sometimes even in tree rows extending from the forest. In my own study, 52% of the nests were located within 25 m of the forest edge, which is hardly less than the expected 59% if nests were randomly distributed over the available forest (Matthysen, unpubl.). Nevertheless, very few census studies have addressed edge effects in a rigorous way. Among these, Blana (1978) found no difference between 'edge' and 'interior' plots, but Moskát and Fuisz (1994) found no Nuthatch pairs in a 10-ha edge plot while three and four pairs were breeding in two 10-ha interior plots in Hungary.

TWO ALTERNATIVE HYPOTHESES

In the rest of this chapter I explore two general hypotheses that may explain why animals of a particular species should be less abundant in habitat fragments. The first possibility is that fragments represent habitat of lower quality, because there may be differences in food availability, nest sites, predation or even parasitism. Such differences, if they exist, may in turn result from changes in the abundance of other species that serve as competitor, prey, predator or parasite. This idea can be tested by comparing breeding success and survival in different forests. The second hypothesis, which has attracted the most attention from theoretical ecologists and landscape planners, is that dispersal between habitat fragments is reduced to such an extent that the population persists at a lower equilibrium level. One can approach this hypothesis at the level of individual patches or at the level of the individual bird. In the former case, a population of Nuthatches in a highly fragmented landscape may be considered a 'metapopulation', that is, a collection of local populations that are linked by dispersal events resulting in the exchange of individuals. Such a metapopulation may persist only if the rate of patches being 'colonized' exceeds the rate of patches going 'extinct'. At the level of individual birds, a crucial question is whether the success of dispersal is affected by fragmentation.

Since much of the following discussion is based on results from my own study in a 'fragmented' population, I will briefly sketch the study area and the essential methods that were used in this study, insofar as they differ from those described in Chapter 5.

THE 'FRAGMENTS' STUDY

From 1990 to 1994 I studied Nuthatches in an area of 200 km^2 just south of Antwerp (actually extending into the suburban area) with many

Aerial view of the 'fragments' study area, with an oak-wood fragment of c. 30 ha in the foreground, and one of 6 ha in the right background. Both fragments contained at most two territorial pairs during 4 years of study. (Photo: Frank Adriaensen).

small forest patches and wooded parks, which nevertheless account for less than 2% of the total area (Fig. 71). One-third of the fragments were largely stands of mature oak, usually with an understory of hazel, alder or black cherry. Slightly more than half of them were public parks or park-like remnants of old estates, characterized by a high diversity of tree species (mainly deciduous) including ornamental species, usually some very old trees, and a variety of lawns, small ponds and buildings. The landscape 'matrix' in between these patches consisted partly of traditional agricultural land and partly of rapidly growing residential areas with a dense road network. Most of the suitable 'fragments' were between 1 and 10 ha in size, and contained at most three breeding pairs. About 50 of these fragments had at least one pair of Nuthatches in at least a single season of observation.

In contrast to my earlier study in the Peerdsbos, only a small sample of the Nuthatches in the fragments were ringed. The main field work was devoted to two annual censuses, one in spring and one in late summer, with two main objectives: determining the number of individuals in each fragment, and locating ringed birds to study survival and movements. Since some of the fragments were inaccessible and the number of visits per area was limited, I estimated that between 75 and 85% of all the birds present were found on each census. In addition to these censuses I also collected data on breeding success by checking nests weekly by observation from a

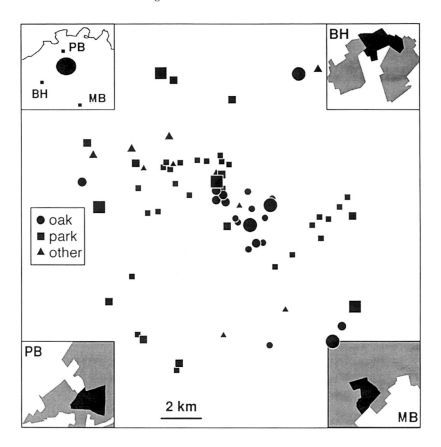

Fig. 71 *The areas used in the study on habitat fragmentation near Antwerp, 1990–1994. The main map shows the 'fragments study area' with forests and parks represented by symbols according to three size classes (<5, 5–10, >10 ha). The insets (representing 2 × 2 km squares) show the study plots, in black, within three large forests: PB = Peerdsbos, MB = Meerdaalbos, BH = Buggenhoutbos. The top left inset shows the location of the plots within northern Belgium.*

distance, which gave a crude estimate of success rate and timing of fledging. I also visited most of the accessible nests (i.e. those below 12 m) about 10 days after the presumed date of hatching. If the young were between 10 and 15 days old they were ringed. Nestlings were pulled out by means of a 'noose', i.e. three loops of nylon string protruding from a flexible tube that can be inserted into the cavity (Jackson, 1982). A total of 220 nestlings were ringed, 17 of which were observed later. Another 87 Nuthatches were trapped as full-grown individuals.

In parallel with these observations in forest fragments, I collected data in three study plots of 25–28 ha within large forests of c. 150, 300 and 1500 ha, respectively (Fig. 71). Although these forests all contained mixtures of oak,

This fragment of oak-wood of 1.5 ha was one of the smallest that ever supported a breeding pair during the 'fragments' study.

One of the smallest fragments of oak-wood (0.8 ha) where Nuthatches were observed during the 'fragments' study.

beech and coniferous stands, the actual study plots were dominated by mature oaks. One of them was immediately adjacent to the study plot in the Peerdsbos that had been used for the 1982–1987 study (see Chapter 5). These large forest plots were studied in the same way as the fragments, except that no late summer census was undertaken. This was because the limited number of colour-ringed birds made it impossible to determine the number of territorial pairs accurately.

BREEDING SUCCESS

Judging from data collected on almost 250 nests in forest fragments, parks and two large forests, there was no indication that laying date, breeding success and nestling weight were affected by forest size (Matthysen and Adriaensen, unpubl.). Broods tended to be smaller in parks than in oak forest, but this was compensated by a slightly higher nest success (Table 22). The only difference between fragments and large forest plots was that failed nests in fragments were more likely to be taken over by Starlings (29% *vs.* 4%). A similar (but non-significant) trend was found for take-over by Great Spotted Woodpeckers (19% *vs.* 8%). This may suggest a higher level of competition for nest sites in the forest fragments, which in turn may be caused by the more abundant Starlings. This bird breeds in large numbers in some of the small woodlots while foraging on nearby agricultural land, but is nearly absent from the large forest plots. Nevertheless, the increased competition for nest sites does not appear to have influenced overall breeding success.

SURVIVAL ESTIMATES

The methodological problem of distinguishing between mortality and emigration has already been mentioned in the previous chapter, and it becomes particularly relevant in a comparison between study areas differing in size and degree of isolation. Given the size of the fragments study area as a whole, and given the fact that birds moved very rarely between fragments once they had settled (see below), the observed local survival of *c.* 60% in adults (Table 22) is probably close to true survival. Unfortunately, the same cannot be said of the large forest plots where the estimate of survival is only 32%, probably because of dispersal (see discussion in previous chapter). Therefore, the data merely suggest that adult survival in the fragments is relatively high and probably not lower than the unknown true survival rate in large forests. There was no relationship between survival and fragment size (Matthysen, unpubl.). For juvenile survival as well, the estimate for the fragments turns out relatively high (54% survival from summer to breeding) and quite close to an indirect estimate for the Peerdsbos population (55%; see Appendix III). The estimate for juveniles in large forest plots was quite low (14%) but based on a very small sample (only seven birds ringed).

DISPERSAL

Natal dispersal distances

In the fragments study, I hoped that, because of the large study area, I would be able to locate the majority of dispersers. This expectation was based on the relatively short median dispersal distances in the order of 1 km known from ringing data (Chapter 8). However, as three consecutive field seasons yielded only 15 recoveries from over 200 colour-ringed nestlings, it became clear that the majority of these nestlings had either died or moved beyond the boundaries of the 200 km^2 study area. Using estimates of adult survival and reproduction, and allowing for the possibility that I missed up to a quarter of the breeding birds in the study area, I estimated that I should have found 16% of all the ringed nestlings if they remained within the study area. Since in reality I found only 4%, the inevitable conclusion was that the majority had moved out of the study area, and even a very cautious guess suggested the median dispersal distance should be around 5 km (Matthysen *et al.*, 1995). Those birds that were found again had moved up to 9 km, close to the maximum distance I was able to detect (Fig. 72). The obvious

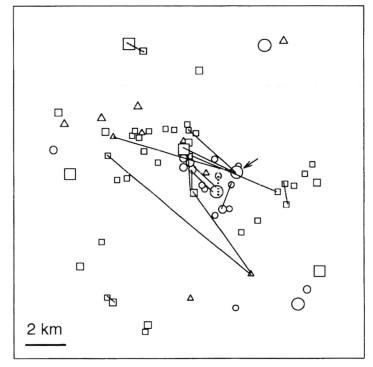

Fig. 72 *Movements by individually marked Nuthatches within the fragments study area. Forest fragments are represented by symbols (cf. Fig. 71). Solid lines indicate movements between hatching and settling, dashed lines are movements from one territory to another. Two young birds settled inside the fragment where they were born (indicated by arrow). From Matthysen et al. (1995).*

alternative explanation is that many locally hatched nestlings died, and their places were taken by immigrants, since the population apparently remained stable in numbers, but this would still imply that the average breeding bird travelled several kilometres before breeding, so the main conclusion would be unaffected. Furthermore, observations in the post-fledging period suggested that, at least prior to dispersal, there was no obvious difference in mortality between fragments and large forest plots (see below).

Thus, dispersal distances appear to be larger in a highly fragmented landscape, as if the movements track the 'dilution' of suitable habitat within that landscape. In fact, if dispersal is scaled to the density of territories within the landscape, the proportion of nestlings breeding within an area of 10–15 territories from the natal territory no longer differs between fragments (2.6%) and large forests (3.0%). The latter value was compiled from six small-scale studies with over 600 nestlings ringed in total (Matthysen *et al.*, 1995). This result agrees with an earlier analysis of the Frankfurt ringing data (Matthysen and Schmidt, 1987), showing that birds hatched in isolated woods were not more likely to breed within the same nestbox plot. In the Frankfurt study the 'fragments' were much larger, however, up to a few hundred hectares. In the same data set there was no indication that Nuthatches dispersed preferentially through more densely forested parts of the study area (Matthysen, 1994).

Timing of natal dispersal

Dispersal distances only reveal the outcome of the dispersal process, but give little information on what actually happens to the birds. With the help of Dave Currie I collected some more information on both the 'leaving' and 'arrival' phases of dispersal. For the 'leaving' phase, we found that the rate of disappearance of young birds was very similar in six territories in a large forest plot and eight territories in fragments (Chapter 8, Fig. 57). This suggests that Nuthatches hatched in fragments did not postpone their dispersal. In combination with the large dispersal distances, this argues against a possible barrier effect of open spaces between fragments. Contrary to a suggestion made in a popular article, Nuthatches do not appear to suffer from agoraphobia (Vermeulen, 1991).

We did find a difference in the arrival of dispersers, however, when we compared observations in the fragments with previously collected data from the Peerdsbos study (Fig. 73). In the fragments, settlers arrived at a much slower rate than they did in the Peerdsbos. Several vacancies even remained unoccupied until July or August, which rarely or never happened in the Peerdsbos. In addition, almost half of the newly established territories in the fragments were initially settled by a single bird, whereas in the Peerdsbos most, if not all, settlers were paired from the first observation (Matthysen and Currie, 1996). It is not entirely clear what caused the slower rate of settlement in the fragments, and this question may be satisfactorily answered only by radiotracking dispersers. On the basis of our own data, we can rule out two possibilities, namely that juveniles hatched in fragments remain with their parents for longer, or that fewer of them are produced in

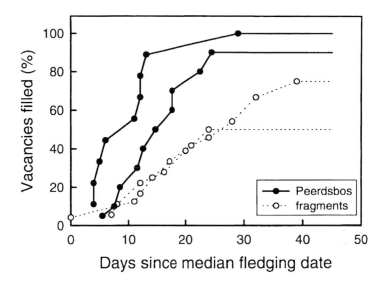

Fig. 73 *The rate of settling in vacant territories in two years in the Peerdsbos and two years in forest fragments, in relation to the median fledging date of that area and year. After Matthysen and Currie (1996).*

the first place. As the most plausible explanation, we suggest that in the fragmented study area many dispersers die before they can settle, which would explain both the low overall recruitment and the slow rate of arrival. The underlying idea is that it is much more difficult to locate suitable vacant territories in a landscape where they make up less than 1% of the total area. Dispersers could easily travel 5–10 km in any direction without even coming near a vacancy. In indirect support of this hypothesis there is no evidence that the best territories are taken up earliest (in contrast to the Peerdsbos study), as we should expect when dispersers cannot easily locate suitable habitat patches. The observations of single males and females settling in neighbouring fragments within the same study area points in the same direction (Matthysen and Currie, 1996). In other words, although dispersal distances appear to be high, the efficiency of dispersal may be much reduced in a fragmented landscape.

Movements after settling

Once they have settled in a particular forest fragment, Nuthatches appear very reluctant to leave it again. Only two of 14 juveniles that survived from summer to breeding moved to another territory (14%), and none of 43 surviving adults did so. One adult female did move but failed to survive to the next breeding season, and two adult males moved a few kilometres in spring but returned before summer. Both males had apparently lost their mate prior to moving. As a consequence of the juveniles' reluctance to leave a fragment, most of them ended up breeding in the same territory where they had settled the preceding summer (71%), which contrasts with the much smaller proportion in the Peerdsbos study (29%, see Chapter 7). Possibly it is too costly for birds in low-quality patches to monitor the availability of vacant territories in other fragments.

Pairing status

Given that up to half of the territorial birds may disappear over winter, that many fragments have only one or two territories, and that owners are unlikely to move between fragments, one should expect a high proportion of unmated birds in fragments in spring. In my own study I observed some males that appeared to be unpaired, but my census data are not sufficiently precise to make good estimates. In the Netherlands, however, Alex Schotman (pers. comm.) found that 12% of all territories ($N = 321$) were occupied by unpaired males. As expected, single males were more commonly found in the most isolated fragments (19% *vs.* 9% in less isolated areas), either because solitary males are more reluctant to leave these areas, or because new females are less likely to arrive, or both. This difference was pronounced only in the medium size class (1.5–5.4 ha) but not in smaller or larger fragments, and there was no overall relation with fragment size either. Single females were only rarely observed, but may of course be hard to detect since they do not advertise their presence the way males do.

IS A FRAGMENTED LANDSCAPE A POPULATION SINK?

As mentioned above, summer recruitment of territorial birds was much lower in the fragmented study area than expected, and this phenomenon was observed in each of the 5 years of study. Whereas the Peerdsbos population on average doubled in size each summer (average increase = 87%; see Nilsson, 1987, for a similar pattern in Sweden), this increase was only 30% in the fragments, and in some years there was hardly any recruitment at all, despite the presence of many suitable vacancies (Matthysen, unpubl.). Nevertheless, the population appeared to be thriving and showed no decline throughout the study. Calculations showed that in most years, the breeding population was considerably larger than what could have been expected from the survival of previous breeding birds and recruits of the previous summer population (survival rates estimated from a colour-ringed sample of the population). The 'deficit' was between 14 and 26% for the three best-studied years. The only explanation seems to be that between autumn and spring there is some immigration from outside the study area. This is in agreement with numerous cases of colour-ringed birds that disappeared over winter and were replaced by unringed birds. The aforementioned calculations suggest that many of them were real immigrants and not 'shifters' from nearby fragments.

The occurrence of post-summer immigration is not surprising given the fact that in the more intensively studied Peerdsbos population, 15–30% of the breeding population consisted of such immigrants (Chapter 8). There is a major difference, however. In the Peerdsbos study this immigration was, at least to some degree, balanced by an unknown but considerable emigration of established birds. Many of these immigration or emigration events may have been short-distance movements covering a few territories only. In the much larger fragmented study area, immigrants or emigrants must have covered much larger distances. Most importantly, frequent emigration from the area seems unlikely given the scarcity of movements between neighbouring fragments.

Therefore, I suggest (Matthysen, unpubl.) that the fragmented study area may act as a population 'sink' (*sensu* Pulliam, 1988) for Nuthatches, which can be sustained only by net immigration from outside. The sink is caused by the poor success of young birds searching for a territory and mate in summer. It seems plausible that the extra immigrants originate from larger forests or parks where summer recruitment is higher and – as in the Peerdsbos – a surplus of young birds lives in low-quality territories or as non-territorial residents. During the late summer/early autumn dispersal phase these birds may move away to areas where vacancies are still available, such as the fragmented study area. This scenario remains largely hypothetical, but is supported by the decrease in population density from north to south in the fragmented study area, that is, a decrease away from the nearest large Nuthatch populations (Matthysen, unpubl.).

The implications of this interpretation are quite important, since they imply that the fragment population would not be viable without the presence of larger forests nearby. The source–sink hypothesis also provides an attractively simple explanation for recent increases in Nuthatch populations in regions where the availability or quality of forest fragments has not really changed (*cf.* Opdam *et al.*, 1993). A sink population depends critically on the surplus produced in a source, and may therefore increase in size solely because of habitat changes in other areas. This may have happened in the fragmented study area itself, where Nuthatches are known to have been much less abundant a few decades ago.

THE METAPOPULATION APPROACH: FRAGMENTS AS LOCAL
POPULATIONS

Recent work by Jana Verboom and others (Verboom *et al.*, 1991a,b, 1993) illustrates how a Nuthatch population in a fragmented landscape can also be viewed as a metapopulation, in which each forest fragment – even if it contains only one territory – has the properties of a local population: it can go extinct and be recolonized from other populations. Evidently, the chance that a fragment goes 'extinct' depends on the survival of the inhabitants, and colonization depends on both the dispersal behaviour and the survival of young birds. Verboom *et al.* (1991a) documented the characteristics of three such metapopulations in different parts of the Netherlands, with a comparable distribution of Nuthatch habitat as in the Antwerp study area. The majority of patches contained only one or a few potential territories, with a total density of 0.3–0.7 suitable territories per km² (Antwerp: *c.* 0.8). In these study areas, fragments were more likely to contain Nuthatches if they were less isolated and if they were larger. 'Extinctions' occurred quite often in fragments with only one territory (*c.* 50%) but only rarely in fragments large enough to accommodate two to three territories (*c.* 10%), and never in larger patches. In other words, extinction rarely implied the disappearance of more than one pair. Empty fragments were more likely to be colonized again if they were less isolated. The fate of such a metapopulation can be predicted quite simply: as mean fragment size decreases or mean isolation increases, the proportion of fragments that support a population goes down, and below a particular threshold the entire metapopulation may go extinct (Verboom *et al.*, 1991a).

Although the original data set showed no evidence that the degree of isolation also influenced extinction in fragments, later analyses showed that this might be the case in some years (J. Verboom, pers. comm). This effect can arise if overwinter immigration is sufficiently frequent to reduce the chance of local extinction (the so-called 'rescue effect'). In the previous section I discussed evidence from the Antwerp study suggesting the importance of overwinter immigration. In the north-eastern part of the Netherlands, Opdam *et al.* (1993) found that in a period of increased colonization rate of fragments between 1972 and 1987, presumably caused by a greater influx of immigrants from 'source' areas, the extinction rate also decreased, possibly because of the 'rescue' effect.

One may question the validity of applying the metapopulation concept to the kind of Nuthatch populations considered here, since this implies – and Verboom *et al.* (1991a) were well aware of this problem – that a single pair of Nuthatches in a 1-ha woodlot is considered a 'population', and the deaths of two single individuals should be regarded as a population 'extinction'. It has been suggested that this kind of population, where the majority of individuals reproduce in a different fragment from the one where they were originated, should rather be called a 'patchy population' (Harrison, 1991). Nevertheless, the merit of the metapopulation concept is that simple presence/absence data per fragment can be used to predict the dynamics of the total population, regardless of the population sizes involved (Verboom *et al.*, 1991a). More complex models are needed, however, to simulate the effects of fragmentation in a more realistic way and to incorporate variation in demographic parameters such as reproduction or mortality. Such a model has also been developed by Verboom *et al.* (1991b), and I briefly present some of the model's characteristics and results in the next section.

A simulation model of a Nuthatch population

Verboom *et al.*'s (1991b) 'structured metapopulation model' is a true population model, where the dynamics of local populations follow from the fate of individuals which reproduce and die. Their paper was intended to illustrate the usefulness of this kind of model in landscape planning. I discuss it in some detail, since it illustrates how ecological and behavioural data on Nuthatches can be incorporated in a model with all the inevitable simplifications and difficulties of providing estimates for largely unstudied parameters. Figure 74 shows a diagram of how the model works at the level of a single forest fragment. The model simulates changes in the numbers of adult and first-year birds for each fragment, but works with males only, and assumes that all males acquire a partner and reproduce successfully. Later developments of the same model show that if the model contains females as well as males, extra dispersal is required to keep the population viable, since otherwise many males remain unpaired (Verboom and Matthysen, unpubl.). Fragments in the model contain a varying number of territories which may differ in quality (A, B or C). Changes in the number of birds within a fragment are governed by several processes. Some of these are deterministic (e.g. Nuthatches always move to a neighbouring higher quality territory if this becomes available), others are modelled as stochastic events with fixed probabilities. For instance, all adults have an 80% chance of surviving winter, whereas juvenile survival in winter depends on territory quality.

Each run of the model works with a particular set of fragments charac-terized by a number of territories and their quality, and the distances between them. The link between the fragments is provided by dispersal of newly fledged birds. Juveniles prefer to settle in the natal fragment, but if no territory is available they settle in a territory within 1 km, then in the next 1-km distance class, and so on. Within each distance class they prefer the highest quality territory (A > B > C). If no suitable terri-

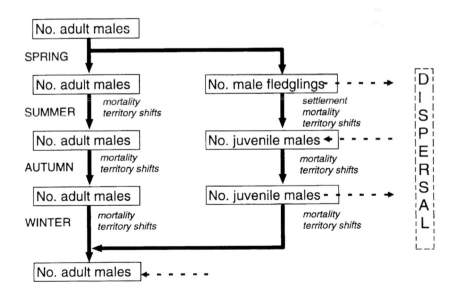

Fig. 74 *Diagram of a simulation model for nuthatch populations in habitat fragments. The diagram represents the dynamics of a single habitat patch within the model. After Verboom et al. (1991b).*

tory is found within 6 km, they die. During dispersal there is a 20% chance of dying per km travelled. Although these rules and parameter values are little more than educated guesses, and we now know that a dispersal limit of 6 km is far too small, the model did capture the essential conclusions of the dispersal study that was later carried out in the Antwerp study area: as fragments become more and more separated, dispersal distances increase but at the same time mortality of dispersers increases. The main weakness may be that the dispersal rules are too deterministic, since in reality juveniles rarely settle in the natal fragment, even if a vacancy is available, and they appear to have difficulties in locating the best available territories. Once juveniles have settled, no more dispersal occurs between fragments during autumn. If owners of high-quality territories die, they are replaced – again according to strict rules – by owners of lower quality territories in the same fragment. During winter there is again mortality and shifts between territories, but also an extra dispersal phase allowing birds from the lowest quality territories to leave a fragment. These territories were intended to simulate marginal territories unsuitable for breeding, or even the presence of non-territorial birds (cf. Chapter 7).

The model was initialized with parameters for mortality and reproduction from field studies (in particular from the Peerdsbos study), but subsequently these were modified so as to achieve a stable population with a proportion of occupied fragments similar to field data (Verboom *et al.*, 1991a). Figure 75 presents some results comparing the fate of metapopulations differing in sizes and configuration of forest fragments. It shows

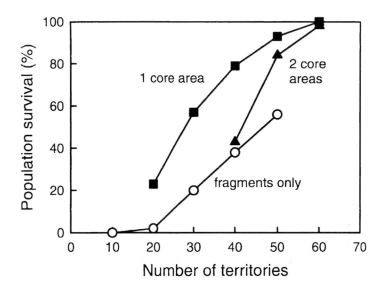

Fig. 75 *The proportion of simulated model populations surviving during 250 years as a function of habitat size, expressed as the number of suitable territories. Data from Verboom et al. (1991b, Fig. 3). Populations contain 0, 1 or 2 'core' areas with 20 territories, and a variable number of fragments each containing two territories.*

that the chance of the population persisting for a long time (250 years or more) increases rapidly with the total forest area. Populations with only 10–20 territories are almost certain to go extinct, while those with 50 territories or more are likely to persist. This is a straightforward demonstration of the vulnerability of small populations. Given a certain amount of habitat, the fate of the population also depends on its spatial configuration. For instance, if the habitat consists of 10 fragments with room for two territories each and a 'core area' of 20 territories, the population has an 80% chance of persisting, whereas this chance is only 40% if the same 40 territories are divided over 20 fragments. Another set of simulations presented by Verboom *et al.* (1993) show that if a 40-territory area is fragmented into progressively smaller patches containing 10, 4, 2 and 1 territories each, the size and longevity of the population decrease particularly strongly once the patches become smaller than 4 territories (which is a common size range in field studies both in Antwerp and the Netherlands). Another interesting result is that, although the occupation of patches decreases away from the core areas, this effect is very gradual and extends far beyond the maximum dispersal distance. For instance, in a particular simulation, 55% of the fragments at 20 km from the core area were occupied (Verboom *et al.*, 1991b). This indicates the importance of fragments functioning as 'stepping stones' for dispersal over larger distances.

Planning landscapes for Nuthatches

Perhaps because of their role as a test-case in the development of models such as the one described, Nuthatches have become popular as a reference species for predicting the effects of fragmentation, estimating minimal area requirements or even evaluating landscape planning scenarios. A simple approach is to predict their presence or abundance from forest fragment characteristics by means of a regression model based on other study areas. In one case study, Schotman and Meeuwsen (1994) predicted that the addition of some large forests to a highly fragmented landscape would increase the Nuthatch population by a factor of 5.7, even though the forest area increased by less than a factor of 3. In a more complex study, Harms and Opdam (1989) evaluated different scenarios for planting new forests in the densely populated Randstad area in the middle of the Netherlands, and used the Nuthatch as an example, considering it a 'key indicator of a species-rich forest bird community as well as a biogeographically favourable locality'. They first identified the existing Nuthatch populations in and around the study area from census data, and then evaluated the accessibility of possible locations for new forests to colonists from these source areas. For this purpose they divided the landscape into grid cells and assigned to each cell a resistance value to dispersal. Cells with high resistance (such as built-up areas or large lakes) were assumed to represent areas that dispersing Nuthatches were unlikely to traverse. The result is a map where the accessibility of a given spot in the landscape is the combined result of the distance from possible source areas and the accumulated resistance values of the cells in between. This provides a way to evaluate both the chance of new forests becoming occupied, and the effect of new forests on the presence of Nuthatches in previously existing forests.

Part III: Nuthatches of the World

Courtship-feeding in Corsican nuthatches.

CHAPTER 11

The Mediterranean Nuthatches

Il s'ensuivit une campagne de presse d'une ampleur sans précédent dans les annales de la zoologie
('This was followed by a press campaign unequalled in the history of zoology')

Vielliard (1978) describing the publicity that followed the discovery of the
Algerian Nuthatch

As we turn our attention from the single best-studied nuthatch species, we also change our geographical perspective. In the present chapter we start our world tour of the nuthatch family by examining three closely related species living in mountain forests around the Mediterranean Sea. All of them have been discovered remarkably late given their proximity to European ornithologists. It seems difficult to imagine now that in 1850, when half of the nuthatch species of the world had already been described, there were still three to be discovered within a thousand kilometres of the European mainland: Krueper's Nuthatch in Asia Minor in 1863, the Corsican Nuthatch in 1884 and the Algerian Nuthatch in 1975. A possible explanation for this curious fact is that all three species have limited geographical distributions and all live in rather under-explored habitats, i.e. montane or submontane forest.

The large distributional gaps between the three species are all the more peculiar because of their obvious close relationships, although these have recently been questioned to some extent (see Pasquet, 1998, and later in this chapter). When the Algerian Nuthatch was discovered in 1975 it was considered a relic of a hypothetical pan-Mediterranean ancestral species, and even hailed as a 'living fossil'. The history of the discovery is worth retelling here (mainly after Vielliard, 1978), especially since some of the earliest conclusions about the species' status and ecology need to be revised completely in the light of more recent data.

THE DISCOVERY OF THE ALGERIAN NUTHATCH

On 5 October 1975, the Belgian botanist Jean-Pierre Ledant and two companions reached the top of the Djebel Babor mountain (1995 m) in the Petite Kabylie mountains of northern Algeria, only 25 km from the coast. Their aim was to study an endemic tree species, the Algerian fir, in the isolated forest remaining on the top and higher slopes of the mountain. To their surprise they saw a nuthatch, clearly not a Eurasian Nuthatch but rather similar to the Corsican species. After his return Ledant came into contact with the French ornithologist Jacques Vielliard, who happened to be engaged in a study of the evolutionary history of the genus *Sitta*. Though sceptical about Ledant's report, Vielliard also realized the potential significance of the discovery, and encouraged him to collect more information. Ledant returned to the Djebel Babor in December 1975 (without success) and again in April and July 1976, on the last occasion together with Vielliard. They observed the new species over a few days and collected a breeding male and female. In the meantime the species had been discovered independently by a Swiss ornithologist, Eric Burnier, in June 1976. On 28 July the discovery of a new bird species was announced in the French newspaper *Le Monde*, on 13 September the formal description was presented to the French Academy of Sciences, and on 30 September it was published in the French journal *Alauda* (Vielliard, 1976), followed in December by Burnier's report in *L'Oiseau*. The discovery was widely publicized, being the first new description of a Palearctic bird species for 16 years (Vuilleumier and Mayr, 1987), and this almost at the border of Europe.

In the following years there were several more expeditions to the Djebel Babor, and the initial population estimate of only 12 pairs was gradually raised to about 80 (Fig. 76). This was attributed partly to a better exploration of the area, and partly to an apparent increase in numbers within the optimal habitat (Ledant *et al.*, 1985). It was nevertheless still believed that the species was confined to the 1200 ha of forest on this single mountain top and depended for its survival on this mixed forest of cedar, oak (mainly xen oak) and fir trees.

Fifteen years after the first observation, on 16 June 1989, a second population of the Algerian Nuthatch was discovered in deciduous and evergreen oak at an altitude well below the Babor forest, and appeared to be several times larger than the Babor population. It was found in Taza National Park,

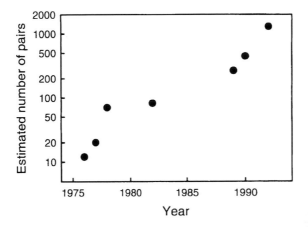

Fig. 76 *Changes in the estimated number of Algerian Nuthatches since their discovery in 1975. Note the logarithmic scale. The first four data points are estimates for Djebel Babor by Ledant and Jacobs (1977), Vielliard (1978), Gatter and Mattes (1979) and Ledant et al. (1985). The 1989 point is the 1982 estimate (82 pairs) plus Bellatreche and Chalabi's (1990) estimate for the newly discovered Taza population (182 pairs). The 1990 estimate assumes that the newly discovered populations (Bellatreche, 1990, 1991) are of comparable size to Taza (c. 180 pairs). The 1992 estimate assumes that breeding densities in Taza and other oak forests are higher than previously assumed, using an intermediate value (c. 2 pairs per 10 ha) between the estimates of Bellatreche and Chalabi (1990: 0.6 pairs per 10 ha) and Doumandji and Kisserli (1993: 3.2–3.5 pairs per 10 ha).*

part of the Guerrouch forest, about 20 km from the Djebel Babor and only 9 km from the Mediterranean coast (Fig. 77) (Chalabi, 1989; Bellatreche and Chalabi, 1990). The preliminary estimate of 364 individuals, though considered no more than indicative, implied a threefold increase in the known population size. Again the discovery became international news. The scientific implication was that the Algerian Nuthatch was not particularly associated with the relict forest on the Babor mountain, and less threatened than had been thought before. In hindsight, the fact that Gatter and Mattes (1979) found a clear preference for foraging in oak, and not for any other tree species, should have pointed out the possibility that the species was not closely linked to the cedars and firs on the Djebel Babor. Earlier attempts to find other populations may have failed because they were limited to cedar and cedar–fir forests on nearby mountain tops (Ledant *et al.*, 1985).

The discovery in Taza prompted more searches in nearby oak forests, and in 1990 two new populations were found within 5–10 km of the known sites (Fig. 77) (Bellatreche, 1990, 1991). Their size has not been estimated, but since the new sites are even larger than the Guerrouch forest (about 11 000 ha) and contain similar habitat, they may easily contain 100–200 breeding pairs. A prudent estimate at that time would have been about 400

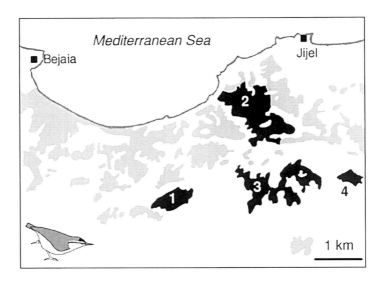

Fig. 77 *Geographical range of the Algerian Nuthatch. Black and grey areas indicate forests with and without nuthatches. 1. Djebel Babor, 2. Guerrouch, 3. Tamentout, 4. Djimla. From Bellatreche (1991).*

pairs for the total population, based on a modest breeding density of *c.* 0.6 pairs per 10 ha. Since, in the meantime, a much higher density has been reported from Taza (Doumandji and Kisserli, 1993; see below), the current population size may be anywhere between 500 and a few thousand breeding pairs (Fig. 76).

It is remarkable that the Algerian Nuthatch had been overlooked for so long and was first discovered in what turned out to be one of its smallest populations, and probably the least representative in terms of habitat. In their handbook of the birds of north-western Africa, Heim de Balsac and Mayaud (1962) cited some alleged, but poorly documented, reports of Eurasian Nuthatches in Algeria in the 19th century, one of them only 300 km west of the present Algerian Nuthatch populations. If these reports are true, either the Eurasian Nuthatch once occurred much nearer to the Algerian Nuthatch, or the latter once had a more extensive distribution and was mistaken for the former.

MORPHOLOGY

The three Mediterranean nuthatch species are smaller and more compact than the Eurasian Nuthatch, with a comparatively large head, short bill and short tail. They share the typical nuthatch colour pattern with bluish-grey upperparts and white to buff underparts, and all have subterminal white spots on the outer tail-feathers. All males and some females have

TABLE 23: *General characteristics of the Mediterranean nuthatches and their closest relative, the Chinese Nuthatch (mainly after Harrap and Quinn, 1996)*

	Krueper's Nuthatch	Corsican Nuthatch	Algerian Nuthatch	Chinese Nuthatch
Discovery	1863	1883	1975	1865
Distribution	Asia Minor, Caucasus	Corsica	Algeria	China, Korea
Range size (km²)	500 000	5000	800	1 000 000
Main habitat	Conifers	Conifers	Evergreen	Conifers, mixed
Male wing length (mm)	71–81	70–76	80–83	63–75
Weight (g)	10–14	12–15	16–18	8–11
Bill length (mm)	18–20	16–19	16–17	14–18
Colour of male cap	~Black	Black	~Black	Black
Colour of female cap	~Black	Grey	Black or grey	Grey
Female cap feathers	Almost wholly black	Almost wholly grey	Black with variable grey fringe	Black with broad grey fringe
Undertail-coverts	Bi-coloured	Uniform	Uniform	Uniform
Breast-patch	Yes	No	No	No
Juveniles separable	Yes	No	Only from adult male	Only from adult male

NOTES:
~ forecrown only.
Range sizes are crude approximations.

black caps, a prominent black eye-stripe and white superciliary. Table 23 summarizes the main morphological characteristics of the three species as well as those of their closest relative, the Chinese Nuthatch (see also Harrap and Quinn, 1996; Pasquet, 1998). Krueper's Nuthatch has a diagnostic rufous crescent-shaped breast-patch, and is the only species with bi-coloured undertail-coverts (rufous and white). There is considerable varia-tion in the width of this breast-patch (13–27 mm at the widest point in the few dozen specimens I examined myself). The sexes differ slightly (females are on the whole slightly duller) but the age dimorphism is the most pro-nounced of all nuthatches, juveniles having grey caps and a much fainter breast-patch (Löhrl, 1962). The Corsican Nuthatch is the smallest of the three with a remarkably short tail and greyish white underparts with only traces of buff. Males in particular have a more prominent head pattern than the two other species, the black cap extending down to the nape rather than ending halfway down the head. The Algerian Nuthatch is larger than the other two, has a moderate degree of sexual dimorphism (see below) and pale buff underparts as in Krueper's, but no breast-patch. The bill appears to have an upward inflection which is caused by a distinct bend halfway along the lower mandible. A similar but weaker tendency can be seen in Krueper's, but apparently not in the Corsican Nuthatch (Cramp and Perrins, 1993). All three species are geographically monomorphic, which is not surprising given their small ranges.

The sexual dimorphism in the Algerian Nuthatch is not yet well under-stood (see Vielliard, 1980 for a detailed discussion). The female forecrown appears to be covered by black feathers with variable grey fringes, which tend to wear off in spring and early summer. Thus, some females (perhaps adults) have some black on the forehead, whereas others (presumably first-year breeders, or birds in autumn) have uniformly grey caps (Cramp and Perrins, 1993). Female Algerian Nuthatches are therefore closer to Krueper's females with their small black caps than to Corsican females with entirely grey caps, though even the latter sometimes have black feather bases on the forehead (also in the Chinese Nuthatch, Fiebig, 1992). The juvenile plumage of the Algerian Nuthatch is also an unresolved problem: most sources describe juveniles as having grey caps, but Vielliard (1978) observed juveniles with black caps, and Fosse and Vaillant (1982) saw one with black and two with grey caps in a single brood. Perhaps this reflects a sexual dimorphism in the nest, as in the Corsican Nuthatch.

PHYLOGENY

Until halfway through the 20th century, all small 'black-capped' nuthatches (Corsican, Krueper's, Chinese, Red-breasted) and the grey-capped Yunnan Nuthatch were regarded as a single species, *Sitta canadensis*, despite their wide geographical separation throughout the Holarctic (Fig. 78) (e.g. Hartert, 1910–1922; Voous and van Marle, 1953). Other authors (notably Kleinschmidt, 1928, and Vaurie, 1957) doubted whether Krueper's and Yunnan Nuthatch should be included in this species, and

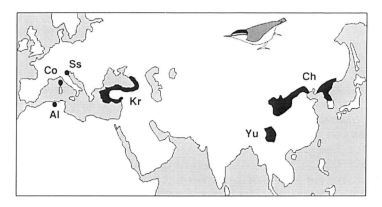

Fig. 78 *Geographical ranges of the* Sitta canadensis *superspecies in the Old World (Chinese, Corsican, Algerian, Krueper's and Yunnan Nuthatches). Ss =* Sitta senogalliensis, *known from fossil remains only.*

ever since the publication of Vaurie's *Birds of the Palearctic Fauna* (1959) all forms have been regarded as closely related species within a superspecies (cf. Sibley and Monroe, 1990). This is partly due to Löhrl's (1960, 1961) detailed comparison of the behaviour and ecology of Corsican and Red-breasted Nuthatches (see Chapter 14). When the Algerian Nuthatch was described in 1976 its close relationship with the others was immediately recognized.

Vielliard (1978) suggested that the three Mediterranean species evolved from an ancestral 'Mesogean' nuthatch – probably close to the Chinese Nuthatch – which colonized the Mediterranean basin in the early Miocene, and whose range was subsequently fragmented by climatic and tectonic changes. This early colonization is supported by the presence of a fossil nuthatch, *Sitta senogalliensis*, in the Italian Miocene (see Chapter 1). In Vielliard's view, the Corsican Nuthatch became isolated first (6–7 million years ago) and Krueper's and Algerian Nuthatch more recently, *c.* 600 000 years ago. In his phylogenetic tree (Vielliard, 1978) the three Mediterranean species and the Chinese Nuthatch are somewhat separated from the other two members of the superspecies, the Red-breasted and Yunnan Nuthatches (Chapter 1, Fig. 2). The Red-breasted is indeed markedly different from the Old World species with respect to voice and nest-building behaviour (Löhrl, 1988; see also Chapter 14). Leonovich *et al.* (1996a) suggested that this species has diverged to the largest degree (particularly in voice) from the Chinese Nuthatch, and conclude that its ancestor probably colonized N America from Europe, not from Asia. On the other hand, recent evidence, notably the presence of the typical harsh call, underlines the probable close relationship between the Yunnan Nuthatch and the remaining species (Chapter 13).

A recent study using mitochondrial DNA by Pasquet (1998) supports a rather different view, which involves a more complex biogeographical history, but appears to fit better with the morphological patterns, and supports the earlier views of Vaurie (1957). The DNA sequences suggest that Krueper's and Algerian Nuthatch are indeed closely related, but that a second subgroup is formed by the Red-breasted, Corsican and Chinese Nuthatches, of which the latter two are closest. This fits well with the general plumage patterns, but it implies that the evolution of behavioural differences in the Red-breasted Nuthatch has taken place in a relatively short time. The separation of the two subgroups is approximately dated to 5 million years ago, whereas the Chinese/Corsican and Krueper's/Algerian species pairs were separated 1–2 million years ago (Pasquet, 1998). It would be interesting to examine the molecular relationships with the last member of the superspecies, the Yunnan Nuthatch.

DISTRIBUTION AND HABITAT

Krueper's Nuthatch is confined mainly to Asia Minor and occupies a broad zone along the eastern and southern coast of the Black Sea and the north-eastern coast of the Mediterranean Sea, though it is absent from the central Anatolian plateau (Fig. 78; for details see Neufeldt and Wunderlich, 1984). To the north-east its range extends into Georgia and the western Caucasus. The south-eastern limits are not very well known but do not seem to reach the Turkish–Syrian border (Neufeldt and Wunderlich, 1984). The only European population of this bird, on the Greek island of Lesbos, within view of the Turkish coast, was discovered as late as 1960 by Watson (Löhrl, 1965a). Krueper's Nuthatch is found between 500 and 2400 m altitude, down to 200 m on Lesbos, but most often between 1200 and 1700 m (Kumerloeve, 1961; Neufeldt and Wunderlich, 1984). At lower altitudes it inhabits pine forest, at higher altitudes also spruce, fir, cedar and occasionally juniper forest. In Turkey it is most abundant in Turkish pine, and the high degree of coincidence of both species' ranges led Frankis (1991) to speculate on the possible dependence of pines on nuthatches for the dispersal of their seed. In the Caucasus region they live mainly in spruce forest between 1000 and 2000 m, and in pine forest only when there is a significant admixture of fir or spruce, though one pair has been found breeding in an aspen stand (Polivanov and Polivanova, 1986). Density estimates are not available but most accounts suggest relatively low population densities (Neufeldt and Wunderlich, 1984). The smallest territory found on Lesbos was 3–4 ha in size (Löhrl, 1988).

The Corsican Nuthatch is a typical inhabitant of the endemic high-altitude forest of the endemic Corsican pine or laricio, at altitudes between 800 and 1800 m but mainly between 1000 and 1500 m (Thibault, 1983; Cramp and Perrins, 1993). Estimates of population densities along transects through different habitat types are provided by Brichetti and di Capi (1985, 1987). The optimal habitat consists of old and unmanaged pure laricio stands in the central and upper parts of the altitudinal range, and

Laricio pine forest near Evisa, optimal habitat for the Corsican Nuthatch.

supports a mean density of 1.1 pairs per 10 ha. Even higher densities of 1.7–3.0 males per 10 ha may be found (Beck, 1992). These stands typically have wide clearings and many standing dead and decaying trees (Fig. 79). Suboptimal habitat comprised four categories: lower elevation laricio stands mixed with cluster pine, beech or fir; young even-aged stands; heavily managed stands lacking standing dead trees; and the highest elevations where trees are scattered and often stunted (Fig. 79). Densities were also higher where suitable dead trees for nesting were abundant (Brichetti and di Capi, 1985). Territory sizes are between 7 and 10 ha with a core of 4–6 ha (Brichetti and di Capi, 1985). A small-scale study in winter suggests that this species may at least locally reach high densities in the cluster pine zone as

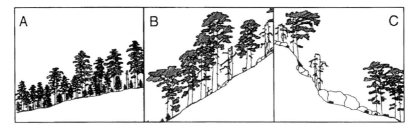

Fig. 79 *Vegetation profiles of laricio pine forest on Corsica at different elevations. Lower (A) and upper (C) parts represent suboptimal habitat, the central part (B) is optimal habitat. After Brichetti and di Capi (1985).*

well, at around 800 m (*c.* 2 pairs per 10 ha in February; Matthysen and Adriaensen, 1989c). These birds were probably not just winter visitors, as suggested in the editorial note to our 1989 paper, since some of them were still present the next summer. Ledant (1978) also reported having found the species just as easily in laricio as in cluster pine stands. Other occasional observations outside laricio forest include a stand of fir surrounded by maquis (Payn, 1927), evergreen cork oak forest at 300 m altitude (Payn, 1931) and sweet chestnut plantations and villages in winter (Payn, 1927; Thibault, 1983).

The Algerian Nuthatch is known from four forests only within an area of less than 1000 square kilometres. The largest known population in Taza is found from 50 to 1120 m altitude, mainly in deciduous xen oak and Algerian chestnut-leaved oak, and at lower altitudes in cork oak as well (Chalabi, 1989; Bellatreche and Chalabi, 1990). This area has a humid Mediterranean climate with a mean annual temperature of 18°C and hardly any frost, quite different from the high-altitude coniferous forests inhabited by Corsican and Krueper's Nuthatches. A preliminary census suggested a relatively low density increasing with altitude (on average *c.* 0.6 pairs per 10 ha; Bellatreche and Chalabi, 1990) but a more recent study suggests more than 3 pairs per 10 ha (Doumandji and Kisserli, 1993). The Tamentout and Djimla populations are also found in deciduous forest but at somewhat higher altitudes (900–1400 m), and here they appear to be rare in the cork oak forest below 1000 m. The best-studied, but also the smallest and perhaps least representative, population is the one on Djebel Babor. The summit of this mountain (up to 1995 m) is one of the few areas in N Africa with a montane Mediterranean climate, markedly cold and humid, with up to 4 m of snow in winter. The forest is unique, has several endemic plant species, and is the only Algerian site for many plants common to European deciduous forests (Ledant *et al.*, 1985). The highest nuthatch density (3–4 pairs per 10 ha) is found in the mixed cedar–oak–fir forest at the top, which is rich in epiphytic mosses and lichens (Gatter and Mattes, 1979; Ledant *et al.*, 1985). The more extensive lower elevation forest of xen oak and cedar supports lower densities (*c.* 0.5 pairs per 10 ha) but still contains more than half of the estimated population (Gatter and

Mattes, 1979; Ledant *et al.*, 1985). Intermediate densities of 1–1.5 pairs per 10 ha are found in the upper parts of the oak forest where cedar and fir become increasingly numerous.

FORAGING AND FOOD

The diet of none of these nuthatches has been studied in detail, but probably all three feed mainly on invertebrates in summer and rely at least partly on seeds in autumn and winter. The food of Krueper's Nuthatch contains the same insect groups that are typical of the Eurasian Nuthatch's diet, such as aphids, caterpillars, Tipulidae and Plecoptera, and seeds of pine and fir (Cramp and Perrins, 1993). Nestlings in the Caucasus area were fed with 44% caterpillars, 17% winged insects, 6% spiders and 30% various small arthropods (Polivanov and Polivanova, 1986; observations on 70 nest visits). Algerian Nuthatches have been seen to take insects and spiders as well as nuts and seeds of oak, maple, cedar and fir, and to feed their young mainly with caterpillars but also beetles, earwigs and spiders (Gatter and Mattes, 1979). Fly-catching seems a common practice in summer in Corsican and Krueper's Nuthatch (Ledant, 1978; Löhrl, 1988), but appears to be rare or absent in the Algerian Nuthatch (Ledant, 1978).

Several authors have commented on the Corsican and Krueper's Nuthatches' agile, titlike foraging behaviour in spring and summer, on the extremities of branches amidst the twigs and cones rather than on the trunk and large branches (Löhrl, 1960, 1988; Ledant, 1978). Löhrl (1960) suggested that free-living Corsican Nuthatches climbed less well and used their wings more frequently than Eurasian Nuthatches, but saw no differences in the aviary. Perhaps the scaly pine bark offers a less suitable climbing substrate than the rough bark of oaks. In any case, Corsican Nuthatches do spend most of their time foraging on trunks in winter, in true nuthatch fashion (60–70%; Matthysen and Adriaensen, 1989a). Foraging on trunks was particularly noticeable in the cluster pines, while birds in the laricio stands more often frequented the tree crowns. Algerian Nuthatches also foraged mainly on trunks and branches in late winter (often 'scaling' bark), but in June one pair spent most of its time looking for caterpillars amongst the twigs and developing leaves (Gatter and Mattes, 1979). A seasonal shift from foraging on trunks and large branches in winter to smaller branches, twigs and needles in summer is also well documented in the ecologically similar Red-breasted Nuthatch (Chapter 14). Male Corsican Nuthatches tended to forage more often in the tree crown and less on bare trunks than females, but this difference was not statistically significant (Matthysen and Adriaensen, 1989a; see also Chapter 4 for other species).

All three species are known to hoard food (Gatter and Mattes, 1979; Löhrl, 1988). Even in June, young Algerian Nuthatches were fed partly on seeds (about 20% in one particular brood). Ledant *et al.* (1985) suspected that yew was particularly important for the species, since in September they found many birds feeding in or near yew trees, probably searching for seeds. Krueper's Nuthatches also spent a lot of time extracting seeds from pine cones in September–October (Frankis, 1991).

BREEDING BIOLOGY

The nests of the Mediterranean nuthatch species differ from those of the Eurasian Nuthatch in two aspects: they are often self-excavated, and the entrance is never reduced with mud. Corsican Nuthatches usually excavate their own nest, but often start from an unfinished woodpecker hole. The majority of nest trees are dead laricios 200–300 years old with little bark left. The majority of entrances (58%) face downhill, only 17% uphill. Many entrance holes are overlooked by a branch stump which may serve as a roost-site or perhaps to camouflage the entrance (Brichetti and di Capi, 1985). Previous nest holes may be re-used, but not commonly. Krueper's Nuthatches, on the other hand, seem to prefer pre-existing holes to excavating their own (8 out of 9 nests in Löhrl, 1988; between 6 and 9 out of 11 in Polivanov and Polivanova, 1986). This preference may depend, however, on the presence of dead wood and thus on forest management. One nest was built in a seemingly natural pile of twigs, needles and bark flakes in the fork of a pine tree (Löhrl, 1988). Algerian Nuthatches probably excavate as a rule, but the large entrance holes (35–50 mm) suggest that many start from unfinished woodpecker holes (Ledant and Jacobs, 1977; Gatter and Mattes, 1979). The majority of the described nests (all from Djebel Babor) have been found in fir or cedar trees, often at the edge of clearings (Gatter and Mattes, 1979). Curiously, one deserted nest had the entrance reduced with mud (Gatter and Mattes, 1979). Since this can hardly be the work of

Krueper's Nuthatch.

Eurasian Nuthatches, which have never been observed in Algeria, the implication is that Algerian Nuthatches might occasionally use mud, but this certainly needs confirmation.

Nest-building is done mainly by the female. The contribution of the male is limited in Krueper's Nuthatch, and although he may provision nest material he does not enter the cavity. Corsican Nuthatch males, on the other hand, take a considerable share in excavation and also assist in lining the nest (Löhrl, 1988). No information is available on Algerian Nuthatches. All three species' nests have a course foundation of bark chips, rotten wood or pine seed wings. The lining is more typically passerine and includes bark fibres, moss, feathers, hair and fur (Ledant and Jacobs, 1977; Polivanov and Polivanova, 1986; Löhrl, 1988; Cramp and Perrins, 1993). Copulations have not been observed. Bundy (1971) saw courtship chases and courtship feeding in Krueper's Nuthatch, with the female soliciting in a crouched posture accompanied by wing-shivering.

First-egg dates for the Corsican Nuthatch vary from 28 April to 13 May, though replacement clutches are possible at later dates (Brichetti and di Capi, 1985). Timing does not seem to vary with altitude in this species. In Krueper's Nuthatch on Lesbos, however, there was a delay of 2–3 weeks in pairs at 600–700 m compared to those at 200–300 m (Löhrl, 1965a). A similar difference was noted in Turkey at higher altitudes (Danford, 1878). Clutches of this species have been found from early April to mid-May (Cramp and Perrins, 1993). Laying dates of Algerian Nuthatches are not known with certainty and may vary considerably between years. In 1976 the majority of nests on Djebel Babor fledged on 8 July while in 1977 the latest fledged on 18 June, and in Taza (at lower altitudes) all nests had probably fledged by 20 June in 1989 (Vielliard, 1978; Bellatreche and Chalabi, 1990). Assuming incubation and nestling periods of *c.* 15 and 22 days, this would imply first-egg dates from early to late May. Second broods have not been identified with certainty for any of the three species, but an observation of dependent juveniles in Corsica on 13 August may suggest a second or very late replacement brood (Löhrl, 1960).

Clutches of Krueper's and Corsican Nuthatches usually contain five to seven eggs (Cramp and Perrins, 1993). Brichetti and di Capi (1985) estimated the Corsican's incubation time at 14 days (slightly less than the Eurasian Nuthatch), and the time between hatching and fledging at 23 days. The latter is in agreement with the comparable development rates of hand-reared young of the two species (Löhrl, 1988). Polivanov and Polivanova (1986) give a similar incubation time for Krueper's Nuthatch (14–17 days) but only 16–19 days for the time to fledging, which would be rather short compared with other nuthatch species. On the other hand, Löhrl (1988) mentioned a nest with young of 22 days old. Breeding success has not been estimated in any species, but Brichetti and di Capi (1985) believe that nest predation by Great Spotted Woodpecker is an important cause of failure in Corsican Nuthatches. Algerian Nuthatches either lay very small clutches or have a poor success rate, since the most commonly observed brood size is two with a maximum of four (Ledant and Jacobs, 1977; Vielliard, 1978).

SOCIAL BEHAVIOUR

The little information that is available suggests that all Mediterranean nuthatches are sedentary, if not permanently territorial. In winter I observed a number of Corsican Nuthatch pairs with well-defined home-ranges, but considerable variation in their response to playback (Matthysen and Adriaensen, 1989c). These pairs generally foraged close together (within 20 m) and maintained contact with soft calls. The suggestion of permanent territoriality agrees with Brichetti and di Capi's (1985) observation of strong response to playback near former nest sites in autumn. However, I also occasionally observed three or four individuals together without obvious signs of aggression, and similar observations have been reported by others (Löhrl, 1960; Guillou, 1964). Jacques Blondel (pers. comm.) saw up to six Corsican Nuthatches in a mixed flock in winter. Thus, Löhrl's (1960) conclusion that territoriality is less pronounced than in the Eurasian Nuthatch may apply to both the winter and breeding seasons. The same may be true for the other two species as well. For instance, Polivanov and Polivanova (1986) stated that territories of Krueper's Nuthatch were probably taken up in autumn by pairs of young birds, and more so by adults, while Kumerloeve (1958) and Frankis (1991) saw small groups in late summer and early autumn. Gatter and Mattes (1979) suggest that Algerian Nuthatches have sharply demarcated breeding territories but nevertheless saw neighbouring pairs foraging peacefully within sight of each other. Ledant and Jacobs (1977) mention 'territorial individuals' in November but without further details.

Displays are not well described, except those of the Corsican Nuthatch. In a fight the male typically ruffles all his body feathers, giving him the appearance of a small feather ball. The tail is raised but not spread and the bill is held low (Löhrl, 1960, 1988). A moderate threat posture with ruffled feathers, in one case with raised wings, has also been observed in Krueper's Nuthatch (Löhrl, 1988). Gatter and Mattes (1979) saw an Algerian Nuthatch displaying with ruffled back and belly feathers, drooping wings and the bill pointing upward. It is difficult to tell whether these anecdotal accounts reflect true interspecific differences. Wing-flicking as a sign of excitement has been observed in Krueper's and Corsican Nuthatch (Löhrl, 1988; Matthysen, pers. obs.). All Mediterranean nuthatches participate in mixed flocks, notably with Coal Tits and Goldcrests (Payn, 1927; Jacobs *et al.*, 1978; Polivanov and Polivanova, 1986). Corsican Nuthatches appear to do this more often in laricio than in cluster pine forest (Matthysen and Adriaensen, 1989c).

VOCALIZATIONS

Song

Mediterranean nuthatches conform to the general pattern of the family in their song, which is a stereotyped repetition of identical notes usually maintained for a few seconds (Vielliard, 1978; Cramp and Perrins, 1993; Fig. 80). However, the songs differ clearly and audibly between species. The song of the Corsican Nuthatch is a repetition of simple notes with some variation between songs; one of the note types is quite similar to the Eurasian Nuthatch's Trilled song (see Chappuis, 1976 for more details). The tempo is variable but generally quite high with 10–15 notes per second. A similar song type is known from the Chinese Nuthatch, which further supports their close relationship (Chapter 13). The songs of Krueper's Nuthatch include a similar (but slower) Trilled song, but also a song based on two alternating notes (Fig. 80). The tempo of both song types is four to eight notes per second. Two variations of the two-note type are shown by Vielliard (1978). According to Vittery and Squire (1972), the two-note song is delivered by pairs in duets, but this has not been mentioned by any other author and remains to be confirmed. Löhrl (1988) found Krueper's Nuthatch to be markedly more vocal than the Corsican Nuthatch. The only song known from the Algerian Nuthatch is based on a repetition of three notes, the first two of which are short and the second linked to a longer third note (Fig. 80). The harmonic structure is much more pronounced than in the other two species. The tempo is rather slow with three to four units or 'strophes' per second. The song has been described as sonorous, sometimes fluty or more nasal, and reminiscent of the Wryneck (Burnier, 1976), and is quite distinct from the trill of the other species. The Algerian Nuthatch does not respond to playback of Corsican or Krueper's song, but the Corsican Nuthatch does respond to Algerian song (Ledant, 1978; Vielliard, 1978).

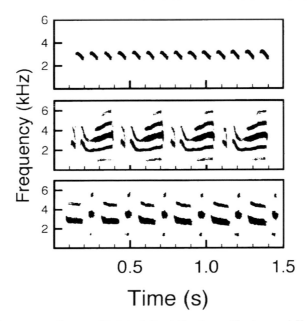

Fig. 80 *Sonagrams of songs of (top to bottom) Corsican, Algerian and Krueper's Nuthatch. Reproduced from Chappuis (1976, Corsican) and Vielliard (1978, Algerian and Krueper's).*

Calls

All three Mediterranean nuthatches, as well as the Chinese and Yunnan Nuthatches, share a harsh, rasping 'schrä' call, often compared to a distant Jay calling. On a sonagram this call appears as a broad band of noise lasting 0.2–0.4 seconds (Vielliard, 1978; Matthysen *et al.*, 1991; Cramp and Perrins, 1993). Algerian Nuthatches responded weakly but noticeably to the calls of the other two species, which underlines the similarities between these species (Vielliard, 1978). Vielliard (1978) hypothesized that this call represents a characteristic ancestral to all nuthatches. Outside the *Sitta canadensis* group – where it is absent only in the Red-breasted Nuthatch – similar calls are known from the more distantly related Kashmir and rock nuthatches. All Mediterranean nuthatches have softer contact calls as well, which have not been described in detail, and a few other calls which may be used in specific contexts only (Cramp and Perrins, 1993). The 'most striking' call of Krueper's Nuthatch (Löhrl, 1988) is a Greenfinch-like 'doid', and according to Harrap and Quinn (1996) resembles the 'quuwee' of the Algerian Nuthatch. Another distinct call of Krueper's is a 'hick' or 'pwit' like a soft Great Spotted Woodpecker call (Cramp and Perrins, 1993). Male Corsican Nuthatches use a low-intensity trill when they approach their rivals in a conflict (Löhrl, 1960).

MOVEMENTS

Hardly anything is known about dispersal in any of the three species, except that Ledant and Jacobs (1977) observed young Algerian Nuthatches outside their natal territory after 7 days, which may suggest early dispersal as in the Eurasian Nuthatch. There is some evidence for altitudinal movements, however, in at least two of the species. Corsican Nuthatches are sometimes recorded well below their breeding altitudes in winter, descending to chestnut plantations and gardens at 300–600 m (Thibault, 1983). Brichetti and di Capi (1985) believe that these are mainly or exclusively first-year birds driven by harsh conditions, given the strong attachment to the breeding territory found in adults. Regular altitudinal movements are also reported for Krueper's Nuthatch in the Caucasus, into coastal areas with dense deciduous forest and some pines. These movements may be related to local failure of seed crops. One observation of a large group of 15–20 birds in September suggests a flock of wandering birds (Neufeldt and Wunderlich, 1984).

POPULATION STATUS AND THREATS

All three Mediterranean nuthatch species are potentially threatened because of their small geographical range and rather specialized habitat requirements. The Algerian Nuthatch is the most vulnerable, being confined to about 20 000 ha of forest and with a population probably below

1000 pairs. Grazing and cutting represent potential threats to the regeneration of its forest habitat, and could endanger the species in the long run. Nevertheless, its prospects appear relatively healthy since two of its populations occur within national parks (Djebel Babor and Taza) and the species is legally protected in Algeria (Bellatreche and Chalabi, 1990). The Corsican Nuthatch occurs over a wider geographical area than the Algerian Nuthatch and may reach higher breeding densities, but the total amount of habitat is not much larger (24 000 ha of montane pine forest). Population densities are negatively affected by intensive forest management but seem relatively insensitive to disturbance by houses, traffic or powerlines (Brichetti and di Capi, 1985, 1987). With an estimated population of 2000 pairs and no evidence for overall decline, the species is not acutely threatened but should be considered vulnerable (Tucker and Heath, 1994). Krueper's Nuthatch has the widest range, but may suffer from the same threats of deforestation and intensive management (Neufeldt and Wunderlich, 1984). The small population on Lesbos is probably stable (Tucker and Heath, 1994) but nothing is known about the main population in Turkey and the Caucasus. The total lack of knowledge about demography and population dynamics of any of the three species remains a serious impediment to the evaluation of threats and their possible remediation.

Western Rock Nuthatch

CHAPTER 12

The Rock Nuthatches

Wo tot und erstorben erscheint alles Leben in der tollen Felswildnis,
wo nichts sich regt als flimmernd die Luft über zerfetzte Felsgerate
und kein Laut vernehmbar als das Summens des durchglühten Gesteins:
dort erschallt plötzlich heraus das höhnische, weithintönende,
langgezogene Gelächter der Felsenspechtmeise
('Where all life appears to be dead in the mad wilderness of rock,
where nothing stirs but the shimmering air over the broken cliffs
and nothing is heard but the buzzing of the sun-drenched stone:
there suddenly the scornful, far-reaching,
long-drawn-out laughter of the rock nuthatch is heard')

Roháček (1919)

It is difficult to imagine a nuthatch spending most or all of its life away from trees. Yet two members of the nuthatch family live almost exclusively on the ground, having exchanged the forest habitat for rocky hill slopes often devoid of any trees. The so-called rock nuthatches are often described as very noisy and lively even in comparison with the Eurasian Nuthatch, and they share some remarkable behaviours such as the construction of mud nests and song duetting. However, they have become notorious mainly for their contribution to ecological theory on geographical variation and speciation, being featured in many textbooks as the classic example of character displacement (the divergence in morphology, behaviour or any other trait of two closely related species where they occur together, as opposed to greater similarity where they live separately). This phenomenon was first described by Vaurie in 1951 but later critically re-examined by Grant (1975) and Panov (1989). Grant's study, though aimed specifically at examining the ecological and evolutionary explanations for geographical variation, still remains one of the major information sources on the ecology and behaviour of both species.

MORPHOLOGY AND AFFINITIES

The basic plumage patterns of the Eastern and Western Rock Nuthatch are quite similar to that of the Eurasian Nuthatch, except for the paler upper parts, the lack of subterminal white tail-spots, the absence of chestnut markings on the flanks and the uniformly coloured undertail-coverts (Harrap and Quinn, 1996). The rock nuthatches also appear to have larger heads and longer necks, an impression further enhanced by their upright posture and the prominent eye-stripes in some populations. Neither species shows any marked sexual dimorphism in plumage or size (Grant, 1975). First-year birds are slightly paler than adults and may have pale rufous tips on the greater coverts, especially in fresh plumage (autumn and early winter) (Vaurie, 1950).

There is no single good criterion to separate the two species throughout their combined ranges, but they are clearly distinct where they meet. Where they overlap in Iran the Western Rock Nuthatch is smaller (by *c.* 1 cm wing length) with a markedly shorter bill (by about 6 mm), and weighs only about half as much as the other species (20–27 g compared to 42–55 g). The most prominent difference is the short eye-stripe, however (*c.* 1 cm, versus *c.* 3 cm in most Eastern Rock Nuthatch populations). All of these differences diminish towards the extreme opposites of both species' ranges, although the Eastern species remains slightly larger with a longer eye-stripe throughout its range.

The rock nuthatches are obviously closely related to the Eurasian Nuthatch and its sister taxa. Voous and van Marle (1953) argued that the uniform undertail-coverts point to a close affinity with the Kashmir Nuthatch. Although this view was not shared by Vaurie (1957), it is supported by two more recent behavioural findings. The first is the harsh call which is shared by the Kashmir Nuthatch, the rock nuthatches and the more distantly related Mediterranean nuthatches. Second, the Kashmir Nuthatch has a habit of building short tunnel-like extensions to the nest, similar to the entrances of the rock nuthatches' mud nests (Löhrl, 1988).

Western Rock Nuthatch. (Photo: Bernard Casteleyn).

DISTRIBUTION

Although the existence of two distinct species of rock nuthatch was recognized as early as 1911, their exact delimitation has remained a matter of confusion for several decades (Table 24). For instance, Buxton (1920) and Hartert (1910–1922) lumped all races except for two forms from the overlap zone, *rupicola* and *tschitscherini*, which are now regarded as belonging to the Western Rock Nuthatch. Even in more recent works such as Dement'ev and Gladkov (1954) all rock nuthatches are considered to be one species. On the other hand, Leonovich *et al.* (1996b) have recently advocated the recognition of a third species, *Sitta obscura*, on the western end of the distribution of the Eastern Rock Nuthatch. The presently accepted classification goes back to von Jordans (1923) (Table 24).

The Western Rock Nuthatch occurs from the Balkan peninsula throughout Asia Minor to Transcaucasia and the western half of Iran (Fig. 81). Within the Balkans it is found mainly along the Croatian coast and throughout Greece, Macedonia and southern Bulgaria, where it has increased in recent decades (Tucker and Heath, 1994). It is absent from most of the larger Greek islands (Reiser, 1905) except Corfu (Böhr, 1962). Its main range in Asia comprises most of Turkey, Armenia, Azerbaijan and southern Georgia just south of the main Caucasus range, and the Elburz mountains along the southern Caspian Sea border. From the centre of its range two 'fingers' extend to the south, one along the

TABLE 24: *The history of classification of the rock nuthatches, largely based on Stresemann (1925) (which see for references)*

Source	Subspecies											
	neu *	tsc *	syr *	zar *	rup *	par *	plu *	obs *	tep *	irra *	dre *	kur *
Hellmayr, 1903	NEU		NEU						NEU=	NEU		
Hartert, 1905	NEU	NEU	NEU		NEU=			NEU=	NEU			
Hellmayr, 1911	NEU	TSC	NEU	NEU	NEU			NEU=	NEU		NEU	
Sarudny, 1911	NEU	NEU	SYR			NEU		SYR	SYR		SYR	
Buxton, 1920	NEU	RUP		RUP=	RUP=	RUP		NEU		NEU	NEU	
Hartert, 1910–1922	NEU	RUP	NEU	NEU	RUP			NEU	NEU	NEU	NEU	
von Jordans, 1923	NEU	NEU		NEU	NEU			TEP	TEP	TEP	TEP	
Stresemann, 1925	NEU	NEU	NEU	NEU	NEU=	NEU		TEP	TEP	TEP	TEP	TEP
Greenway, 1967	NEU	NEU	NEU=	NEU	NEU=	NEU	NEU	TEP	TEP=	TEP	TEP=	TEP

NOTES:
1. The modern classification by Greenway (1967) is added for comparison.
2. Entries in capitals indicate to which species a particular (modern) subspecies was assigned by a particular author.
3. = indicates that neighbouring subspecies were lumped.
4. Subspecies marked * occur in sympatry with a race of the other rock nuthatch species.
5. Full names of taxa are: *neumayer, tschitscherini, syriaca, zarudnyi, rupicola, barvi, plumbea, obscura, tephronota, iranica, dresseri* and *kurdistanica*.

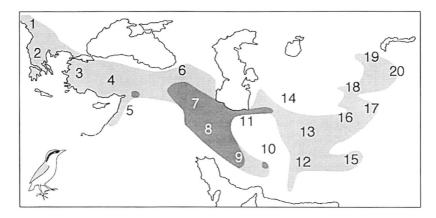

Fig. 81   *Geographical distribution of Western and Eastern Rock Nuthatches, after Wunderlich (1986) and Loskot et al. (1991). Overlap areas are shaded darker. Numbers correspond to populations represented in Figs 83 and 84.*

Mediterranean coast down to northern Israel, one south-east into the Zagros Mountains of Iran and beyond. Of the five subspecies listed by Greenway (1967), the three westernmost ones differ only slightly in plumage and size, and have been lumped by Harrap and Quinn (1996). These are *neumayer* in the Balkans, *syriaca* in Asia Minor (including *zarudnyi*, listed by Cramp and Perrins, 1993) and *rupicola* in the Caucasus and northern Iran. Further south-east, two smaller subspecies occur with a strongly reduced eye-stripe that is virtually absent in some specimens. Of these, *tschitscherini* in Kurdistan and the Zagros mountains is paler than *plumbea* in south-eastern Iran, while the latter has an ashy rather than white throat and breast (Vaurie, 1959). These two races and *rupicola* live in sympatry with the Eastern Rock Nuthatch.

The most westerly populations of the Eastern Rock Nuthatch are found in eastern Turkey, including an isolated population near Gaziantep, less than 100 km from the Mediterranean coast (Eggers, 1977; Beaman, 1986), and in the Zagros and Elburz mountains in sympatry with the Western Rock Nuthatch. It is absent from large parts of central Iran, but occurs in most of Afghanistan and parts of Pakistan, and further north through Tajikistan and Kyrgyzstan (Fig. 81). There are three subspecies: *tephronota* in north-eastern Iran, Afghanistan and further east; *obscura* in Turkey, north-western and central Iran; and *dresseri* in south-western Iran (Vaurie, 1950; Greenway, 1967). The subspecies *tephronota* is the smallest with the least developed eye-stripe, and does not overlap with the Western Rock Nuthatch; *obscura* is intermediate in size and eye-stripe and has darker upperparts than the other two, and *dresseri* is the largest with the most prominent eye-stripe and heaviest bill, and the palest upperparts (Vaurie, 1959).

HABITAT

As suggested by their name, bare rock is an essential requirement for the habitat of both species. They live in arid or semi-arid regions, mainly on rocky outcrops in hills and mountains, on crags, cliffs, steep sides of ravines and valleys; but also on rocky slopes with scrub or scattered trees, and occasionally in very open woodland with scattered rocks (Wunderlich, 1986; Loskot *et al.*, 1991). In southern Iran the Eastern Rock Nuthatch lives in xerophilous pistachio forest on stony ground (Desfayes and Praz, 1978). Both species may descend to lower elevations in severe winter conditions and can be observed in gardens, woodland and orchards (Dement'ev and Gladkov, 1954; Wunderlich, 1986; Loskot *et al.*, 1991). Both are found over a wide altitudinal range. The Western Rock Nuthatch occurs from almost sea level (e.g. on Corfu; Böhr, 1962) to near the snow-line at *c.* 2500 m in several parts of its range (Wunderlich, 1986). In northern Israel it may even be observed above the snow-line outside the breeding season (Paz, 1987). In Turkey it is most abundant between 700 and 1500 m (Kumerloeve, 1961). The Eastern Rock Nuthatch is typically observed from 500–600 to 2000–2300 m, exceptionally up to 3300 m (Loskot *et al.*, 1991). In southern Iran it is very abundant between 2000 and 3000 m (Desfayes and Praz, 1978). In the north the upper limit is lower, e.g. 2000 m in the Tien Shan and 1000 m in the Dzhungarskiy Alatau in easternmost Kazakhstan.

In most areas where they co-occur, the two species seem to occupy similar habitats, for instance in southern Iran (Grant, 1975) and southern Turkey (Eggers, 1977). In the Elburz mountains, however, the Western Rock Nuthatch is found at higher altitudes, mostly from 1300 to 3500 m, while the Eastern occupies the lower, warmer and drier elevations, mainly from 1000 to 1400 m (Grant, 1975).

Breeding densities of the Eastern Rock Nuthatch appear quite variable, from *c.* 0.03 to almost 1 pair per ha, probably typically around 0.1 (Loskot *et al.*, 1991). In ravines, typical densities are 1–2 pairs per km, but in one case seven occupied nests were found along 600 m of a gorge (Loskot *et al.*, 1991; Cramp and Perrins, 1993). Densities are lower in the upper mountain ranges, and possibly also in pistachio forest (*c.* 1 pair per 20 ha) (Loskot *et al.*, 1991). I found no density figures for the Western Rock Nuthatch.

CHARACTER DISPLACEMENT

It was Vaurie (1950) who first drew attention to the marked differences between the two rock nuthatches where they occur together and the much larger similarity between the allopatric populations (Fig. 82). He explained this pattern by a model of natural selection that became known as 'character displacement'. This model proposes that when two closely related species evolve independently in allopatry and then meet again, each becomes subjected to directional selection driving its characteristics away from the other species. In the case of the rock nuthatches, differences in bill size would be selected for to alleviate competition, and differences

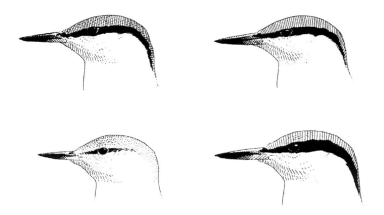

Fig. 82 *'Character displacement' in the rock nuthatches. Bottom left and right: Western and Eastern Rock Nuthatch in overlap zone (south-western Iran), top left: Western Rock Nuthatch at western range limits (Dalmatia), top right: Eastern Rock Nuthatch at eastern range limits (easternmost Uzbekistan).*

in plumage attributes such as the eye-stripe would be favoured to facilitate the recognition of conspecifics and avoidance of hybridization.

So much for the theory. When Grant (1975) undertook a more detailed analysis of variation in the rock nuthatches, he concluded that the story was probably more complicated. Rather than comparing sympatric and allopatric populations as such, he looked more closely at neighbouring populations within and outside the other species' range. He found some evidence for character displacement in eye-stripe size in southern and western Iran, particularly in the Western Rock Nuthatch (populations 8 to 10 in Fig. 83), whereas in northern Iran (populations 7 and 11) the eye-stripes are more similar to the allopatric populations. There is less evidence for character displacement in body size or bill size. For instance, bill sizes of both species diminish gradually from west to east, and there are no marked changes where one species crosses the border of the other species' range (Fig. 84). Furthermore, the sizes of males and females do not converge in sympatry, which would have been expected if competition was driving natural selection on size (Grant, 1975). Grant did some further experiments to test the idea that divergence in plumage facilitates species recognition in areas of sympatry. Indeed, individuals from sympatric populations were able to recognize their own conspecifics when presented with models (museum specimens), some of these having been painted to manipulate the appearance of the eye-stripe. However, Western Rock Nuthatches in Greece – which had no experience of the other species – did not discriminate, and actually reacted more strongly to the largest models with the largest eye-stripes. Of course, these experiments did not test for recognition by voice or displays. So far no hybrids have been described between the two

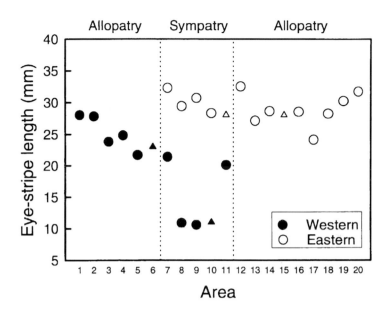

Fig. 83 *Geographical variation in postocular eye-stripe length in Western and Eastern Rock Nuthatches. Populations are arranged from west to east, corresponding to numbered areas in Fig. 81. Triangles indicate means based on fewer than five individuals. Data from Grant (1975)*

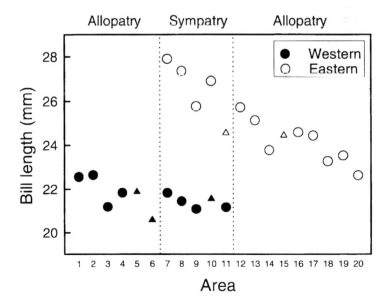

Fig. 84 *Geographical variation in bill length in Western and Eastern Rock Nuthatches (legend as in Fig. 83).*

species, and there also is no noticeable aggression between them except close to the nest (Grant, 1975).

In conclusion, Grant (1975) suggested that the two rock nuthatches had become fully ecologically isolated before they came into contact, as an adaptation of each species to a different climatic environment: a Mediterranean climate in the western species, a more continental climate with hotter summers and colder winters in the eastern species. This selected for a larger body size and bill in the east, and also led to differences in diet and habitat selection. Once the two species made contact in Iran – probably caused by a climatic change – they underwent directional selection on eye-stripe size and body size until full reproductive isolation was attained, and the species were able to coexist at the same altitudes. The effect of this divergent selection may have 'spilled over' through gene flow beyond the area of sympatry. For instance, the large eye-stripes in the allopatric population 12 (Eastern Rock Nuthatch) may be caused by gene-flow from sympatric populations further to the west. The northern contact zone in the Elburz may represent more recent contact, where the species have not diverged sufficiently to allow coexistence at the same altitudes (Grant, 1975). More recently, Panov (1989) rejected the character displacement hypothesis altogether, arguing that the differences between the allopatric populations are larger than they have generally been recognized, in morphology (particularly size), vocalizations and timing of breeding (based on an English summary by E. Panov, pers. comm.).

FORAGING AND FOOD

Rock nuthatches search for their food on bare rock faces, cliffs and boulders, looking for invertebrate prey in the cracks and crevices. They also forage on the ground or occasionally on fallen logs and branches. Grant (1975) also saw them digging with the bill, perhaps in search of beetle larvae, and fly-catching or hovering to take insects from flowers. He found no conspicuous differences in foraging technique between the two species. They do not often fly but generally hop and jump about (rather than run), and show a remarkable ability to hold on to small ledges, more than other nuthatches. They easily cling to vertical surfaces, even head downwards, but do not climb down, and they avoid smooth vertical walls (Löhrl, 1988). During foraging they are very attentive and scan the surroundings frequently, and 'freeze' when in danger. Löhrl (1988) found them extremely wary compared with other hand-reared nuthatches, and very suspicious of unfamiliar human visitors. This wariness may be connected to the dangers inherent in living on the ground.

Although Grant (1975) rarely observed rock nuthatches in trees and never saw them feeding there, the Eastern Rock Nuthatches living in the open xerophilous forests of southern Iran do sometimes forage on trees (Sarudny and Härms, 1923; Desfayes and Praz, 1978) and there are anecdotal reports of tree foraging in Western Rock Nuthatches as well, behaving in a similar way to Eurasian Nuthatches (e.g. Moore and Boswell, 1956).

Roháček (1919) even reported a simultaneous observation of a Western Rock Nuthatch exploring tree trunks and a nearby Eurasian Nuthatch climbing rocks!

The diet has been studied best in the Eastern Rock Nuthatch, but may be very similar – depending on local conditions – in the other species (Loskot *et al.*, 1991; Cramp and Perrins, 1993). In spring and summer the main food is large insects, particularly beetles, grasshoppers, butterflies and their larvae, and also spiders and snails. Snails are typically wiped on the ground before being swallowed (Löhrl, 1988). A Western Rock Nuthatch has been seen eating a small lizard (Roháček, 1919). In the winter various seeds are taken including pine, various fruits (cherry, almond, apricot), pistachio, hawthorn, and various grasses and weeds. Captive Eastern Rock Nuthatches also took earthworms, slugs and bees (Löhrl, 1988). Both species wedge snails and seeds in small rock crevices to hammer them. In late summer they start hoarding food, mainly seeds and snails, in crevices and cover them with stones or soil, or store them behind loose bark. There is little evidence for differences in diet between the two species, except that in southern Iran the larger-billed Eastern Rock Nuthatches used larger seeds, which required much more force to be opened (Grant, 1975). It seems plausible that they would take larger arthropods as well. Young Eastern Rock Nuthatches in Turkmenistan were fed mainly on arthropods (caterpillars, beetles, grasshoppers, Hymenoptera, woodlice, spiders and centipedes) but 11% of them had pistachio nuts in their stomachs as well (Rustamov, 1958, cited by Cramp and Perrins, 1993). In Armenia, 81% of the food brought to nestling Western Rock Nuthatches consisted of caterpillars (Adamyan, 1965).

THE NEST

Rock nuthatches have taken the art of plastering a considerable step further than their arboreal relatives, by building a nest entirely of mud. The most typical nest site (particularly in the Western Rock Nuthatch) is an exposed vertical face, often where the rock has recently broken off, or somewhat sheltered by an overhanging rock. In this case the nest is half-conical in shape, directly attached to the vertical wall, and with the entrance usually perpendicular to the wall. Such nests may weigh a few hundred grams to several kilograms (Cramp and Perrins, 1993). The walls are usually several centimetres thick, but sometimes only 1 cm near the entrance. The entire construction is so strong that a human may need an axe to open it (Peus, 1954). Some nests have also been found on buildings (Western, Cramp and Perrins, 1993) or on steep river banks (Eastern, Loskot *et al.*, 1991). The second most common site is a wide crack or crevice among the rocks or occasionally in a tree, which is almost completely walled up and the cavity further reduced with mud. Huge amounts of mud may be used, sometimes weighing 10–40 kg and covering an area up to 1 m^2 (Cramp and Perrins, 1993). This type of nest is more typical of the Eastern Rock Nuthatch, which usually nests near the ground (Ali and Ripley, 1973).

Eastern Rock Nuthatch.

Both species also occasionally nest in smaller holes which are reduced in the typical nuthatch manner. Eastern Rock Nuthatches have been recorded in abandoned holes of woodpeckers, but also those of bee-eaters and rollers (Sarudny and Härms, 1923; Desfayes and Praz, 1978). The nest entrance typically has a short protruding tunnel 5–10 cm long and 3–5 cm wide, but these are less common and also shorter in the Eastern Rock Nuthatch. The inner diameter of the entrance is a good indicator of the birds' size and may even be used for species identification (Grant, 1975). A few nests have been found with a double entrance (Sarudny and Härms, 1923; Löhrl, 1988).

Both sexes contribute to nest-building (Löhrl, 1988). Some authors suggest that the male builds more than the female (Adamyan, 1965), but the female may specialize on the inside work (Sarudny and Härms, 1923; Kipp, 1965). Old nest sites are often re-used (Panov, 1989). For example, six out of eight Western Rock Nuthatch nests were still in use after 7 years, and one after 19 years (Löhrl, 1965b, 1988). Even after complete destruction, a nest may be re-built on the same site (Roháček, 1919), or a new nest may be built next to an old one (Dement'ev and Gladkov, 1954), suggesting that nest sites are carefully chosen. Re-building may take as little as 5 days. Old or incomplete nests are sometimes used by other birds including owls and wheatears (Loskot *et al.*, 1991). In winter the nest may be used for roosting (Loskot *et al.*, 1991), and refurbishing has been observed in autumn (Cramp and Perrins, 1993).

The main building material is mud, which is smeared out with the bill, in contrast to the Eurasian Nuthatch which fixes little lumps on to the nest with blows of the bill (Löhrl, 1988). A large variety of other materials is mixed with the mud (but no saliva), some of them quite conspicuous, including dung, pistachio resin, feathers, hair, raptor pellets, bone and skull fragments, plant fibres, pieces of cloth, snail shells, small stones, wings of butterflies and bugs, beetle elytra, berries, bees' wax, and even tin-foil or cellophane wrappers of sweets (compiled from various sources). Some nests are almost completely covered on the outside with beetle elytra or butterfly wings, giving them a bright appearance that can be spotted from afar (Roháček, 1919; Sarudny and Härms, 1923). Some of the nest material, for instance feathers, may be scattered around the nest as well, and some observers have interpreted this as decoration (Géroudet, 1963) or even as mimicking an old deserted nest (Rokitansky, 1962). Whether such decorated nests have a display function as, for instance, those of the bowerbirds, remains an unanswered question. Löhrl (1988) believes that the various additives serve to strengthen the construction and that the scattered material has been dropped for later use or because the birds were disturbed. Such storing of nest material for later use is known from other nuthatches as well (Löhrl, 1958; Fiebig, 1992). The nest chamber is lined with a felt-like loose mass (reminding Peus (1954) of the contents of a vacuum cleaner bag) mainly consisting of hair, often from raptor pellets, wool and feathers, but also thistle seed wings, cloth, plant fibres and pieces of shed snakeskins (Roháček, 1919; Peus, 1954; Géroudet, 1963; Kipp, 1965; Löhrl, 1988).

BREEDING BIOLOGY

Roháček (1919) described a copulation by Western Rock Nuthatches as follows: the female solicited by means of begging calls and wing-shivering, after which the male adopted a vertical posture with the bill pointing upwards, approached the female sideways with the head moving slowly from side to side, and mounted. This sequence was repeated 3–4 times, and appears very similar to the Eurasian Nuthatch's courtship behaviour. Löhrl (1988) observed a similar copulation once, but also other less ritualized cases. In the Eastern Rock Nuthatch he also observed rapid wing-shivering and the male vertical posture, but no Pendulum movements. The copulation call of the latter species is described as a repeated 'jib jib jib', rather different from that of the Eurasian Nuthatch (Löhrl, 1988).

The Western Rock Nuthatch lays its eggs mainly in late March and April, occasionally in early May, apparently with little geographical variation (Reiser, 1905; Dement'ev and Gladkov, 1954; Panov, 1989; Cramp and Perrins, 1993). However, Roháček (1919) reports considerable altitudinal variation, with complete clutches between 25 April and 5 May below 800 m, around 15 May between 800 and 1200 m, and in the last week of May at higher altitudes (region unspecified). In the Eastern Rock Nuthatch there is more variation between regions and habitats. The earliest clutches are laid in late February and mid-March in southern Tajikistan, south-eastern Iran and Pakistan (Loskot *et al.*, 1991; Roberts, 1992). In most other areas (Transcaucasia, Turkmenistan, Tajikistan, Tien Shan) eggs are laid in late

March and April, and in the northernmost populations of the Dzhungarskiy Alatau (easternmost Kazakhstan) even in late April (Loskot *et al.*, 1991). Late first clutches can be found in early May in many areas (Loskot *et al.*, 1991). Complete clutches range from four to eight eggs in the Eastern Rock Nuthatch, occasionally nine (Cramp and Perrins, 1993). In Pakistan clutches typically contain only four eggs (Roberts, 1992). The Western Rock Nuthatch lays markedly larger clutches from six to 13 eggs, with a mean probably around eight or nine (Cramp and Perrins, 1993).

Incubation and nestling care appear to be largely comparable to those of the Eurasian Nuthatch. Incubation is said to last 12–14 days in the Eastern and 14–18 days in the Western Rock Nuthatch (Roháček, 1919; Rustamov, 1958, cited by Cramp and Perrins, 1993; Adamyan, 1965). Females start incubating before the end of laying, which may result in asynchronous hatching, at least in the Eastern species (Cramp and Perrins, 1993). The male feeds the female on the nest during incubation. Eggs are covered with nest material in the egg-laying period but not during incubation breaks (Löhrl, 1988). Löhrl (1988) twice observed that a raptor pellet was stuck in the entrance of a nest (Western) and suspected that it might serve as camouflage. A similar observation has been reported of an Eastern Rock Nuthatch blocking the entrance with a plug of soft nest material before going to roost (Kanevski, 1978, cited by Cramp and Perrins, 1993). Young rock nuthatches leave the nest after 23–26 days (Cramp and Perrins, 1993), in one case almost 30 days (Western Rock Nuthatch, Armenia; Adamyan, 1965). Newly fledged young do not fly very well, but climb around and hide in crevices when the parents give alarm calls. They stay with the parents for about a month, perhaps longer (Loskot *et al.*, 1991; Cramp and Perrins, 1993).

Second broods have been reported in both species, and fresh clutches can be found until early July in the Western Rock Nuthatch (Roháček, 1919). Nevertheless, Roberts (1992) doubts the existence of second broods (in the Eastern) because of the long period of dependence of the young. Adult rock nuthatches start their annual moult relatively late, from mid-July (Loskot *et al.*, 1991) or mid-August (Vaurie, 1950) to mid-September. This could be in agreement with a frequent occurrence of second broods, since Eurasian Nuthatches are known to start moulting immediately after the young fledge in late May or early June (Chapter 2).

BEHAVIOUR AND DISPLAYS

Rock nuthatches have two typical behaviours used in several contexts, particularly when excited. The first is bobbing with the entire body, reminding Schüz (1957) of a Redstart; the second – often accompanying the first – is wing-flicking, well known from other nuthatch species (Schüz, 1957; Kipp, 1965; Löhrl, 1988). The latter is less commonly shown by female Western Rock Nuthatches (Kipp, 1965), and both behaviours are said to be less common in Eastern Rock Nuthatches (Cramp and Perrins, 1993). Also typical is the upright stance, especially when the birds are alerted.

Grant (1975) studied the displays of both species, mainly in response to models, and found few differences between these two. The main displays of the male towards model intruders were Bobbing with horizontal bill and a Vertical display with drooping wings and erect tail (Fig. 85). The latter was once accompanied by sideways Pendulum-like movements. Löhrl (1988) described the same Vertical display as the main aggressive posture, sometimes including spreading the tail and directing it to the opponent. During these displays the male's eye-stripe becomes more conspicuous (Grant, 1975). Females approaching a model adopted a crouching posture, repeatedly switching their orientation to the model (Horizontal display, Fig. 85). Vertical postures and Bobbing were also used in mutual displays of mated birds away from the model. Other displays include a submissive Hunched posture, observed in females being chased by their mates, and a Gaping display used in aggressive interactions (Fig. 85). The remarkable Head-curving display was observed only twice by Grant (1975). In this display the male raised its head slowly sideways and upward, assumed the Vertical posture for a few seconds, lowered the head with bill pointing downwards, and made a similar curve to the other side (Fig. 85). In contrast to these observations, Panov (1989) observed few stereotyped displays and no overt aggression in natural conflicts, and suggested that the reactions to the models were to some extent artefacts, appearing as 'a nervy and chaotic version of everyday behaviour' Nevertheless, the displays of the rock nuthatches share many similarities with those of the Eurasian Nuthatch, for example the Vertical posture, Pendulum-like sideways movements and Gaping (see Chapters 5

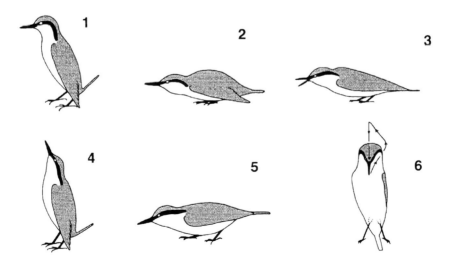

Fig. 85 *Postures and displays of the Western Rock Nuthatch, reproduced with permission from Grant (1975). 1. Bobbing, 2. Hunched, 3. Gaping, 4. Vertical, 5. Horizontal, 6. Head-curving.*

and 6). Grant (1975) speculated that the displays communicate information on characteristics such as body length (Vertical display) and eye-stripe length (Horizontal display and sideways movements), and in this way would facilitate species recognition in areas of sympatry.

VOCALIZATIONS

Rock nuthatches are usually described as very noisy with a loud and rich voice (Sarudny and Härms, 1923; Löhrl, 1988) which has given them the name 'Rock-Nightingale' in Turkish (Danford, 1878). Kipp (1965) could hear a Western Rock Nuthatch singing up to 400 m away, and Löhrl (1988) was forced to move his hand-reared birds out of the living-room because of their loud and constant calling. Both species have highly varied repertoires but share the same basic elements (Panov, 1989). Grant (1975) recognized no fewer than 29 types of vocalization by syllable structure. The distinction between song and calls seems even more difficult than in other nuthatches. Kipp (1965) discriminated between a fast trill or 'tschitscherstrophe' that both sexes use, and 'male song' which includes the majority of the other vocalizations. Löhrl (1988) also interpreted the fast trill as a contact-alarm call rather than song, since both sexes use it and young birds as well. Trills can have up to 20 notes per second, often dropping noticeably in pitch towards the end. This phenomenon is known as 'drift' and has been well studied in songs of other birds (e.g. Lambrechts and Dhondt, 1988). The notes that make up the trill can also be delivered singly or in series of two, and serve as the main contact-call.

Löhrl also noticed a softer contact-call used during foraging, which reminded him of the calls of the Chestnut-bellied and Velvet-fronted, but not Eurasian, Nuthatches. Both rock nuthatches also share the harsh call known also from the Mediterranean and Kashmir Nuthatches (Chapters 11 and 13). This call appears to be used in antagonistic as well as heterosexual encounters (Cramp and Perrins, 1993), but less so by the Eastern species (Löhrl, 1988). Some other calls that have been described include a 'clicking' call, a nasal or hoarse call often heard near the nest, a copulation call, and a loud whistle (Cramp and Perrins, 1993).

The songs of the rock nuthatches consist of one, occasionally two and rarely three syllables that are repeated a number of times. Many of the monosyllabic song types are quite similar to those of the Eurasian Nuthatch, in the same general frequency range between 2 and 4 kHz. The length and tempo (number of notes per second) are highly variable, though the majority of songs have four to seven notes per second. Panov (1989) remarks that intermediates may be heard between different vocalization types or 'compound' vocalizations, such as a combination of song and trill. The songs of the two species appear to be quite similar, and share many song types. Grant (1975) recorded 14 from both species, 11 from the Western and four from the Eastern Rock Nuthatch only. Although the larger repertoire in the former species may in part reflect sampling effort (Grant, 1975), a more variable song in this species has also been suggested by Mörike (1964) and Eggers (1977). Not surprisingly, the larger Eastern Rock Nuthatch sings at a lower pitch (Eggers, 1977; Panov, 1989) and is also louder than the other species (Mörike, 1964; Grant, 1975). The difference in pitch appears to be more pronounced in sympatry, but the evidence for this is not conclusive (Grant, 1975). While in Turkey the two species did not respond to each other's song playback (Eggers, 1977), in Iran the Eastern Rock Nuthatch did respond to the Western's song (Grant, 1975). Since there is no evidence that songs are more similar in sympatry, a possible alternative explanation is that Grant's playback experiments were done closer to the nest, and evoked a stronger response. Some information on vocal differences between species and subspecies can be found in Leonovich *et al.* (1996b).

In contrast to the majority of their congeners, female rock nuthatches do sing, but why is still somewhat unclear. While Grant (1975) simply states that no particular vocalizations were limited to one sex, Panov (1989, pers. comm.) describes the female's song as more stereotyped than the male's, perhaps even limited to one song variant per individual, and used particularly for duetting (Fig. 86). The female's song often, but not always, starts as a typical trill, but then changes to true song notes. She starts after or shortly before the male has finished. Duetting pair members are often widely spaced, suggesting that the behaviour may in part serve to maintain contact. On the other hand, it is rarely heard outside the breeding season, and sometimes birds duet within 20–30 cm of one another (Panov, 1989). While Panov considers this behaviour common in both species (from his studies in Transcaucasia, Turkmenistan and other areas), however, it has not been mentioned by other authors.

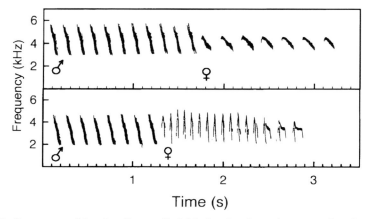

Fig. 86 *Sonagrams of duetting Eastern Rock Nuthatches. In each case a series of notes by the male is immediately followed by female song. Reproduced with permission from Panov (1989).*

SOCIAL ORGANIZATION AND MOVEMENTS

Rock nuthatches are not very social, but the extent of their territorial behaviour remains to be clarified. The immediate vicinity of the nest is vigorously defended, and neighbouring nests are usually at least 150 m apart. Some conflicts have been observed away from the nest where neighbouring pairs meet (Grant, 1975; Panov, 1989). Pair members usually move around together (Cramp and Perrins, 1993). Outside the breeding season they live in pairs or solitarily, and tend to stay near the nest site (Roháček, 1919; Löhrl, 1988; Panov, 1989; Roberts, 1992). Löhrl (1988) mentions territorial fights in September (Western). Pair members may roost together in the nest in winter (Panov, 1989). On the other hand there are reports of groups of adult and young birds, probably families, staying together as late as August or September (Dement'ev and Gladkov, 1954; Cramp and Perrins, 1993), and large groups have been observed as late as October in Pakistan (Roberts, 1992). Reports of altitudinal movements also suggest that not all birds stay in or near the breeding territories in winter. These movements are probably a response to severe winter conditions (Roháček, 1919; Wunderlich, 1986; Loskot *et al.*, 1991). In Tajikistan regular vertical migration may start in September–October in the highest regions, while birds from lower altitudes descend only during the heaviest snowfalls (Abdusalyamov, 1973, cited by Cramp and Perrins, 1993). In contrast, Western Rock Nuthatches in Israel are said to move up Mount Hermon after the breeding season until the first snow falls (Paz, 1987).

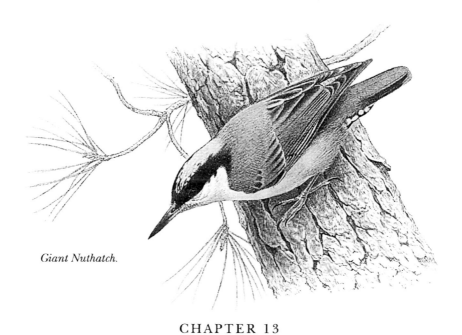

Giant Nuthatch.

CHAPTER 13

Oriental Nuthatches

This pretty little species, which recalls to the wanderer in the wilds of Ceylon the familiar little Nuthatch of England, (. . .) is one of the most active birds imaginable

Legge (1983) on the Velvet-fronted Nuthatch

In this chapter we move away from Europe and the Middle East, and into the world's 'hot spot' of nuthatch diversity in the subtropical zone of eastern Asia. No fewer than 14 species are confined to the region between the Himalayas and the Philippines, and up to seven species can be found together in some parts of continental SE Asia. None of them has been studied in much detail, which is why it is possible to treat them all in a single chapter. In many cases even the most basic knowledge is lacking on habitat, vocalizations and breeding biology, except for largely anecdotal information in handbooks or expedition reports. For some species, Löhrl's (1988) observations on captive birds represent the major source of information on

217

behaviour. Some of the Oriental nuthatches have already been mentioned in other chapters. Notably, the Chestnut-vented, Chestnut-bellied and Kashmir Nuthatches are close relatives of the Eurasian Nuthatch and have been discussed briefly in Chapter 2, while the Chinese Nuthatch is closely linked to the Mediterranean nuthatches (Chapter 11). The Yunnan Nuthatch is also, but less clearly, related to the latter group, and the White-cheeked Nuthatch is obviously close to the N American White-breasted Nuthatch. The affinities of the other Oriental species are less obvious.

ECOLOGICAL RELATIONSHIPS

The co-occurrence of so many closely related species has inevitably attracted attention with respect to their ecological relationships (Ripley, 1959; Lack, 1971). I deal with these mainly in the following species accounts, but a summary of Lack's conclusions is given in Table 25. Out of 28 geographically overlapping species pairs, 15 are separated by habitat and altitude, four or five by habitat alone, and only seven species pairs occur in the same forests in the same geographical region. The altitudinal zonation in different parts of Asia is summarized in Fig. 87. Thus, niche separation on a local scale is expected in only a few cases. For example, the Beautiful Nuthatch co-occurs with several of the smaller nuthatches (in different parts of its fragmentary range) which are only about half its weight, and it seems logical to assume that they use different foraging substrates or foraging techniques, and actual competition for food is limited. However, none of these relationships has been investigated in any detail.

TABLE 25: *Probable means of ecological segregation of nuthatches in the mountains of central and SE Asia, modified after Lack (1971)*

	1	2	3	4	5	6	7	8	9	10	11
1 Chestnut–bellied	1										
2 Eurasian	–	2									
3 Beautiful	–?	–	3								
4 Velvet–fronted	F?	–	F?	4							
5 White–tailed	Ha	–	F?	Ha	5						
6 White–cheeked	Ha	–	–	Ha	Ha(–?)	6					
7 Giant	Ha	–	–?	Ha	–?	–	7				
8 Chestnut–vented	Ha	–	F?	Ha	–(H)	H	F?	8			
9 White–browed	Ha	–	–	Ha	H(–?)	–	–	Ha	9		
10 Snowy–browed	–	Ha	–	–	–	–	–	–	–	10	
11 Yunnan	–	–*	–	–	H	H	–	HF?	–	–	11
12 Yellow–billed	F?	–	F?	Ha	Ha	–	–	H?	–	–	–

NOTES:
* following maps of Harrap and Quinn (1996) and Tso-Hsin (1987), I considered Yunnan and Eurasian Nuthatch not to overlap (*contra* Lack, 1971).
– no geographical overlap, H segregation by habitat, Ha where this is closely linked with altitude, F probably by foraging sites.

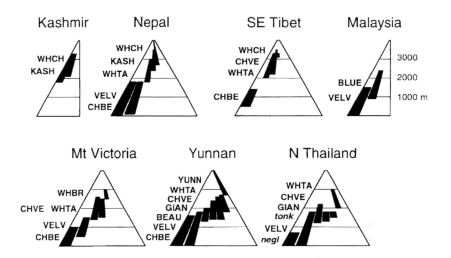

Fig. 87 *Altitudinal zonation of nuthatch species in different parts of Asia. Species plotted together do not necessarily coexist on a single mountain. Sources. Ali and Ripley (1973) for Kashmir, Inskipp and Inskipp (1985) for Nepal, Ludlow (1944, 1951) for south-eastern Tibet, Stresemann and Heinrich (1940) and Ludlow (1944) for Mt Victoria (range of White-tailed is hypothetical), de Schauensee (1984) for Yunnan (range of Chestnut-bellied is hypothetical), Deignan (1945) and Lack (1971) for northern Thailand, and MacKinnon and Phillips (1993) for Malaysia. Species are Kashmir (KASH), White cheeked (WHCH), White tailed (WHTA), Chestnut-bellied (CHBE), Velvet-fronted (VELV), Chestnut-vented (CHVE), White-browed (WHBR), Yunnan (YUNN), Giant (GIAN), Beautiful (BEAU) and Blue Nuthatch (BLUE). In northern Thailand two different subspecies of Chestnut-bellied are represented* (tonkinensis *and* neglecta).

CHINESE NUTHATCH

This species was regarded by Vielliard (1978) as the most primitive living nuthatch. It is confined mainly to China and North Korea (Fig. 78), with some recent observations from Sakhalin and southern Ussuriland (Fiebig, 1992; Harrap and Quinn, 1996). Its closest relative is probably the Corsican Nuthatch from which it differs mainly by the darker underparts (pale buff to orange-cinnamon, rather than greyish-white) and the less well-defined eye-stripe (Harrap and Quinn, 1996; see also Chapter 11). At less than 10 g it is one of the smallest nuthatches. There are two subspecies completely separated in range: the nominate *villosa* in north-eastern China and Korea, and *bangsi* in western China (Greenway, 1967). The latter is more rufous below, has somewhat longer wings and more white on the tail-feathers (Dunajewski, 1934; Harrap and Quinn, 1996).

Like the other species in the *canadensis* group, the Chinese Nuthatch lives mainly in coniferous forest: pine (particularly red pine), spruce and larch, sometimes mixed with birch, at altitudes up to 2600 m (Gao, 1978;

Mauersberger, 1989; Matthysen *et al.*, 1991; Fiebig, 1992). In southern Ussuriland the birds are present only from mid-May (Nazarenko, 1988, cited by Fiebig, 1992) while in north-east China they are said to be present for about 200 days (Gao, 1978), suggesting that this species migrates, perhaps altitudinally. Nevertheless, the majority of populations are probably largely resident (e.g. Fiebig, 1992). The species' range overlaps considerably with that of the Eurasian Nuthatch which prefers deciduous habitats (Fiebig, 1992), and also slightly with that of the White-cheeked Nuthatch in north–central China (Gansu and Sichuan provinces). Here both species have been observed in pine forest (Stresemann *et al.*, 1937), but nothing further is known about overlaps in habitat or foraging niche.

Anecdotal observations suggest that Chinese Nuthatches forage in a similar way to the closely related Mediterranean nuthatches (Mauersberger, 1989; Fiebig, 1992). They feed on pine cones in September, and otherwise search for their food on branches and twigs and more occasionally on trunks, or even practise fly-catching. Stomachs collected from April to August contained almost exclusively insects, mainly beetles and Hemiptera in adults, and caterpillars in nestlings (Gao, 1978). Pine seeds have been found in stomachs collected as late as May (Mauersberger, 1989). Food hoarding appears to be common (Fiebig, 1992).

In north-eastern China and Korea nest-building may start in late March and clutches appear to be produced mainly in April, though no exact dates are reported (Gao, 1978; Fiebig, 1992). Chinese Nuthatches either excavate their own hole or use an existing cavity such as old or unfinished holes of Lesser Spotted, White-backed and Japanese Pygmy Woodpeckers (Fiebig, 1992). Both partners contribute to nest-building (Gao, 1978). The nest is lined mainly with villous plant material, feathers and grass. The males also bring mud to the female during incubation, which she uses to repair the inside wall of the nest hole (Gao, 1978). Mud was found within all 13 nests examined. This behaviour might represent an ancestral state to 'true' plastering behaviour (see Chapter 1). There is one intriguing observation of a pair in Sakhalin using a nest hole with a reduced entrance, but the possibility cannot be excluded that this hole had previously been occupied by a Eurasian Nuthatch (Leonovich and Veprintsev, 1986). Curiously, two independent observers have seen females removing nest material after the young had already fledged (Fiebig, 1992). A copulation event observed by Fiebig (1992) was reportedly comparable to that of the Eurasian Nuthatch, with wing-shivering by both sexes, but it is not clear whether a Pendulum-like behaviour was included. Only one clutch is laid, containing four to nine eggs, usually five or six (Gao, 1978), though Fiebig (1992) suspected that one pair initiated a second clutch. Eggs are incubated for 15–17 days and nestlings fledge after 17–18 days. This is comparable to the nestling period for Krueper's Nuthatch, but considerably less than in the Corsican Nuthatch (Chapter 11). Gao (1978) presents some further data on provisioning rates and nestling weight changes.

The few observations outside the breeding season all suggest that Chinese Nuthatches live in pairs throughout the year, but the winter home-ranges may be larger than the breeding territory. Pairs are often found in mixed flocks with several species of tits, treecreepers, goldcrests, tit-warblers and woodpeckers (Stresemann *et al.*, 1937; Mauersberger, 1989; Fiebig, 1992). The vocalizations have been relatively well described (Matthysen *et al.*, 1991; Fiebig, 1992). The most striking element in the repertoire is the harsh call which this species shares with the Mediterranean nuthatches. The fast trilling song is also quite similar to that of the Corsican Nuthatch, with some variation in note structure between individuals (Fiebig, 1992). Singing birds adopt an erect posture which is probably comparable to that of the Eurasian Nuthatch (Fiebig, 1992). However, the main calls as recorded in autumn in Sichuan are quite different from those of the Mediterranean nuthatches (Matthysen *et al.*, 1991). They consist of long irregular series of single notes, typically four to five per second, made up of seven different note types, one of them with a clear harmonic structure. It has been suggested that this represents an ancestral vocalization which has subsequently been lost (or modified into a lower-intensity contact-call) in the other members of the *canadensis* group (Matthysen *et al.*, 1991).

YUNNAN NUTHATCH

The next species is another member of the *canadensis* superspecies from coniferous forests in China. It was once considered a subspecies of the Red-breasted Nuthatch but restored to specific status by Vaurie (1957, 1959). It resembles the Chinese and Corsican Nuthatches in size and plumage, but without the black cap in the male. The underparts are dirty pinkish grey, the eye-stripe rather broad and long with a well-defined supercilium, and there is some pale mottling on the nape. Juveniles are dull compared with the adults, with the supercilium and eye-stripe reduced or absent (Harrap and Quinn, 1996). This species' phylogenetic position remains unclear. Voous and van Marle (1953) considered it the most primitive member of the *canadensis* group, but Vaurie (1957) suspected – without stating specific arguments – that it was more closely related to the White-tailed Nuthatch. It has even been considered a subspecies of the Eurasian Nuthatch (Hellmayr, 1903, cited by Vielliard, 1978). These divergent views were somewhat reconciled by Vielliard (1978) who placed the Yunnan Nuthatch between the Red-breasted and White-tailed Nuthatch in his phylogenetic tree, at the origin of the 'plastering' nuthatch line (see Chapter 1, Fig. 2).

The Yunnan Nuthatch lives in high-altitude coniferous forest from 2700 to 4500 m in the Chinese provinces of Yunnan, Tibet, Sichuan and Guizhou (de Schauensee, 1984; Harrap and Quinn, 1996). It hardly overlaps in altitude with other nuthatches, except for the White-cheeked Nuthatch which it meets in extreme south-eastern Tibet (Fig. 87). Here it seems to prefer dry pine forest while the White-cheeked prefers fir and spruce at higher altitudes (Schäfer and de Schauensee, 1938). In winter some birds may descend down to 1200 m. Collar *et al.* (1994) considered it a threatened species because of the widespread disappearance of mature pine forest on which it depends. Very little is known about the behaviour and breeding biology of this nuthatch. It seems to have a variety of calls, including several with a nasal quality, as well as the harsh scolding call common to all Palearctic members of the *Sitta canadensis* superspecies (Harrap and Quinn, 1996).

WHITE-CHEEKED NUTHATCH

The White-cheeked Nuthatch presents yet another biogeographical enigma within the nuthatch family. Although confined to the mountains of central Asia, it is obviously closely related to the N American White-breasted Nuthatch. The two were even considered conspecific by Kleinschmidt (1933). Some authors have suggested a close relationship to the *canadensis* group as well, but without much supporting evidence (Dunajewski, 1934; Voous and van Marle, 1953).

The diagnostic features of this moderate-sized nuthatch are the missing eye-stripe (hence 'white-cheeked'), the black cap shared by both sexes, and the contrasting rufous rear flanks. There are two clearly recognizable sub-species, separated by a large gap of about 1000 km. In the west, *leucopsis* occurs from north-eastern Afghanistan to Nepal, while the eastern *przewalskii* lives in the Chinese provinces of Tibet, Gansu, Sichuan and Qinghai (Harrap and Quinn, 1996). The latter has a markedly thinner and shorter bill, slightly shorter wings, and darker underparts (rich cinnamon rather than buffish white). There are some isolated records from south-eastern Tibet which may indicate that the geographical separation is not as complete as previously believed (Harrap and Quinn, 1996).

The White-cheeked Nuthatch is usually described as rare and shy except by Roberts (1992) who found it 'plentiful' in northern Pakistan. It is mainly confined to various types of coniferous forest: spruce, fir, cedar, pines and occasionally juniper (Stresemann *et al.*, 1937; Schäfer, 1938; Ripley, 1959; Löhrl and Thielcke, 1969) but also in moist, mixed temperate forest in Pakistan (Roberts, 1992). Throughout its range it occupies the highest altitudes of all nuthatches, between 2000 and 4000 m (Fig. 87), with some altitudinal movements down to 1800 m in winter (Ali and Ripley, 1973). It overlaps in range and altitude with a few populations of Kashmir, White-tailed and Yunnan (q.v.) Nuthatches only. Kashmir Nuthatches have a broader habitat range, and where they overlap with the White-cheeked the latter lives at lower densities (Löhrl and Thielcke, 1969; Wunderlich, 1988). In Nepal there is some overlap in altitude with the White-tailed Nuthatch, but hardly any in range.

Paludan (1959) and Macdonald and Henderson (1977) observed this species most often high in the canopy foraging on small branches, while Löhrl and Thielcke (1969) observed it mainly on the underside of lichen-covered branches in the lower parts of cedar trees. In the same areas, the sympatric Kashmir Nuthatches foraged higher up (Macdonald and Henderson, 1977) or more often on trunks and the upperside of branches (Löhrl and Thielcke, 1969). The preference for undersides contrasts markedly with the avoidance of these substrates by the closely related, but larger, White-breasted Nuthatch (Chapter 4, Fig. 21). Fly-catching has also been observed (Jamdar, 1987). Captive birds were very agile and tit-like in their foraging behaviour, inspecting small twigs as well as branches and trunks, and hoarding food as well (Löhrl, 1988).

Nest sites include large crevices in dead trees, natural and excavated holes, and one hole between the roots of a fir (Löhrl and Thielcke, 1969; Roberts, 1992). Whether self-excavation occurs is unknown. The nest is lined with moss, hairs, grass or leaves, but no mud is used (Ali and Ripley, 1973; Löhrl, 1988; Roberts, 1992). Löhrl (1988) observed the same bill-wiping behaviour around the nest site that has been described in more detail in the White-breasted Nuthatch (see Chapter 14). Clutches of four to eight eggs have been found in late May and June (Whistler, 1930, cited by Wunderlich, 1988; Ali and Ripley, 1973; Roberts, 1992). However, in one case young were being fed in late May, thus egg-laying must have occurred in April (Wunderlich, 1988).

Social behaviour and song are virtually undescribed. Löhrl (1988) describes a threat display, with wings and tail spread, from a captive bird. Wing-flicking is mentioned by Ali and Ripley (1973). The main call is described as a nasal, bleating sound delivered in series, similar to the White-breasted Nuthatch's calls, and the song a faster repetition of similar notes. However, a disyllabic 'ti-tuï ti-tuï ti-tuï' may represent another song variant (P. Alström, cited by Harrap and Quinn, 1996).

CHESTNUT-BELLIED NUTHATCH

This bird replaces the Eurasian Nuthatch in most of southern Asia with no overlap in range (*fide* Vaurie, 1957; *contra* Voous and van Marle, 1953) (see Chapter 2). Its distribution comprises large parts of the Indian low-lands, the southern foothills of the Himalayan range from northern Pakistan to easternmost India, and large parts of SE Asia including a few sites in China (Harrap and Quinn, 1996). At the western and eastern edges of its range it is replaced at higher altitudes by the Kashmir and Chestnut-vented Nuthatches, two other representatives of the *Sitta europaea* super-species. Voous and van Marle (1953) believe this species to be a relatively old (late Tertiary) offshoot of an ancestral Eurasian Nuthatch population

The Chestnut-bellied Nuthatch has the most pronounced sexual dimorphism of all nuthatches. The male's underparts are entirely dark chestnut to mahogany red, while the female's colour is best described as cinnamon to beige, comparable to Kashmir Nuthatch males (see below) but darker than Eurasian males. The contrasting white chin and cheeks are diagnostic, particularly in males. Juvenile males are intermediate between adult males and females, but juvenile and adult females hardly differ. Sometimes a faint supercilium is present. The undertail-coverts are grey or white with brown edges, the exact pattern varying between subspecies (see Harrap and Quinn, 1996).

The four montane subspecies *almorae* (western Himalayas), *cinnamoven-tris* (eastern Himalayas), *koelzi* (northern Myanmar) and *tonkinensis* (north-ern Thailand, Vietnam and Laos) are also known as the 'foothills group'

(Greenway, 1967). They have generally larger and heavier bills than the lowland forms *castanea* (large parts of India) and *neglecta* (lower elevations in SE Asia). The four montane races differ only slightly in the colour of the underparts and the undertail-coverts (Voous and van Marle, 1953; Harrap and Quinn, 1996); *neglecta* has a less well-developed eye-stripe and markedly lighter underparts (especially the males) than the other subspecies. The apparent lack of intermediates between lowland and montane forms has drawn the attention of many authors (Vaurie, 1950; Voous and van Marle, 1953; Ripley, 1959). In particular, Harrap and Quinn (1996) raised the question whether *tonkinensis* and *neglecta* might be reproductively isolated and, in that case, whether one of them merits specific status.

This species is found in tropical and subtropical forest, usually open woodland, in lowlands and foothills up to 1600 m, but generally below *c.* 1300 m (Lack, 1971). Only *tonkinensis* is found at higher altitudes, from 1200 to 2200 m (Beaulieu, 1944; Lack, 1971) (see also below). The habitat seems varied and includes deciduous, semi-evergreen and pine forest, dry lowland forest, mango groves, orchards and other plantations, scrub, gardens, bamboo clumps and even masonry walls (de Schauensee, 1934; Ripley, 1959; Ali and Ripley, 1973). In Thailand Deignan (1945) found them mainly in parklike forest with large and well-spaced trees, and in the lower plains of Myanmar they were very common in gardens (Stanford and Ticehurst, 1938). Hellmich (1968) found them to be quite common in Nepal, comparable to Eurasian Nuthatches in Germany, which suggests densities of several pairs per 10 ha.

Lack (1971) suggested that most populations are held below 1300 m through competition with the higher-altitude Chestnut-vented and White-tailed Nuthatches (Fig. 87). The subspecies *tonkinensis* is an exception to this pattern, since its geographical range interdigitates with that of the Chestnut-vented Nuthatch (q.v.), and where it meets the White-tailed Nuthatch their altitudinal separation lies higher, around 2000 m. The Chestnut-bellied Nuthatch also overlaps widely in range with the Velvet-fronted Nuthatch, but lives in drier and more open woodland and forages on larger substrates and more often on the ground (Lack, 1971). Although some aggression between the two has been noted, they are often found in the same flocks (Ali and Ripley, 1973).

Foraging behaviour and diet appear very similar to those of the Eurasian Nuthatch, with a preference for foraging on trunks and large branches. In mixed forest in Nepal this species has been seen foraging on both broadleaved trees and long-leaved Indian pine (Löhrl, 1988). Occasionally it forages on the ground, for instance on ants and termites, and seems to be fond of foraging on walls, banks and cliffs (Voous and van Marle, 1953). It should be noted, however, that the latter tendency has been described for the montane subspecies only, while others have described the species as 'purely arboreal' (Whistler, 1963). Stomach contents yield the typical mixture of various insects and spiders (see Ali and Ripley, 1973; Roberts, 1992). Seeds and nuts are probably taken in at least some populations. Löhrl's (1988) captive birds (of the nominate subspecies) did not hoard seeds and rarely ate them, but their captive-bred offspring hoarded occasionally.

The montane subspecies seem to have a particular preference for nest sites on the ground in stony walls and steep banks (Voous and van Marle, 1953; Whistler, 1963) but otherwise typically nest in holes in trees, which may or may not be reduced with mud, sometimes mixed with resin (Ali and Ripley, 1973; Löhrl, 1988). Males participate actively in plastering, and in one case a male and female started nest-building each in a different part of the aviary (Löhrl, 1988). Some nests contain the familiar woody material (bark chips, seed wings) while others have a more extensive foundation of moss and bark, and a lining of fur. Whistler (1963) suggests that the former type is more common in tree cavities, and the latter in nests in holes in the ground or a wall. This may reflect a geographical, or even subspecific, variation in nesting behaviour, but needs to be investigated further. There is an intriguing parallel here to the rock nuthatches, which nest in holes in or among rocks, and use soft lining material as well (Chapter 12).

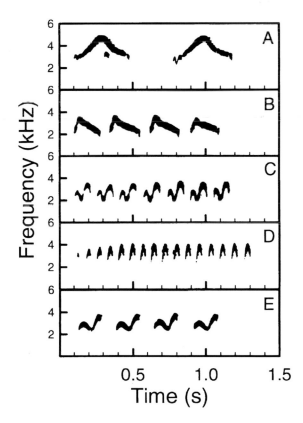

Fig. 88 *Some songs of Chestnut-bellied and White-tailed Nuthatches. A = Chestnut-bellied subspecies* almorae *from Nepal; B,C = Chestnut-bellied subspecies* castanea *in captivity; D, E = White-tailed Nuthatch. Sonagrams reproduced from Löhrl (1988).*

Breeding and social behaviour are not well studied, but again are probably quite similar to those of the Eurasian Nuthatch. A defensive posture with drooping wings and erect tail is common in both species. Pre-copulation behaviour (as seen in captivity) includes Pendulum movements but appears to lack the rigid male vertical posture (Löhrl, 1988). Breeding seasons are given as February–March for the lowland races and April–May for the montane forms. Clutches contain two to seven eggs (mainly five to seven) and incubation apparently lasts only 11–12 days (Ali and Ripley, 1973). Chestnut-bellied Nuthatches are usually said to live in 'pairs or loose family parties', typically associated with mixed flocks of small insectivorous birds. Rand and Fleming (1957) found them usually in pairs in winter in Nepal.

Löhrl (1988) described a number of vocalizations, with sonagrams, of captive birds (subspecies *castanea*) and free-living birds of the *almorae* subspecies (Fig. 88). Though obviously related to the Eurasian Nuthatch's sounds, the song as well as the calls are distinctly different to the human ear (Hellmich, 1968) as well as to the birds themselves, since the song of captive birds did not elicit a response from nearby Eurasian Nuthatches (Löhrl, 1988). The alarm call is said to be a 'loud raucous note', but different from the Kashmir's harsh call (Roberts, 1992). Löhrl (1988) recognized two different contact-calls, one of which reminded him of the Eurasian's 'tsit'. He found no obvious contextual differences between the two calls, and even newly fledged birds used both. The 'threatening call' described by Löhrl (1988) may be quite similar to the Eurasian's alarm call (see also Harrap and Quinn, 1996).

KASHMIR NUTHATCH

The relationships of this species have already been discussed in Chapter 2. In short, it is intermediate between the Eurasian and Chestnut-bellied Nuthatches in the colour of the underparts, which are also somewhat darker in males, and by having a less clearly demarcated white throat and cheeks than the Chestnut-bellied. On the other hand, the uniform under-tail-coverts and the harsh alarm call may suggest some affinity to the rock nuthatches (see Chapter 12). It is found mainly in Kashmir but also in north-eastern Afghanistan, northern Pakistan and north-western Nepal (Harrap and Quinn, 1996). It lives at altitudes from 1500 to almost 4000 m, mainly between 2000 and 3000 (Fig. 87), with limited altitudinal migration down to *c.* 1800 m in winter (Ali and Ripley, 1973). Like the Eurasian Nuthatch it inhabits a variety of forest types such as holm oak, deodar, fir, spruce, cedar and pine, although it may require the presence of at least some broadleaved trees. Paludan (1959) also found it in hazel scrub and poplar groves in a valley bottom. In coniferous forests it may co-occur with the White-cheeked Nuthatch (q.v.). In Nepal there is a sharp geographical separation with the White-tailed Nuthatch (Inskipp and Inskipp, 1985) which may suggest competitive exclusion. The Kashmir Nuthatch is said to forage mainly in the lower parts of trees, sometimes on the ground (Ripley,

1959). Paludan (1959) and Löhrl and Thielcke (1969) observed it mainly on trunks and the uppersides of branches (see also White-cheeked Nuthatch). Its diet contains a high proportion of nuts and seeds in winter, including hazel, walnut, pine and yew (Roberts, 1992) and it hoards seeds in captivity (Löhrl, 1988).

The few descriptions of breeding biology suggest only minor differences from the Eurasian Nuthatch. In some nests (three out of nine examined) the mud wall extends slightly into a tunnel-like protrusion from the entrance (Löhrl and Thielcke, 1969). A mating between a Kashmir male and a female Eurasian Nuthatch in captivity was characterized by the absence of a copulation call, a less rigid male posture, and less marked Pendulum movements compared with Eurasian Nuthatches (Löhrl, 1988). Nests with five to seven eggs have been found in April and May (Löhrl and Thielcke, 1969; Ali and Ripley, 1973; Roberts, 1992). The song is similar to the Eurasian's but less loud and delivered less frequently (Löhrl, 1988). The calls are probably also similar (see sonagrams in Löhrl, 1988) except for the Jay-like harsh alarm call which it shares with most members of the *canadensis* superspecies and the rock nuthatches. The 'territorial call' is described as a rapid squeaking whistle 'pee-pee-pee-pee-pee' (Roberts, 1992). Not much is known about social behaviour, except that Löhrl (1988) found it more social and less aggressive in captivity than the Eurasian Nuthatch. Free-living birds participate in mixed flocks with *Phylloscopus* warblers and Crested Black Tits (Macdonald and Henderson, 1977).

CHESTNUT-VENTED NUTHATCH

The fourth species in the *Sitta europaea* superspecies complex (cf. Chapter 2) is characterized by grey rather than buff or brown underparts, with highly contrasting brownish flank regions. Also notable is some pale mottling on the nape, especially in juveniles. The centre of its distribution is in the mountains of north-eastern Myanmar and south-west–central China, extending into India and Tibet (Harrap and Quinn, 1996). In addition three rather isolated populations occur in the Naga and Chin Hills on the Indian–Myanmar border, the Da Lat plateau in southern Vietnam, and the Chinese province of Fujian. The sympatric occurrence with Eurasian Nuthatches (at lower altitudes) in Fujian has been an important argument for the taxonomic separation of the two species. However, further east in central Sichuan, the two species also meet but may show signs of intergradation (details in Harrap and Quinn, 1996). Apart from the nominate subspecies, Greenway (1967) recognizes only *grisiventris* which comprises two of the most southerly, but mutually widely separated populations on Mount Victoria (Myanmar) and in southern Vietnam. The subspecies *nagaensis* has more buff on the underparts (particularly in fresh plumage) which is expressed most in the north-eastern populations (also known as *montium*; Vaurie, 1957). These are therefore more similar to sympatric Eurasian Nuthatches.

The Chestnut-vented Nuthatch lives at altitudes between 1200 and 2100 m, occasionally up to 4000 m (Lack, 1971). It is found in deciduous, ever-green and coniferous forests, with a preference for more open forest (de Schauensee, 1929; Stresemann and Heinrich, 1940; Stanford and Mayr, 1941; Deignan, 1945). In south-eastern Tibet it is quite commonly seen on poplar and walnut trees near villages (Schäfer and de Schauensee, 1938). In this area and in neighbouring Chinese provinces it appears to avoid conifers, perhaps because of competition with the Yunnan Nuthatch (Lack, 1971). In northern Myanmar it occurs alongside the White-tailed Nuthatch (q.v.), from which it is separated by altitude and partly by habitat. In north-ern Thailand, northern Vietnam and Laos it is replaced by Chestnut-bellied Nuthatches (q.v.) of the *tonkinensis* subspecies.

The behaviour is again very similar to that of the Eurasian Nuthatch, but some differences in vocalizations are known. An alarm note is described as 'chit-it-it-it-it', recalling a Wren (Harrap and Quinn, 1996). Another call-note is variously described as a nasal 'tjäb' (Stresemann and Heinrich, 1940), a 'yank' similar to that of the White-breasted Nuthatch (de Schauensee, 1929), or a whining 'quir', 'kner' or 'mew' (Harrap and Quinn, 1996). The song is similar to the Eurasian Nuthatch's Trill, fast and monotonous, but not as loud (Stresemann and Heinrich, 1940; Harrap and Quinn, 1996). Chestnut-vented Nuthatches seem to breed quite early in some areas, since nest-building has been seen in Thailand and Vietnam in January (Harrap and Quinn, 1996), fledged young have been observed on Mount Victoria on 31 March (Stresemann and Heinrich, 1940) and moulting birds were collected in southern Vietnam from May to July (Voous and van Marle, 1953). Harington (1914) reported a clutch in April and observations on nest-building in June, which suggests possible double-broodedness. Clutches number two to five eggs (Harrap and Quinn, 1996). The birds are usually seen singly, in pairs or small groups, and often in mixed flocks with various tits, woodpeckers, babblers and warblers (Schäfer and de Schauensee, 1938; Stanford and Ticehurst, 1938; Stanford and Mayr, 1941).

WHITE-TAILED NUTHATCH

The name of this central Asian nuthatch betrays its single diagnostic fea-ture: the white basal halves of the central tail-feathers, appearing as a clear white patch which is not always easily visible in the field. Other notable traits are the uniform undertail-coverts and the pale neck patch made up of white feather bases. Otherwise the colour pattern is similar to that of the *Sitta europaea* group, in particular the Kashmir Nuthatch, but the body appears relatively small and 'dumpy' with a short bill (Harrap and Quinn, 1996). A white tail patch and pale neck patch are also found in the closely related White-browed Nuthatch (see below) and the dwarf nuthatches of N America. Some authors have therefore included them in a separate sub-genus *Mesositta* (Glutz von Blotzheim, 1993). Others consider the White-tailed Nuthatch a relatively old offshoot of the Eurasian/rock nuthatch lineage (Voous and van Marle, 1953; see also Fig. 2). Eastern and southern populations of this bird are slightly more rufous, and the size decreases gradually from west to east (Vaurie, 1950), but no subspecies are recognized by Greenway (1967).

White-tailed Nuthatches occur from the Punjab across the Himalayas to northern Myanmar and western Yunnan (China), and further south-east have a very fragmentary distribution in Myanmar, Thailand, Laos and Vietnam. They live chiefly in broadleaved temperate forest, also in broadleaved evergreen (including rhododendron forest), and more rarely in pure coniferous forest, and are further said to prefer oaks and lichen-rich forests (Stanford and Ticehurst, 1938; Ripley, 1959; Hellmich, 1968; de Schauensee, 1984). The altitudinal distribution is mainly from 1400 to 3400 m, occasionally down to 1000 m in winter (Ali and Ripley, 1973; Lack, 1971). In northern Myanmar it overlaps with the Chestnut-vented Nuthatch in range and altitude, but here the latter is found in more open broadleaved forest and in pine forest as well, and at somewhat lower altitudes (1200–2100 m, compared to 1400–2700 m for White-tailed) (Lack, 1971). Ripley (1959) suggested that character displacement (cf. Chapter 12) had occurred in these populations leading to stouter bills and a reduced eye-stripe in the White-tailed, and an opposite trend in the Chestnut-vented Nuthatch, but this idea was not supported by later authors (Vaurie, 1957; Lack, 1971).

While foraging, White-tailed Nuthatches seem to prefer lichen-covered branches to trunks, perhaps because of the smooth bark of oak trees in their habitat (Hellmich, 1968; Löhrl, 1988) and sometimes forage in the undergrowth (Ripley, 1959) but little else is known. Insects and spiders are probably the main food, some of it caught on the wing (Stanford and Mayr, 1941; Löhrl, 1988) but seeds may also be taken (Ali and Ripley, 1973). In winter they are seen in mixed flocks of tits, babblers or tit-babblers (Stanford and Ticehurst, 1938; Ali and Ripley, 1973).

The nest entrance is reduced with plastering, in some cases with a mixture of mud, dung, berries and crushed insect larvae. Whether the larvae were used on purpose or because of a shortage of other material is unknown (Löhrl, 1988). Only the female was seen building, the male occasionally provisioning her with nest material. Besides the familiar lining of wood and bark fragments (Löhrl, 1988), moss and rhododendron leaves have also been found (Ali and Ripley, 1973). Clutches of four to seven eggs (Ali and Ripley, 1973; Etchécopar and Hue, 1983) are laid relatively early, in late March or early April (Vaurie, 1950; Löhrl, 1988).

Sonagrams of the calls and songs of this species (Löhrl, 1988) appear rather different from those of the Eurasian Nuthatch, but correspond to the general structure of nuthatch vocalizations. Löhrl's recordings show three song types including a fast trill and two other fast songs with six and 10 notes per second, respectively (Fig. 88). Another trill is described as an excitement call in aggressive contexts. Harrap and Quinn (1996) describe some call notes as 'chak', 'chik' or nearly disyllabic 'ts'lik' that may be repeated in short bursts or longer rattles. They also mention 'long, shrill, squealing or quavering' notes of up to 1.5 seconds long. Finally, Harrap and Quinn suggest that pairs may engage in rapid duetting using notes that differ in pitch. If this behaviour is confirmed, it would represent a remarkable correspondence with the dwarf nuthatches, and would strongly support the systematic relationship that was mentioned earlier.

WHITE-BROWED NUTHATCH

Definitely one of the rarest and least well known of all nuthatches, this species is restricted to Mount Victoria in the southern Chin Hills in Myanmar. It has been seen and collected only a few times since its discovery in 1904, and was not seen at all between 1938 and an expedition in April 1995. This expedition found the nuthatches not only on Mount Victoria but also at another locality 22 km to the north-west (Harrap and Quinn, 1996). Another recent expedition to Myanmar failed to find the species in supposedly suitable habitat in the northern Chin Hills, less than 300 km to the north (P. Rasmussen, pers. comm.). With a poorly known but obviously very limited range and population size, it has been included in the World List of Threatened Birds (Collar *et al.*, 1994), but fortunately there seems to be no immediate threat (C. Robson, pers. comm.).

The White-browed and White-tailed Nuthatches are obviously closely related and have formerly been considered conspecific. A major argument for taxonomic separation was the collection of a single White-tailed Nuthatch from Mount Victoria (Ludlow, 1944), but recent information suggests that it does not breed there but occasionally strays south from the northern Chin Hills (C. Robson, pers. comm.). In any case the two species are quite distinct: the White-browed has broad white supercilia meeting above the bill, a much shorter (by 1–2 mm) and especially narrower bill (4–5 instead of 6 mm), and more contrasting underparts with greyish-white throat and breast and bright orange-rufous flanks and undertail-coverts. Juveniles have less rufous on the sides of the neck (Harrap and Quinn, 1996).

The White-browed Nuthatch has been found from 2500 to 3000 m in lichen-rich deciduous and evergreen broadleaved forest, but not in coniferous forest (Stresemann and Heinrich, 1940; Ludlow, 1944; C. Robson, pers. comm.). For a montane species it breeds remarkably early (cf. Chestnut-vented Nuthatch) since fledged young were observed on 28 April by Stresemann and Heinrich (1940). Vocalizations are poorly known but include an insistent 'pee, pee, pee' call and a fast song similar to the White-tailed and Eurasian's (Harrap and Quinn, 1996). Nothing else has been recorded about its ecology or behaviour.

GIANT NUTHATCH

With a wing length of 113–122 mm this bird surpasses all other nuthatches in size, but otherwise shares the same general colour pattern. It has a broad and long black eye-stripe, blue-grey upperparts, white subterminal tail-spots and chestnut-and-white undertail-coverts. The underparts are soft blue-grey with whitish chin and throat. The crown is marked to a highly variable degree with black streaks, from immaculate grey to almost solid black (Deignan, 1938). The streaks are formed by grey crown feathers with black fringes on both sides, a unique plumage feature among nuthatches (Fig. 89). Females and juvenile males have decidedly duller markings, and the underparts are suffused with buff. The flight is said to be 'bold and dipping, like that of a woodpecker' (Smythies, 1986). The phylogenetic position is unresolved. Wolters (1975–1982) included the Giant Nuthatch in the subgenus *Sitta* with the Eurasian Nuthatch, while Vielliard (1978) placed it in the branch leading to the brightly coloured tropical nuthatches.

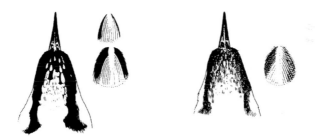

Fig. 89 *Top view of the head of male (left) and female (right) Giant Nuthatch, with some crown feathers magnified.*

The Giant Nuthatch lives in east–central Myanmar, northern Thailand and south-west China (Yunnan, Guizhou and Sichuan). It is rare and local in Thailand, its status in Myanmar is unknown, and there is only one recent record from China. Recently the species was missed in north-west Thailand on a site where it was known to occur before 1945 (Round, 1984). It is included in the Threatened Birds list (Collar *et al.*, 1994). Two subspecies have been described: *ligea* in Yunnan, with a markedly short and slender bill (Deignan, 1938), and *magna* in the rest of the range (Greenway, 1967). This nuthatch lives mainly in dense evergreen forests but also in more open pine forest, at altitudes between 1400 and 1800 m (Lack, 1971). In Thailand it is said to be restricted to the more open evergreen hill forests of oak and chestnut where khasi pine are frequent. This habitat is under severe threat of cultivation and cutting for fuel (Round, 1988).

The birds are generally observed singly or in pairs (de Schauensee, 1929; Deignan, 1945), and are described as silent and difficult to observe (Deignan, 1945). The commonest call is a trisyllabic, Magpie-like chattering call (described as 'get-it-up') and variations on it. Other calls include a toy trumpet-like 'naa' and a tree frog-like 'kip' or 'keep' note, which is repeated irregularly and may represent song (Harrap and Quinn, 1996). Two nests have been described, both from early April and containing three young (Livesey, 1933; Round, 1983). In the former case there was no sign of plastering, but the hole was barely large enough for the birds to enter anyway. Löhrl (1988, p. 193) considers the Giant Nuthatch one of the plastering species, but gives no reference.

BEAUTIFUL NUTHATCH

A 'large, showy, black and blue nuthatch' (Ali and Ripley, 1973), this bird deserves its name from the brilliant blue and white streaks on the dark upperside, quite uncharacteristic for a nuthatch. Other features are the lilac edges to the outer primaries, two white wing bars, an inconspicuous eye-stripe and pale supercilium. The creamy white throat, cinnamon–buff underparts (with uniform undertail-coverts) and white tail-spots are more typically nuthatch-like. The sexes are very similar, but juveniles tend to have dark blue streaks, rather than light blue or white, on the back and head. The species is somewhat smaller than the Giant Nuthatch but appears bulkier and more solidly built (Hopkin, 1989). Its actions and behaviour are also slower and more woodpecker-like than that of other nuthatches (Ali and Ripley, 1973). Wolters (1975–1982) found it sufficiently dissimilar to include it in a separate genus, *Callisitta*.

The known range is fragmentary, comprising the Eastern Himalayas (Sikkim, Bhutan, north-east India), northern and south-west Myanmar, southern Yunnan (one record only), northern and central Laos, northern Vietnam and one recent record from northern Thailand (Hopkin, 1989; Collar *et al.*, 1994). Descriptions of its habitat are rather vague as 'forest', 'deep forest' or 'dense evergreen forest', at altitudes from 1000 to 2000 m (occasionally 2300 m), but it may descend to 300 m in winter (Harrap and Quinn, 1996). It is included in the World Checklist of Threatened Birds (Collar *et al.*, 1994).

The Beautiful Nuthatch is among the shyest, most elusive and least well known of all nuthatches (Ali and Ripley, 1973). It is observed in pairs or small parties of four or five, usually high up in large trees. Beaulieu (1944) encountered a group of about 10 individuals in northern Laos. The few available notes on vocalizations include a 'typical nuthatch call' less harsh than the Eurasian's, and a voice that is 'low and sweet in tone', lower in pitch than that of the other nuthatches (Ali and Ripley, 1973; Smythies, 1986). The nest is lined with leaves, fur and chips of bark, with an entrance that is reduced with mud, and clutches number four to six eggs (Ali and Ripley, 1973).

BLUE NUTHATCH

Despite the name, this tropical nuthatch appears black-and-white rather than blue. The head, back, wings, tail, lower abdomen and flanks are black or blackish blue, with prominent pale blue markings on the wing-coverts and flight feathers. The throat and breast are whitish, the undertail-coverts

Blue Nuthatch.

pale blue, and the conspicuous eye-ring, bill, legs and feet are bluish white. Wolters (1975–1982) includes the species in a separate genus (*Poecilositta*), but Voous and van Marle (1953) suspected an affinity with the White-cheeked/White-breasted group. The conspicuous eye-ring and deep black forehead are reminiscent of the Velvet-fronted Nuthatch (q.v.), as are its agile and highly social behaviour. Three subspecies occur on peninsular Malaysia and the south-western half of Sumatra (*expectata*) and on western (*nigriventer*) and eastern Java (*azurea*): *nigriventer* has a buffish rather than white breast, and *azurea* has bluer upperparts and more violet underparts than *expectata* (Harrap and Quinn, 1996).

The Blue Nuthatch lives in submontane and montane forest between 900 and 2400 m, usually above 1200 m (Lack, 1971; MacKinnon and Phillips, 1993). Löhrl (1988) saw them foraging in broadleaved trees as well as pines in Malaysia. Clutches (containing three or four eggs) and nestlings have been found in April and May (Hoogerwerf, 1949; van Marle and Voous, 1988). The nests have not been described. While foraging, the species is as agile as the Velvet-fronted Nuthatch but may stay longer on the same tree or branch (Löhrl, 1988). In broadleaved trees they used the undersides of branches more than the trunks, but in a pine plantation the trunks were preferred to twigs or needles (Löhrl, 1988). Chasen and Hoogerwerf (1941) observed a flock that kept to the same area for a few consecutive days. T. Luijendijk (pers. comm.) observed a flock of about six birds in April, and Löhrl (1988) found them in mixed flocks in February. The calls are described as high, shrill cheeps and twitters, which may be repeated and accelerated in staccato trills or rattles, or as squeaking single or disyllabic calls (Harrap and Quinn, 1996). T. Luijendijk (pers. comm.) described a continuous contact-call as a soft, rolling 'wrrt-wrrt, wrrrrrr, wrrt-wrrt'.

VELVET-FRONTED NUTHATCH

The last three species to consider in this chapter are obviously close relatives and were formerly treated as conspecific, and by some authors placed in a genus of their own (*Oenositta*; Wolters, 1975–1982). The most widespread is the Velvet-fronted Nuthatch. This bird is violet-blue on top and lilac-buff below, with a whitish chin and throat, a marked red bill, velvety black forehead and (in males only) a narrow black supercilium. The light blue margins to the black outer primaries and the conspicuous eye-ring (orange–yellow) are reminiscent of the Blue Nuthatch. The tail is black with blue central feathers, a blue-grey terminal fringe and pale subterminal spots. Juveniles are relatively distinct with an orangish tone to the underparts, finely barred undertail-coverts and a blackish bill (Harrap and Quinn, 1996).

This species is the most widespread among Oriental nuthatches, occurring from the Himalayan foothills south to Sri Lanka, and throughout most of SE Asia from southern China to Borneo, Java and Sumatra, as well as two Philippine Islands (Palawan and Balabac). Like the Chestnut-bellied Nuthatch, it is very cold-sensitive in captivity; sunbathing postures are described by Löhrl (1988). Three of the subspecies differ only slightly in plumage (*frontalis, saturatior* and *palawana*) but the fourth (*corallipes* from Borneo) has distinct red feet (see Harrap and Quinn, 1996, for more

Velvet-fronted Nuthatch. (Photo: Pam Rasmussen).

details). It is a bird of lower altitudes, occasionally up to *c.* 2200 m but rarely above 1500 m (Ali and Ripley, 1973). It is found in a variety of habitats including tropical evergreen and moist deciduous forest, mixed bamboo forest, mangroves, coffee and tea plantations, orchards and pine forest (Robinson, 1927; Smith *et al.*, 1940; Chasen and Hoogerwerf, 1941; Smythies, 1960; Löhrl, 1988). Many authors comment on its absence in deep and dark forests and preference for clearings and forest edges. It is probably strictly resident throughout its range. It is by far the commonest nuthatch in northern Thailand (de Schauensee, 1934), and may be so in most of its range.

The Velvet-fronted Nuthatch is renowned for its agility and quick movements ('one of the most active birds imaginable . . . it does not remain long in one tree'; Legge, 1983) but in captivity, it is remarkably shy (Löhrl, 1988). According to de Schauensee (1929, 1934) it forages mainly on small branches and hardly ever visits trunks, and this view was shared, or adopted, by Ali and Ripley (1973) and Lack (1971). However, Partridge and Ashcroft (1976) observed them mainly on trunks, less often on small outer branches, and least on the main branches in Sri Lanka. Occasionally they descend to the undergrowth and forage on brushwood or fallen logs, but not on the ground (Ali and Ripley, 1973). Price (1979, cited by Harrap and Quinn, 1996) suggested that they use wing-flicking as a means to flush insects from trunks. In captivity only some individuals took seeds, and none of them hoarded (Löhrl, 1988) which may suggest that seeds are not a regular part of the diet.

These nuthatches are usually seen in small loose flocks of from four to six birds (Whistler, 1963; Legge, 1983) up to 20 (Deignan, 1945; Ali and Ripley, 1973). However, only one to four were seen per mixed flock in Sri Lanka, and two was the most common number in Nepal in winter (Rand and Fleming, 1957; Partridge and Ashcroft, 1976). Typical mixed-flock companions include babblers, white-eyes, flycatchers, tits and Yellow-eyed Bulbuls (Chasen and Hoogerwerf, 1941; Deignan, 1945; Partridge and Ashcroft, 1976).

The nest is usually in a natural hole, occasionally an abandoned barbet or woodpecker hole (Phillips, 1939). If the entrance is large it may be reduced by plastering, but this seems to be a rare habit (Phillips, 1939; Ali and Ripley, 1973). Often the hole is so small that the birds actually enlarge the entrance (Ali and Ripley, 1973; Legge, 1983). The small nest cup is lined with moss, hair, feathers and sometimes a few dead leaves (Phillips, 1939; Ali and Ripley, 1973; Legge, 1983). The breeding season is quite variable, from January–May in southern India to April–June in the north (Ali and Ripley, 1973). In Thailand juveniles have been observed in early May (Deignan, 1945). Full clutches contain three to five, occasionally six eggs (Ali and Ripley, 1973; Legge, 1983; Smythies, 1986).

Löhrl (1988) could not identify true song in this bird, meaning that no vocalization was heard exclusively from males or in any particular season. The main call notes are a hard 'chat' or 'chip' and a softer 'sip' or 'tsit' (with many intermediates), which may be combined in short rattling series such as 'chip chip sit-sit-sit-sit-sit' (Harrap and Quinn, 1996; see also sonagrams in Löhrl, 1988). Legge (1983) also mentioned a 'short little warble'. All calls are relatively high-pitched compared with those of other nuthatches, with the main frequency range above 4 kHz. In southern Asia it is often the latest bird to be heard calling after dusk (Ali and Ripley, 1973).

YELLOW-BILLED NUTHATCH

This species differs from the previous one mainly by its yellow bill. Other differences include the pale 'collar' extending from the back of the neck over the ear-coverts, the somewhat darker upperparts, the white – instead of pale bluish – tail-spots, the slightly longer bill and tail and shorter wings. Three distinct subspecies are recognized, occurring in widely separated areas: *solangiae* in the Fansipan mountains of northern Vietnam, *fortior* on the Da Lat plateau in southern Vietnam (with one unconfirmed record from central Vietnam; Collar *et al.*, 1994), and *chienfengensis* on Hainan island off the Chinese coast: *chienfengensis* is paler with a longer bill and less marked tail-spots, *fortior* has the underparts suffused with violet and is larger, and *solangiae* has a paler crown. Where they occur together with Velvet-fronted Nuthatches (e.g. in the Da Lat region) the latter live at lower altitudes (Eames *et al.*, 1992).

On Hainan the species appears to be fairly common in tropical rain-forests at 750–900 m, but these forests have been cleared extensively over the last few decades (Harrap, 1991). The same is true for southern Vietnam where the species has been observed in primary and logged evergreen for-est between 1700 and 2100 m (Eames *et al.*, 1992). In northern Vietnam the type specimen was collected at 2500 m (Delacour and Jabouille, 1930) and only one further record exists (Harrap and Quinn, 1996). Because of its restricted distribution the species is included in the World Checklist of Threatened Birds (Collar *et al.*, 1994). Very little is known about its behav-iour and ecology, which are probably very similar to those of the Velvet-fronted Nuthatch. Birds are reported singly or in small flocks (Harrap, 1991; Eames *et al.*, 1992). Four collected stomachs contained insects only (Tso-Hsin *et al.*, 1964). Recently fledged juveniles have been observed on 22 April in Hainan (Harrap and Quinn, 1996).

SULPHUR-BILLED NUTHATCH

The Sulphur-billed Nuthatch is found on the Philippine Islands except for Palawan and Balabac, the islands closest to Borneo. It differs from the Velvet-fronted by the yellow bill and lilac neck collar, from the Yellow-billed by the indistinct tail-spots and more richly coloured underparts, and from both by the white spot at the base of the lower mandible (Harrap and Quinn, 1996). It is probably closer (if not conspecific) to the Yellow-billed Nuthatch. No fewer than five subspecies are listed by Greenway (1967): *mesoleuca*, *oenochlamys*, *lilacea*, *apo* and *zamboanga*, and a sixth (*isarog*) was described by Rand and Rabor (1967). They differ moderately in plumage with increasing colour saturation and lilac tones from north to south. The pale collar is especially pronounced in *oenochlamys* and *mesoleuca* (see Harrap and Quinn, 1996). Previously it was thought that different forms occurred at low and high altitudes on Luzon, but this view is not supported by Rand and Rabor (1967).

This species occurs in lowlands as well as mountains up to 2700 m (McGregor, 1920; Hachisuka, 1930; Greenway, 1967) and inhabits pine and evergreen forests and forest edge (Dickinson *et al.*, 1991). On Cebu it was found at 800–1500 m in broadleaved dipterocarp evergreen forest, in large groups (up to 20 individuals) and often with mixed flocks (G. Dutson, pers. comm.). On Negros the birds were found mainly from 700 to 1200 m accompanying mixed flocks of white-eyes and other birds, but on average only two nuthatches per flock. They foraged mainly on trunks and proximal parts of branches, not on distal branch parts and twigs, perhaps because of competition with Elegant Tits. Stomachs collected in summer contained mainly small beetles and spiders (Gonzales and Alcala, 1969). Breeding is reported in mid-June (Dickinson *et al.*, 1991). The voice is very similar to that of the previous two species, but also includes a 'squeaky toy' call and a very fast rattle that are reminiscent of the calls of the Blue Nuthatch (Harrap and Quinn, 1996).

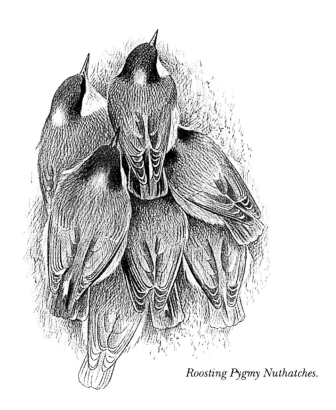

Roosting Pygmy Nuthatches.

CHAPTER 14

New World Nuthatches

The little birds seem so happy, animated, and lively – a playful gathering of talkative, irrepressible, woodland gnomes

Bent (1948) on Red-breasted Nuthatches

There are four species of nuthatch in N America, all of them fairly common with moderate to large geographical ranges. Despite the large number of descriptive notes on various aspects of their natural history, detailed ecological or behavioural studies are few, and only one involved a population

238

of colour-banded individuals followed over several years (Pygmy Nuthatches in Arizona). Except for the two 'dwarf' nuthatches (Pygmy and Brown-headed), the four species are not closely related and may represent three independent colonizations from the Old World (Voous and van Marle, 1953). The White-breasted Nuthatch is the ecological counterpart of the Eurasian Nuthatch (but not closely related to it), and is a common bird in various forests and also a well-known visitor of garden feeders. I have already mentioned the Red-breasted Nuthatch in several chapters as a close relative to several Old World species. It is, among other things, remarkable for being the only truly migratory nuthatch. The Pygmy and Brown-headed Nuthatch are by far the most social of all nuthatches, roosting and breeding communally.

WHITE-BREASTED NUTHATCH

This species is found throughout most of the United States, north to the southern limits of the boreal forest in Canada (50–55° N), and south to about 17° N in Mexico. It is most abundant in the north-east USA and some areas along the Mississippi but is absent from some arid and semi-arid areas in the west and parts of the Great Plains (Root, 1988; Pravosudov and Grubb, 1993; Harrap and Quinn, 1996; Sauer *et al.*, 1997). With body weights up to 26 g it is among the largest of nuthatches, and considerably larger than its close relative, the White-cheeked Nuthatch of central Asia (80–97 compared to 72–81 mm wing length). In comparison with the White-cheeked the underparts are white to grey rather than buff or rufous, and the tail-spots more prominent. Flight feathers and wing-coverts have distinct white or grey edges and tips in some subspecies, and a pale wing bar may be visible in flight. Juveniles are somewhat duller than adults. Females can be sexed accurately in the hand – at least in eastern N America – by the grey tips on some or all crown feathers, but this criterion is not reliable in the field where the crown may vary from grey to wholly black (Wood, 1992). Females with dark crowns are particularly common in the south-east (up to 80% in Florida).

Only one subspecies (nominate *carolinensis*) is found east of the Great Plains, but no fewer than six in the west: *aculeata* along most of the western coast, *tenuissima* from southern British Columbia to the mountains of California and Nevada, *nelsoni* in the main range of the Rocky Mountains from Montana to Mexico, *lagunae* in southern Baja California, *mexicana* in central Mexico and *kinneari* in southern Mexico (following Greenway, 1967). Phillips (1986) and Harrap and Quinn (1996) recognize four more subspecies. The Rocky Mountain and Mexican subspecies have generally darker upperparts and show less contrast than *carolinensis*, while the West Coast populations have darker (more rufous-brownish) vent and rear flanks. The southernmost *kinneari* is markedly smaller (see Harrap and Quinn, 1996 for details).

Habitat and densities

In the north-eastern part of its range the White-breasted Nuthatch prefers deciduous forest and avoids boreal forest (Pravosudov and Grubb, 1993), but in the south-east and the west it is commonly found in pines and mixed forest (e.g. Morse, 1970). It it said to favour relatively open woodland (Pravosudov and Grubb, 1993) with large mature trees, particularly oaks. Its habitat also includes beech–maple–birch forest (Holmes *et al.*, 1979), evergreen oak (Austin and Smith, 1972), orchards, parks and shade trees in towns and cities (Bent, 1948). In the east it is generally a lowland species but in the west largely montane up to the treeline, e.g. up to 3500 m in Nevada (Bent, 1948).

Population densities seem relatively low compared to those of the Eurasian Nuthatch (mainly between 1 and 2 pairs per 10 ha; Table 26), but detailed estimates are scarce and often based on small study plots. This is in agreement with territory size estimates from 5 to 15 ha (Butts, 1931; Ingold, 1981; Kilham, 1981). The presence of cavities or nestboxes seems to have little effect on densities (Brawn and Balda, 1988; Waters *et al.*, 1990). The population as a whole is thought to have decreased in the south-east in the early 20th century, but more recently seems to have expanded in the north-west as well as increasing in abundance in the north-east because of regrowth of the forests (Phillips, 1986; Pravosudov and Grubb, 1993). Indeed the

Deciduous woodland inhabited by White- and Red-breasted Nuthatches in winter in Ohio.

TABLE 26: *Breeding densities of White-breasted Nuthatches in different parts of the USA*

Habitat/state	Pairs per 10 ha	Source
Mature hardwood, West Virginia	4.9	Pravosudov and Grubb, 1993
Ponderosa pine, Arizona	1–2.5	Szaro and Balda, 1979; Brawn and Balda, 1988
Deciduous, West Virginia	1.9	Pravosudov and Grubb, 1993
Beech–maple, Ohio	1.8	Kendeigh, 1944
Maple-beech upland, Wisconsin	1.2–1.5	Ingold, 1977b
Oak–pine, California	0.5–1.5	Waters *et al.*, 1990
Oak–hickory, Ohio	1.0	Kendeigh, 1944
Kentucky (habitat unspecified)	max. 1	Mengel, 1965
Ponderosa pine, Colorado	0.5	Stallcup, 1968
Ponderosa pine, Colorado	0.2	McEllin, 1979

Breeding Bird Survey shows a steady increase by about 2% per year since the 1960s, particularly in the western- and easternmost USA (Sauer *et al.*, 1997). This pattern does not apply everywhere, as shown for instance by a 20–30% decrease from 1960 to 1985 in the New York State Christmas Counts (Yunick, 1988). In Florida, White-breasted Nuthatches were described as 'fairly common' in the 1930s, but are now confined to the northernmost counties and are considered endangered (Rodgers *et al.*, 1995).

Food and foraging

The White-breast's foraging techniques, diet and manner of food handling are quite similar to those of the Eurasian Nuthatch. They spend 70–90% of their time on large trunks and limbs (Morse, 1970; Willson, 1970; McEllin, 1979; Williams and Batzli, 1979a; Grubb, 1982). In mixed scrub pine–oak forest in Maryland there was a particular preference for conifer trunks, while in deciduous trees small branches were used more often (Morse, 1970). Several studies have shown a preference for oaks (Willson, 1970) or for deciduous trees to conifers (Morse, 1970; Szaro and Balda, 1979). With respect to foraging techniques, Holmes *et al.* (1979) recorded 66% gleaning, 31% probing into bark crevices and 3% fly-catching. There is one record of a bird using a bark flake to pry off other pieces of bark (Mitchell, 1993) in the manner of Brown-headed Nuthatches (see below in this chapter).

In a series of studies in deciduous forest in Ohio, Grubb (1982, and other references therein) found that foraging behaviour was relatively insensitive to weather conditions compared with other bark-foraging birds. On cold and windy days the nuthatches used less exposed parts of their home-range and foraged more often on large branches and trunks at lower heights. They also travelled more slowly and made more stops. In ponderosa pine, however, no seasonal differences were apparent (Stallcup, 1968; McEllin, 1979). Contradictory results have also been reported with respect to sexual

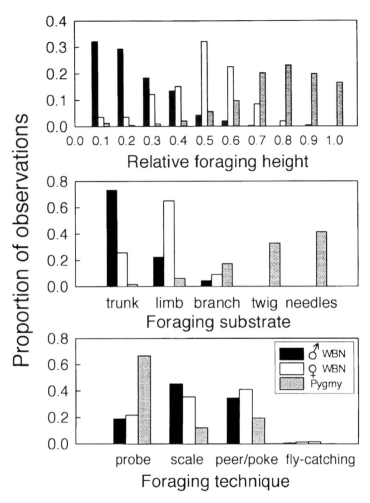

Fig. 90 *Foraging niche differentiation between White-breasted (male and female) and Pygmy Nuthatches in Colorado. Data from McEllin (1979).*

differences in foraging, without a clear explanation (see also Chapter 4). No differences were found in deciduous forest (Grubb, 1982) and in an aviary study (Pierce and Grubb, 1981), but in ponderosa pine forest males foraged more often on trunks at lower heights than females, and used the scaling technique more often (McEllin, 1979; Fig. 90). The foraging niche overlaps considerably with those of several woodpeckers (e.g. Stallcup, 1968; Holmes *et al.*, 1979) but the effects of competition have not been explored further. In areas with beech mast, nuthatches appear to avoid the territories of Red-headed Woodpeckers by which they are frequently chased (Williams and Batzli, 1979a,b). Interactions with Downy

Woodpeckers have also been observed (Willson, 1970; Williams and Batzli, 1979b).

The general diet is quite similar to that of the Eurasian Nuthatch. Stomach contents from Oregon showed a predominance of beetles and earwigs (Anderson, 1976), with larger prey thàn those taken by sympatric Pygmy and Red-breasted Nuthatches. In Michigan the main prey were Hemiptera and beetles (Sanderson, 1898). The contribution of vegetable food (mostly seeds) to the diet varies greatly between studies, from 0 to 24% in spring/summer and from 8 to 74% in autumn/winter (see Table 5 in Chapter 4). In one study about 50% of the diet was maize (Williams and Batzli, 1979c). Suet is readily taken from feeders, and there are occasional reports of feeding on tree sap (Forbush, 1929; Brackbill, 1969b). A remarkable observation is a male that brought large numbers of dragonflies to the nest (Ingold, 1977a).

Hoarding

Hoarding has been well studied in this species (Petit *et al.*, 1989; Woodrey, 1990, 1991; see also Chapter 4). Food items are stored mainly in trunks and large branches, wedged under bark or in a furrow. Petit *et al.* (1989) suggested that it serves to protect food against theft rather than weather, but no information is available on rates of cache robbing. As in the Eurasian Nuthatch, food is hoarded relatively closer to the source than in tits, perhaps because of the nuthatch's dominant status. Typically, only one seed or mealworm is taken per visit. In contrast to the Eurasian Nuthatch, sunflower seeds are not necessarily covered (*c.* 50%) and mealworms rarely or never. Most seeds are shelled before caching. As in Eurasian Nuthatches, the intensity of food hoarding decreases throughout the day (Waite and Grubb, 1988a). Females may avoid competition for caching sites and/or cache robbing by storing in a wider variety of sites (trunks and branches) than males (mainly trunks). Males follow females and try to rob them, and females try to avoid this by often leaving in the opposite direction. Fear of harassment may also explain why females do not lower their vigilance when their mate is present, whereas males do (Waite, 1987).

Social behaviour and roosting

In one of the earliest field studies ever on colour-ringed birds, Butts (1931) noted that White-breasted Nuthatches live in pairs on the same territories year-round. Several studies suggest that winter territories are rather loosely defended with extensive overlap (Butts, 1931; Ingold, 1977b; Larson, 1979; Fig. 91). Birds stray occasionally beyond their territory in search of feeding sites, and neighbouring pairs may feed without aggression at the same feeders (Butts, 1931; but see Kilham, 1981). An aberrant situation was described by Buckingham (1975) who found three different social units in separate home-ranges within a 12-ha woodlot in winter: one male with three females, two males and a female, and a group of two males. Breeding territories correspond closely to winter territories but are more

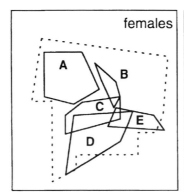
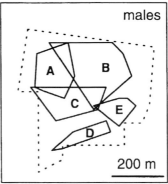

Fig. 91 *Home-ranges of five pairs of White-breasted Nuthatches in a 21-ha woodlot in Ohio, December 1989–January 1990. Data from Matthysen and Cimprich (unpubl.).*

vigorously defended (Larson, 1979). Little is known about the dynamics of the social system. Zirrer (in Bent, 1948) stated that families would stay together until about the end of November, but Ingold (1977a, 1981) saw parents being aggressive towards young about 18 days after fledging and later. Observations of threesomes in winter (Grubb, 1982; Matthysen, pers. obs.) suggest the existence of satellites as in the Eurasian Nuthatch,

A White-breasted Nuthatch caught in a mistnet, marked with leg streamers.

and the rapid replacement of owners that disappeared (Butts, 1931) may also suggest a 'floater' surplus.

Aggressive displays are described in detail by Kilham (1981). They include tail-fanning, wing-flicking (a general sign of excitement in nuthatches), threat displays with drooped wings and erect tail, similar to those of Eurasian Nuthatches, and a 'driving' display (see also Löhrl, 1988). In intensive fights opponents may chase one another or even fight in mid-air. Ingold (1981) also describes a 'free fall' display which sometimes follows a chase. This may be related to the 'floating flight' which Kilham (1972a) could not relate to a particular context.

In winter the nuthatches are usually found in mixed flocks, for instance with Downy Woodpeckers, Tufted Titmice and Carolina Chickadees in Ohio (Grubb, 1982). In captivity they are subordinate to the woodpeckers but dominant over titmice and chickadees (Waite and Grubb, 1988b). They seem to benefit from other flock members by copying their foraging behaviour, for example learning to take mealworms from a hidden site in an aviary (Waite and Grubb, 1988b). Tramer (1994) reported that some nuthatches use alarm calls deceptively to scare other birds, notably House Finches, away from a bird table. Nuthatches often roost in former Downy Woodpecker holes (Kilham, 1971a). A notable habit is the regular removal of faeces from the roosting hole at dawn (Kilham, 1971a). Harvey's (1902, cited by Bent, 1948) description of 29 White-breasted Nuthatches entering the same roosting hole is probably a mistake for the Pygmy Nuthatch, where communal roosting is common (see below).

Vocalizations

Songs and calls have been described in detail by Ritchison (1983; see also Tyler, 1916; Kilham, 1972a). The most commonly used calls, especially in winter, are the 'hit' and 'tuck' notes that apparently serve as contact-calls. 'Hits' are short notes with a downslur, 'tucks' slightly longer ones with a slight upslur. The best-known calls, however, are probably the nasal calls rendered as 'quank', 'kaan', 'kun' etc. (Fig. 92). These can be given as single or double notes, or in series varying in tempo and length, and are probably analogous in function to the Eurasian Nuthatch's Excitement Call. A 'rough quank' is also described with a harsher 'rr' sound. A trilling call was associated with intense aggression and might be related to the aggression call of Eurasian Nuthatches (see Ritchison, 1983, for other less common calls).

The song is rendered phonetically as 'hah-hah-hah' or 'what-what-what' and appears even more stereotyped than that of other nuthatches. A slow variant (on average 6.7 notes per second) has notes with a rising inflection (Fig. 92), while a faster song (on average 11.5 notes per second) has a rising and falling inflection. The main frequency range is relatively low, between 1.5 and 2.5 kHz (Ritchison, 1983) (compared to 2–4 kHz in the Eurasian Nuthatch). Songs contain 1–25 notes, usually around 10 (Brackbill, 1969b). Singing is heard throughout the year, but chiefly from February to May (Tyler, 1916; Ingold, 1981) and most often in the early

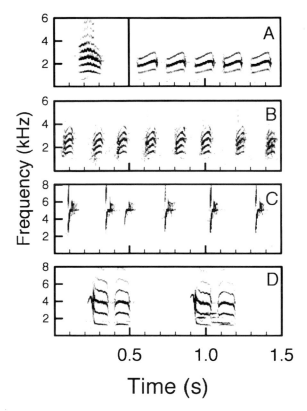

Fig. 92 *Sonagrams of vocalizations of White-breasted Nuthatch (A; left = common call, right = song, both recorded in Ohio), Red-breasted Nuthatch (B, calls, Ohio), Pygmy Nuthatch (C, song, California) and Brown-headed Nuthatch (D, calls, N Carolina). Source tapes are catalogued as BLB 10011, 13214, 12789, 18731 and 4520, © Borror Laboratory of Bioacoustics, Dept of Zoology, Ohio State University, Columbus, OH, all rights reserved.*

morning (Ingold, 1977b). It may be resumed at the time of nest leaving, perhaps as an enticement for the young about to fledge (Ingold, 1981).

Breeding biology

Nests are built mainly in live trees, in natural cavities or woodpecker holes (McEllin, 1979). There are some reports of excavation in decayed wood, but this is probably rare (Forbush, 1929; Oberholser, 1974). The 7% excavation figure given by Martin (1993) may be an overestimate, including observations of cleaning and enlarging (T. Martin, pers. comm.). According to Kilham (1971a) White-breasted Nuthatches prefer roosting and nesting holes with entrances much larger than the bird's size, perhaps to allow them to slip out when a predator tries to reach in. It is noteworthy that Eurasian Nuthatches also prefer larger entrances for roosting than for nesting (Chapter 5). Nest predation rates tend to be higher in nests with smaller entrances (K. Purcell, pers. comm.) but predation rates on adults on nests are unknown. In comparison with other cavity-nesters, relatively low sites are chosen in large, wide trees (Stauffer and Best, 1982). In ponderosa pine, most nests faced south (McEllin, 1979). White-breasted Nuthatches use nestboxes less often than other hole-nesters, but nevertheless up to 60% of pairs used boxes in thinned and open ponderosa pine plots (Brawn, 1988). A typical nest consists of a layer of bark strips, then small, scattered pellets of dried earth and mud, and finally a lining of fur, twigs, grasses and rootlets (Harris, 1927, cited by Bent, 1948). Mud pellets may be stored near the cavity entrance but are not used to reduce the entrance (Duyck *et al.*, 1991; K. Purcell, pers. comm.). The female builds while the male assists in bringing nest material (Forbush, 1929).

The most remarkable aspect of nest-building is the bill-sweeping by the female around the nest hole, with the whole body moving. This is usually done with insects (often taken from storage) in the bill, but also with fur, plant material or nothing at all (Kilham, 1968, 1971b). In one case the insect was identified as a species of blister beetle, which exudes a copious oily fluid when handled. Bill-wiping is performed in- and outside the nest hole, but particularly on protuberances and at branch junctions. The display is most intense at the time of nest-building, less so in the incubation and nestling periods, and especially intense if squirrels are present (Kilham, 1968). Its purpose may be to repel predators or nest competitors (mainly squirrels) or to confuse their scent trails.

In courtship, a prominent element is the 'bowing' that accompanies song in early spring. In this display, the male slowly sways forward during a strophe and then the head and neck shoot up to a vertical position, with crown feathers erect. Sometimes only the bill moves (Bent, 1948; Kilham, 1981). This display is often delivered with the back to the motionless female (see Bevier, 1993 for an aberrant but related display). Another element is mate chases, reminiscent of the Eurasian's zigzag flights (Brackbill, 1969b; Kilham, 1972a). Females solicit copulations with a plaintive 'phee-oo', crouching with drooped and quivering wings (similar to Eurasian Nuthatch) and fluffed flank feathers (Kilham, 1972a). This may be followed by a variant of the Eurasian's Pendulum Display. In a 'most bizarre' (Kilham, 1972a) display, the male moves from her right to her left side, tail erect and head stretched upward, sometimes passing under her bill or hopping on and off her back.

First-egg dates vary between late March in Florida and California, and late June in Oregon and Colorado (Bent, 1948). Full clutches number 5–10 eggs with a mean of 7.3 (50 clutches in the north-east part of the range; Pravosudov and Grubb, 1993). In Florida, Texas and the Rocky Mountains five or six eggs are usual (Bent, 1948; Oberholser, 1974). Incubation is said to last 12–15 days (the latter is probably more typical) and the nestling period 18–26 days (Allen, 1929; Kilham, 1968; Ritchison, 1981; Duyck and McNair, 1991). Despite statements to the contrary (Allen, 1929) only the female incubates. Ghalambor and Martin (unpubl.) compared the incubation behaviour of three North American nuthatches and attributed the higher nest attentiveness and lower rate of incubation feeding in White-breasted Nuthatches (compared with Red-breasted and Pygmy) to their higher risk of nest predation. In the former species, incubation pauses lasted, on average, 3.8 minutes (4.9 in the two other species) and females were fed only 3.6 times per hour, compared with 4.2 in Red-breasted and 6.4 in Pygmy Nuthatches. The higher nest failure rate in White-breasted Nuthatches (40% *vs.* 26% for Red-breasted and 13% for Pygmy Nuthatch; Martin, 1995) may be associated with their use of existing, rather than self-excavated nests. In addition, the cost of nest failure may be higher for a species that does not excavate its own hole (Ghalambor and Martin, unpubl.).

Second broods have not been reported for White-breasted Nuthatches, but lost clutches may be replaced (Bent, 1948; Duyck and McNair, 1991). Feeding rates are roughly similar to those of the Eurasian Nuthatch: *c.* 10 per hour in the first days after hatching (by the male) and 20 per hour (both sexes combined) in the last week before fledging (Ritchison, 1981; see also Ingold, 1977a). Brood parasitism by Brown-headed Cowbirds has

Fig. 93 *Anti-predator display of the White-breasted Nuthatch.*

been reported once (Friedmann, 1934, cited by in Bent, 1948). Broods may also fall victim to woodpeckers such as the Red-headed Woodpecker (Williams, 1918, cited by Bent, 1948).

A peculiar display used to defend the nest has similarities to both bill-sweeping and courtship (Kilham, 1968; Long, 1982). While sitting head downward on a trunk or branch (or sometimes on the ground; Bancroft, 1987) the birds sway slowly from side to side (for 10–20 seconds) with raised feathers, fully extended wings and tail spread, and the bill pointing upward (Fig. 93). The display may be accompanied by calls or song. A similar display, with wings not only spread but raised, was recorded once in the Eurasian Nuthatch (Löhrl, 1988; see Chapter 6). Some tits have a comparable behaviour (Long, 1982). A related threat display may be used to intimidate heterospecific birds at feeders (Whittle, 1926; Allen, 1929; Woods and Bens, 1981).

Ecology and movements

Very little is known about this species' demography or population dynamics. Martin (1995) estimated breeding success at 60%. Karr *et al.* (1990) calculated local survival (winter to winter) as 35% in Maryland, but this seems a dubious estimate since it was based on recaptures in a very small but heavily food-supplemented area. The maximal recorded age is 9 years and 10 months (Pravosudov and Grubb, 1993). White-breasted Nuthatches are rarely found in predator pellets (Liknes, 1994). A few parasitic Diptera were mentioned by Peters (Peters, 1936, cited by Bent, 1948). Winter survival probably depends on food and weather. Birds have slightly more fat in winter and are heavier, but also have an increased metabolic rate and cold tolerance (Liknes and Swanson, 1996). Food supplementation with sunflower seeds and suet had a positive effect on winter condition as measured by tail-feather regrowth (Grubb and Cimprich, 1990) whereas this condition parameter was not affected by low temperatures or wind (Zuberbier and Grubb, 1992) or even pesticides (Herbert *et al.*, 1989).

Dispersal is probably as limited as in other nuthatches. Only one of 109 ring recoveries exceeded 10 km (500 km north-east, Butts, 1931). As early as July, short movements may be recorded (Stewart and Robbins, 1958, cited by Heintzelman and MacClay, 1971). In late August and mid-September there is some migration by birds withdrawing from northern areas and high altitudes (Bent, 1948). A remarkable autumn migration in 1968 involved 297 nuthatches passing at 1600 m in Pennsylvania from early September to mid-October, compared to only 22 the next year (Heintzelman and MacClay, 1971). In the same year migrants were observed in the New York City Area, Maryland, Washington, D.C. (with a peak in October!) and Kentucky. However, the extent of migratory behaviour remains poorly known. On the one hand, Nuthatches have been said to be more common in winter or during migration in most New England states (Forbush, 1929). In Baltimore, for instance, only four of 77 birds banded in winter were permanent residents (Brackbill, 1969b). Judging from dates of first banding and last observation, migrants arrived from August to December and left mainly from February to April (Fig. 94).

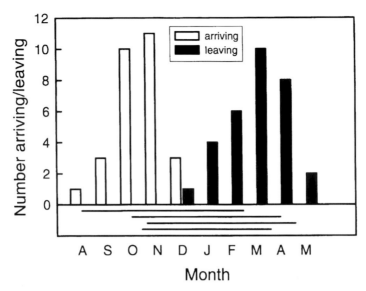

Fig. 94 *The number of migrant White-breasted Nuthatches arriving and leaving per month in Baltimore, USA. Horizontal lines show the presence of four individual nuthatches. Data from Brackbill (1969b).*

In some more northern localities, including upper New York State, Lake Erie, Nova Scotia and New Brunswick, nuthatches are more common in summer (Butts, 1931). On the other hand, many populations are considered as permanently resident, and there is insufficient evidence for a general withdrawal from the northern parts of the range (Butts, 1931). Migrants may be predominantly juveniles, as suggested by Murray's (1966) observations (46 juveniles *vs.* 3 adults passing in autumn 1963 in New Jersey). Vertical migration in the Rocky Mountains is proven by at least one ring recovery in Arizona (Dunning and Bowers, 1984).

RED-BREASTED NUTHATCH

Probably one of the most popular species of nuthatch, this bird has been described as 'a happy, jolly little bird, surprisingly quick and agile' (Bent, 1948) or 'a small boy at play among its elders' (in the company of larger nuthatches and woodpeckers) (Forbush, 1929). For a long time it was thought to be conspecific with a number of Old World nuthatches (see Chapters 11 and 13). It differs from them morphologically by the bright ochre–cinnamon underparts, more pronounced eye-stripe and supercilium, lower weight to wing length ratio, and short bill (Löhrl, 1960; Harrap and Quinn, 1996). The female is duller than the male and has a grey instead of a black cap. Löhrl's (1960, 1961) behavioural studies have been decisive in supporting its specific distinctness from the Old World species, the Corsican in particular.

Breeding populations are found throughout the boreal forest zone with the northern limit varying from 62°N in southern Alaska to 50°N in Ontario–Quebec. The eastern and western populations are hardly in contact except for a narrow junction in Saskatchewan and Manitoba (Harrap and Quinn, 1996). Their separation into subspecies (Burleigh, 1960) has been questioned because some allegedly darker eastern birds may just have a more sooty plumage because of pollution from industrialization (Banks, 1970). The western range extends throughout the Rocky Mountain, Cascade and Sierra Nevada ranges almost down to the Mexican Border. The eastern range includes most of the Great Lakes area, the northern Atlantic Coast and the Appalachians south to Tennessee and North Carolina. The southern boundary is variable and apparently depends on local cone crops. The winter distribution is even more variable. In years with poor cone crops massive irruptions reach down to northern Florida, the Gulf Coast and near to the Mexican border, but migrants are generally absent from the Great Plains (Root, 1988; see below in this chapter). In the 19th century the breeding range extended southwards, especially in the east, possibly because of the spread and maturation of conifer plantations (Phillips, 1986; Root, 1988). The Breeding Bird Survey shows a steady increase by c. 3.2% per year since the late 1960s (Sauer *et al.*, 1997). On the other hand, Christmas Bird Counts in New York State showed no clear trend between 1960 and 1985 (Yunick, 1988).

Habitat and densities

The main breeding habitat is coniferous forest, especially spruce–fir but also hemlock, Douglas fir, pine and incense cedar stands, and also mixed woodland and occasionally pure deciduous stands such as aspen (Bent, 1948; Anderson, 1976; Simpson, 1976; Morrison *et al.*, 1985; Harrap and Quinn, 1996). The southern distribution is largely montane between 1000 and 3000 m (e.g. Simpson, 1976). In the southern Rocky Mountains dry pine and juniper forests at lower altitudes are avoided except by wintering birds (D.A. McCallum, pers. comm.). Estimates of breeding density are not available, but Löhrl's (1961) observations in New Hampshire suggest a density higher than that of Corsican Nuthatches, i.e. more than 1 pair per 10 ha. In winter a wide range of habitats is used: deciduous forest, orchards, scrub, riparian woodland, parks and gardens, but a preference for conifers remains (Sutton, 1967). In good cone crop years the winter distribution in California corresponds well to the distribution of white fir (Widrlechner and Dragula, 1984). Density estimates in winter vary from less than 1 to about 5 birds per 10 ha in various forest types (Morse, 1970; Matthysen *et al.*, 1992).

Food and foraging

Red-breasted Nuthatches forage to a large extent on small branches, twigs and needles in summer (up to 50%; Airola and Barrett, 1985; Morrison *et al.*, 1985). The foraging techniques are mainly gleaning (over

Red-breasted Nuthatch flycatching.

40%), probing (about 25%), pecking (20%) and some fly-catching (Morrison *et al.*, 1987). In winter they forage more often on trunks and less on twigs (Morse, 1970; Morrison *et al.*, 1985), especially at lower temperatures (Fig. 95). Sexual differences in foraging have not been found (Matthysen, unpubl.). Conifer seeds or beech nuts, when available, are eagerly taken from the trees or from the ground (Kilham, 1975a; Matthysen, unpubl.). Löhrl (1961) explained the relatively short bills, compared with the Corsican Nuthatch, by the larger cones exploited by the latter. Several studies have found a preference for coniferous to deciduous trees (Morse, 1970; Airola and Barrett, 1985; but see Morrison *et al.*, 1987). In Maine there was a tendency to forage more on deciduous trees in winter than in summer (Morse, 1970).

Stomachs examined by Anderson (1976) in Oregon contained mainly beetles (Curculionidae and Chrysomelidae) and to a lesser extent Hymenoptera and seeds. In this study the diet overlapped more with Pygmy than with White-breasted Nuthatches, particularly outside the breeding season. The proportion of seeds in the diet was low even in winter, but this obviously depends on the cone crop. Bent (1948) mentions the consumption of various seeds as well as bits of apple, suet and tree sap. Food is hoarded intensely even in the wintering range, with up to 30 mealworms stored per hour (Grubb and Waite, 1987) but food items are not necessarily covered (Kilham, 1975a; Hendricks, 1995). In a clever experiment,

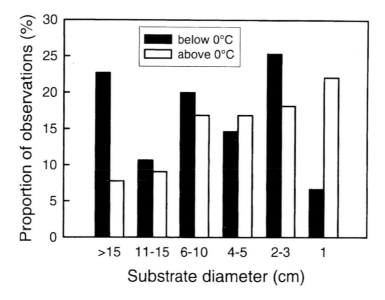

Fig. 95 *The use of foraging substrates by Red-breasted Nuthatches in a deciduous Ohio wood-lot at temperatures above and below 0°C. Data from Matthysen (unpubl.).*

Heinrich *et al.* (1997) demonstrated that the nuthatches prefer heavier sun-flower seeds, even if these were plaster-filled dummy seeds.

The Red-breasted Nuthatch's diet has also attracted attention in a more sinister context. Since nuthatches show a clear foraging response to increasing population levels of spruce budworms – a caterpillar that regularly defoliates conifer stands – Crawford *et al.* (1990) advocate the examination of nuthatch stomachs to detect early stages of budworm outbreaks. But, since most stomachs contain at most a few budworms, a staggering 150 birds would need to be shot to obtain a reasonable estimate!

Social behaviour and roosting

The social organization of this bird is not well known, particularly on the breeding grounds where pairs may stay together during winter (Knight, 1907, cited by Bent, 1948). In the winter of 1989–1990 I studied the home-ranges and social groups of nuthatches wintering in Ohio (Matthysen *et al.*, 1992), and found that the majority had small (1–4 ha) home-ranges and were usually found in stable groups of two to three individuals, but with little aggression between them. These groups included male–female, male–male pairs and threesomes (Fig. 96). This picture is quite different from the typical pair-territories of many other nuthatches, and may be explained by their migratory status, since migrants have no long-term territories or pair-bonds to defend (Matthysen *et al.*, 1992). Other reports confirm that nuthatches are typically seen in groups of one to four birds in winter (De Kiriline, 1952, 1954; Morse, 1970; Kilham, 1975a). In some

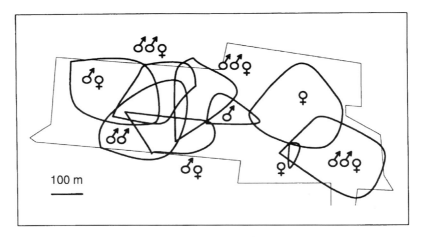

Fig. 96 *Home-ranges of social groups of wintering Red-breasted Nuthatches in a central Ohio woodlot, January 1990. After Matthysen et al. (1992).*

areas, however, larger numbers may concentrate around feeders. In the Adirondacks, for instance, 26 different birds were captured on a single day in January (Yunick, 1980). Red-breasted Nuthatches frequently associate with mixed flocks of tits, other nuthatches, woodpeckers and treecreepers. In central Ohio, 89% of all mixed flocks had Red-breasts in them (Matthysen *et al.*, 1992).

The main threat display is a typical nuthatch posture with drooping wings and erect tail, but the raised crest feathers appear to be unique among nuthatches (Löhrl, 1961; Kilham, 1973). In contrast to the Corsican Nuthatch, the body feathers are not ruffled. In a related threat posture the head is lowered and the bill pointed down. In threats to other bird species the wings are lifted and spread, and the tail fanned, sometimes with a Pendulum-like movement (de Kiriline, 1954). Together with the White-breasted Nuthatch, this is the only species where Pendulum-like displays are known to be used in aggressive contexts and not just in courtship.

No observations are available on roosting in the wild. Captive birds used nestboxes only on very cold nights, and at other times preferred sheltered corners (Mugaas and Templeton, 1970). These authors concluded that Red-breasted Nuthatches have better thermal isolation than similar-sized birds of lower latitudes, and suspected the existence of true hypothermia in colder temperatures. Other thermoregulatory adaptations are panting in warm conditions (with feathers tightly compressed, mouths open, and often hanging upside down) and shivering at low temperatures.

Vocalizations

The typical call has been compared to a toy tin horn, and described as a nasal, bleating or mewing 'yna, yna, yna' or 'knair, knair, knair' (Harrap and Quinn, 1996) (Fig. 92). It may be repeated in rapid series or more slowly with longer notes. When excited, the birds give very rapid series of this call, sometimes recalling an electric buzzer. When in groups, the continuous calling enhances the birds' lively appearance, hence descriptions such as 'talkative' or 'chattering'. As in other nuthatches the distinction between calls and song remains unclear. Kilham (1973) distinguished 'courtship song' from 'agonistic song', the latter conforming to the rapid calls described above. Courtship song is described as a repeated series of plaintive, nasal 'waa-aa-ns', with 2–6 notes in a series and 12–16 series per minute. Unmated males sang more frequently, both in captivity and the field (Löhrl, 1961; Kilham, 1973). Other calls include a soft contact-note, a 'phew' given singly or in short series, and a 'grrr' or 'chirr' in conflict situations. Occasionally a short, high-pitched trill is heard, apparently derived from the nasal notes and thus only superficially similar to the Old World species' trills (Löhrl, 1961; Kilham, 1973). Finally, several authors mention a 'more musical' song, so rapid and variable that it is difficult to describe (Tyler, cited by Bent, 1948; de Kiriline, 1952; Kilham, 1973). Females use it too but far less frequently than the male, and short sequences may be heard in autumn and early winter, probably as contact vocalization. Further investigation is needed to find out if this vocalization is really as different from other nuthatch sounds as it appears. In any case, the repertoire of the Red-breasted Nuthatch is very different from the other members of the *canadensis* superspecies, and in some respects (e.g. the nasal quality of the calls) rather reminiscent of the White-breasted Nuthatch.

Breeding biology

Most nest holes are probably excavated in decaying wood, but occasionally abandoned woodpecker holes or even nestboxes are used (Bent, 1948; Löhrl, 1961; Martin, 1993). A pair may start several excavations before the final nest site is decided (Kilham, 1973). Excavation may last up to 2 months (Knight, in Forbush, 1929) but 3–9 days is more usual (Löhrl, 1961). One pair excavated almost continuously from dawn to dusk, with a total of 6 hours' work by the female and over 4 hours by the male (Löhrl, 1961). Single males may also excavate (Löhrl, 1961). The nest is lined with grasses, roots or bark shreds. The male may bring lining material but does not normally enter (Bent 1948). A unique habit of Red-breasted Nuthatches is the smearing of pitch or resin around the nest entrance, usually by the male. The pitch is taken from spruce, pine or balsam fir trees and carried to the nest in the tip of the bill, but only after the nest is completed, and more pitch is added during the nestling period. A captive male used bits of oranges instead of pitch, which made a sticky, sugary mass reminding Kilham (1975b) of 'the candy houses in German fairy tales'. This behaviour probably serves to defend the nest from small rodents or insects. Apart from being sticky, resin also contains a diversity of monoterpenes known for their antiseptic and toxicological properties (Clark and Mason, 1985; see also Chapter 6). Its effectiveness is sadly documented by two nuthatches found dead, stuck to the pitch of their own nest (Kilham, 1972b; Speirs,

1985). Tiny globules have also been found on feathers of museum specimens (Rand, 1959). However, the nuthatches have several behavioural
adaptations to reduce their own exposure. Adults usually fly straight into
the nest hole, sometimes after hovering before the entrance, and nestlings
deliver their faecal sacs into the nest opening. When the young are about to
fledge, the parents cover the entrance floor with litter or fur, probably to
ensure a safe exit (Kilham, 1972b, 1975b).

Courtship displays have not been described in the same detail as in the
White-breasted Nuthatch, perhaps because the courtship period is shorter.
Kilham (1973) describes courtship as particularly lively, switching from
courtship displays to aggression in a matter of seconds. Prior to copulation
there is a Pendulum-like display close to that of other nuthatches, but
Kilham's descriptions suggest it is very brief. Pendulum displays were also
observed in singing males and in females during pair-formation in midwinter or when soliciting courtship-feeding (de Kiriline, 1954; Kilham, 1973).
Kilham also describes a more aggressive courtship posture resembling a
threat display, courtship flights which are exaggerated slow or floating
flights near the nest, and pursuit flights.

Quantitative breeding data are limited. Clutches of four to eight, typically
five or six eggs, are reported from late April to late June (Forbush, 1929;
Bent, 1948) and are incubated for 12 days (Bent, 1948; Martin, 1995). The
young fledge after 18–21 days with one record of 14 days (Bent, 1948,
Kilham, 1975b). Breeding success in Arizona was estimated as 74% (Martin,
1995). Second broods have been observed in the field (de Kiriline, 1954)
and in captivity (Kilham, 1975b) but their frequency remains unknown.

Movements

The Red-breast is the only true migratory nuthatch, but not on a regular
annual basis. Large-scale movements to the south occur about every 2 or 3
years (Yunick, 1988) and often coincide with failing cone crops in the north
(Fig. 97; see also Widrlechner and Dragula, 1984). These 'irruptions' are
correlated with those of other species such as Black-capped Chickadees
and Pine Siskins (Bock and Lepthien, 1976). However, both relationships
appear to be variable, since between 1920 and 1950 there was less synchrony with other bird species than before or after (Larson and Bock,
1986), and in the same period Ball (1947) failed to find any relationship
with cone crops.

Usually the first observations south of the breeding range occur in
September, more rarely August or even July, and movements may continue
until November. In Quebec large flights were seen in early July (Ball, 1947).
The return passage to the north occurs from March to May (Bent, 1948).
Migrants may stray far outside the normal range, as indicated by observations on the Bermudas, Iceland and England (Harrap and Quinn, 1996)
and on a ship 300 miles off the coast (Audubon, 1844). Most migrants
appear to be juveniles (71% in New Jersey; Murray, 1966). A detailed
account of nuthatch passage on the Gulf of St Lawrence coast was published by Ball (1947). During peak passage on 15 August 1941, 1177 birds

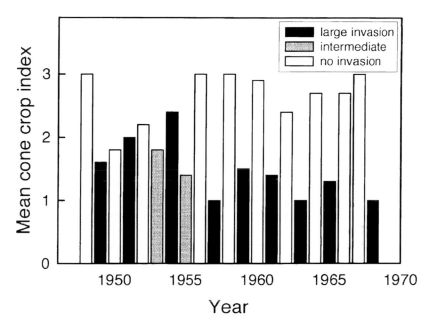

Fig. 97 *Occurrence of Red-breasted Nuthatch irruptions in relation to the cone crops of several boreal tree species. Data from Bock and Lepthien (1976).*

were seen in the first 2 hours of daylight. Ball suspected that they crossed the Gulf mainly at night which probably took them 2 hours or more. Night movements are also documented in other areas by calls recorded at night (B. Evans, pers. comm.). At Gaspé, several 'waves' of migrants were found each year, which probably represented birds from different origins (Ball, 1947).

Migrants are by no means confined to forest habitat. They can be particularly common along the coast, on barren points or islands, climbing on boulders or foraging among beach grass (Brewster, 1906, cited by Bent, 1948). On Fire Island, New York, they were 'everywhere – on the buildings, on trees, bushes, and weeds and even on the ground' and entered houses to catch flies (Dutcher, 1906, cited by Bent, 1948). In the 'superflight' year of 1969–1970 (Bock and Lepthien, 1976) nuthatches were found in the middle of the eastern Colorado prairie, foraging on fence posts miles from the nearest tree (see also Forbush, 1929; Oberholser, 1974).

THE DWARF NUTHATCHES

The Pygmy and Brown-headed Nuthatches are comparable to the *canadensis* group in their small size (60–70 mm wing length), lively behaviour, degree of sociality and link to coniferous forests, but are morphologically quite distinct. The two species are sufficiently similar to be treated together. The major sources for this account are the detailed comparative study by Norris (1958) combining museum and field work, and the recent field studies in Arizona on Pygmy Nuthatches by Russ Balda's team.

Morphology and taxonomy

The two 'dwarf nuthatches' are, at first glance, distinct from other nuthatches because of their brownish or greyish cap which is hardly separated from the slightly darker eye-stripe, without a supercilium. Otherwise they share the familiar nuthatch colours: bluish-grey upperparts, subterminal white tail-spots, white chin and throat, and dull whitish underparts with buff or grey tones and uniform undertail-coverts. The sexes are indistinguishable, and juveniles similar to adults but duller. The two species are mainly separated by the head colour: dark brown (in fresh plumage) in the Brown-head, olive-grey to brownish olive in the Pygmy. The Brown-head also has a clearer whitish neck patch, and the Pygmy a central white tail patch which is rudimentary in the other species (more details in Norris, 1958). As recently as 1970 the two were considered allopatric forms of the same species (Mayr and Short, 1970) under the name of 'Pine Nuthatch' (Sutton, 1967, cited by Phillips, 1986). Norris (1958) suggested, however, that they were reproductively isolated on the basis of numerous differences in morphology, physiology, vocalizations and plumage. The relationship to other nuthatches remains unclear but they are obviously not close to the other N American species. Vielliard (1978) placed them in a separate branch between the Mediterranean nuthatches and the White-breasted/White-cheeked species pair (Chapter 1, Fig. 2). The white neck and tail patches are highly suggestive of a relationship with the White-tailed and White-browed Nuthatches (Chapter 13) but otherwise these two species pairs have rather little in common.

Distribution

Brown-headed Nuthatches are largely limited to the pine belt of the south-eastern USA, and belong to the few species typical of this forest type (Haney, 1981). They occur from southernmost Delaware (with occasional records north to New York State) along the Atlantic and Gulf coasts to eastern Texas, remaining east of the Appalachians except for the recently colonized south-eastern part of Tennessee (Haney, 1981; Harrap and Quinn, 1996). They are also found throughout the Florida peninsula and on Grand Bahama Island. The highest concentrations are found in the Mississippi swamp forests and North Carolina (Root, 1988). Only two subspecies are generally recognized: the nominate on the continent and the longer-billed *insularis* on Grand Bahama (Norris, 1958), but the validity of the latter has been questioned (Smith and Smith, 1994; P.W. Smith, pers. comm.). Continental birds gradually decrease in size from north to south with some minor colour variations. Florida birds have longer bills and darker eye-stripes and have formerly been separated as *caniceps* or 'Grey-headed Nuthatch'. Recent surveys indicate that the Bahama population has greatly declined since the 1960s (possibly over 90%) because of forest logging, and may have to be considered endangered (Smith and Smith, 1994). The Breeding Bird Survey shows an overall decline since the 1960s by *c.* 1.7% per year (Sauer *et al.*, 1997). This decline is least marked along the northern and eastern fringes of the range.

The Pygmy Nuthatch has a wider but more disjunct distribution across the mountains of western N America, from southern British Columbia down to south–central Mexico (about 19°N). The biggest concentrations are found in the north-western USA, Colorado, Arizona, the Nevada–California border and the northern California coast (Root, 1988). Seven subspecies are listed by Greenway (1967) and an eighth (*elii*) was recently added by Phillips (1986). While nominate *pygmaea* is restricted to the coast of California, most of the USA range and northernmost Mexico is occupied by *melanotis* with a darker head and nape and more pronounced eye-stripe. The remaining subspecies differ only slightly in dimensions and head colour (Norris, 1958; Harrap and Quinn, 1996). The Pygmy population has been increasing steadily since the 1960s (1.4% per year) but with wide fluctuations (Sauer *et al.*, 1997).

Habitat and densities

Both dwarf nuthatches are particularly associated with pine trees. The Brown-headed Nuthatch is common in shortleaf and loblolly pines but can also be found, for instance, in ornamental pines in residential areas (Yaukey, 1996) and in Caribbean pine on Grand Bahama (Smith and Smith, 1994). They also occur in mixed and swamp forest (Bent, 1948; Root, 1988). They seem to prefer more open parts of woodland such as clearings, burnt areas and ponds with standing dead trees (Bent, 1948; O'Halloran and Conner, 1987) and also prefer older pine stands to younger ones (Withgott and Smith, 1998). Nests are often found in fence posts, stumps in clearings or stubs in ponds with considerable distances between nest site and foraging area (up to 460 m; Burleigh, 1958, cited by McNair, 1984). The Pygmy's range in the USA coincides roughly with the combined distribution of Jeffrey and ponderosa pine, except that it is absent in Montana where winters may be too cold (Root, 1988). It is also found in other pine species such as Monterey, bishop and lodgepole (Norris, 1958). Many populations live at high altitudes, for instance up to 3300 m on south-facing slopes in New Mexico (Sutton, 1967) and up to 4000 m in Mexico (Harrap and Quinn, 1996).

Pygmies have relatively high population densities with an overall mean of 4.9 pairs per 10 ha (1.3–8 in various studies) compared to 1.5 pairs per 10 ha (1–4) for Brown-heads, and Christmas Counts suggest that Pygmies are about three times as abundant (Norris, 1958). Two more recent studies found low densities for Pygmies as well, around 1 pair per 10 ha (Manolis, 1977; McEllin, 1979). More recent data for Brown-heads (including winter counts) are given by Withgott and Smith (1998), with the same general picture. In Georgia, Yaukey (1996) found the highest densities in residential areas where the nuthatches may benefit from ornamental pine crops.

Both species are much rarer in mixed forest (Norris, 1958; Morse, 1970) although Yaukey (1996) found no particular avoidance of deciduous stands in mixed habitat. Somewhat surprisingly for an excavator (see below), Pygmy densities can be raised dramatically by providing nestboxes. In severely thinned ponderosa pine forest in Arizona, the density increased from 0.7 to 2.5 pairs per 10 ha, and in moderately thinned forest from 3.7 to 6.2 (Brawn and Balda, 1988; see also Bock and Fleck, 1995).

Food and foraging

Both species exhibit the typical foraging behaviour of small nuthatches. They make extensive use of small branches, twigs and cones in their search for insects, practise fly-catching (e.g. on swarming termites) and hovering, and sometimes explore larger branches and trunks as well. When available, pine seeds are an important part of the diet (Morse, 1967). Norris' suggestion that Pygmies are more tit-like in their behaviour, foraging more often on small branches and twigs, is not supported by data on the use of trunks (Table 27), which varies in both species from almost zero to 35%, depending on season and social context (see below). Pygmies use trunks more often in winter than in summer, like the Mediterranean and Red-breasted Nuthatches. They also use a wider variety of trees (species, live/dead) in summer (Stallcup, 1968). Increasing the use of trunks in winter may reduce the degree of competition with the main flock companion, the Mountain Chickadee (Manolis, 1977). Norris (1958) observed a sexual difference in foraging height in Brown-heads, since males foraged lower than females in 18 out of 20 cases when they were closely associated. He mentioned no such difference in the Pygmy but did not provide comparable data (see Chapter 4 for more discussion on sexual niche differentiation).

TABLE 27: *Proportion of foraging time spent on trunks by dwarf nuthatches in summer and winter*

	Summer	*Winter*
BROWN–HEADED NUTHATCH		
Morse, 1967 (monospecific flocks)	–	25–30%
Morse, 1967 (mixed flocks)	–	10%
~Emlen, 1981	<76%	–
~Yaukey, 1997	–	<44%
PYGMY NUTHATCH		
Stallcup, 1968	4%	18%
Bock, 1969	–	35%
McEllin, 1979	3%	<1%
Szaro and Balda, 1979	2–9%	–
Löhrl, 1988	10%	–
Manolis, 1977	c. 15%	c. 30%

NOTE:
~ value for large branches and trunks, combined.

The foraging behaviour of Brown-headed Nuthatches on longleaf pine in Louisiana varied with cone crop abundance and social context (Fig. 98). They spent 20–30% of their time on pine cones when these were abundant. Distal branch parts and cones were used more often in mixed flocks, particularly in poor cone crop years. This effect is mainly due to the presence of Pine Warblers, which forage closer to the trunk than the nuthatches and presumably exclude them from these sites. In another study, however, where warblers were less numerous, the dominance relationship appeared to be reversed (Yaukey, 1997). Aggression between the two species increased about threefold when pine seeds were available, which Morse (1967) explained by the increasing overlap in foraging sites. However, this is not particularly evident from the data, and an alternative explanation is that the warblers tried to rob the nuthatches. Where dwarf nuthatches meet with the White-breasted Nuthatch their foraging niches are well separated (e.g. Fig. 90), but some hostility was observed by Morse (1967). Whether this affects the foraging niche is unclear, particularly since White-breasted Nuthatches are less abundant in coniferous forest. A more beneficial interaction between species has been reported by Jackson (1983), who observed Brown-headed Nuthatches associating with Red-cockaded Woodpeckers

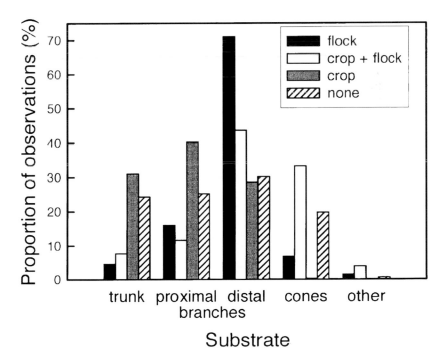

Fig. 98 *The use of foraging substrates by Brown-headed Nuthatches in relation to cone crops and mixed-species flocks. Dark bars = before cone crop, light bars = during cone crop; black/white = in mixed flocks, shaded/hatched = monospecific flocks. Data from Morse (1967).*

and probably picking up food items dislodged but missed by the latter. It is remarkable that no such commensal relationships have been reported from other nuthatch species.

Stomachs collected in winter (October to February) contained mainly pine seeds, while in the rest of the year the vegetable content was only about one half (Pygmy) or a quarter (Brown-head) with considerable variation (Chapter 4, Table 5; details in Norris, 1958). Even large Pygmy nestlings were fed with seeds. The most important insect prey are beetles (in about 50% of stomachs), mainly Curculionidae, Chrysomelidae and Scolytidae; less important groups are spiders, Hymenoptera and Lepidoptera (Norris, 1958). More data are given by Beal (1907, cited by Bent, 1948), Morse (1967), Anderson (1976), Nesbitt and Hetrick (1976) and Withgott and Smith (1998). Both dwarf nuthatches hoard food (Norris, 1958; McNair, 1983) but little is known about its frequency. Sealy (1984) watched two Pygmy Nuthatches pursuing flying ants and hiding some of them under bark flakes. These were apparently 'helpers' (see below) with a breeding pair, which might explain why they indulged in a seemingly uneconomical action at the time of feeding young.

Brown-headed Nuthatches are remarkable for their use of tools, which has been observed in several populations (Morse, 1968; Pranty, 1995; Withgott and Smith, 1998). The 'tool' is a chip of bark held in the bill and used to remove loose bark scales. Each tool is used only a few times at most. This behaviour has mainly been reported from longleaf and slash pines with relatively loose bark, but even then it is rarely seen. Similar behaviour has once been reported from a White-breasted Nuthatch (q.v.). There is one observation of a different kind of tool use by Pygmy Nuthatches: using a twig to probe bark crevices (Pranty, 1995).

Social behaviour

Dwarf nuthatches have the least pronounced territorial behaviour of all well-studied nuthatches. Aggression is often limited to a small area around the nest site, and neighbouring breeding pairs or groups may overlap extensively in their home-ranges (Norris, 1958; McEllin, 1979; Löhrl, 1988). Home-range estimates of breeding pairs vary from 5 to 8 ha for Brown-heads and 2 to 4 ha for Pygmies. In both cases pairs with helpers had larger home-ranges (Norris, 1958). Territories are mainly defended by advertisement (song) but some fights and pursuits have been observed (Norris, 1958). Löhrl (1988) saw a threat display similar to that of other nuthatches, with ruffled feathers, drooping wings and slightly erect tail. Wing-flicking is more conspicuous in Pygmies than in Brown-heads (Löhrl, 1988).

Extra male helpers at the nest are common (*c.* 20%) in both species (Norris, 1958). In a detailed study on Pygmy Nuthatches in Arizona up to three helpers were found per nest with a mean 'unit' size of 2.5 birds (Sydeman *et al.*, 1988; Sydeman, 1989, 1991). Helpers assist in nest construction (according to Norris, but not Sydeman), feed the female and the nestlings, remove faecal sacs, and (in Pygmies) even roost with the pair in the nest cavity. On the other hand, no copulations or courtship-feeding were observed, and helpers did not actively participate in territorial defence and sang much less than breeding males. The majority of helpers in Arizona were first-year males (Sydeman, 1991), and all were closely related (son or brother) to at least one of the breeders, and had been in the same winter group. Some had lost a nest of their own earlier in the season and some returned to help the same pair in a second year. Sydeman estimated that between 76 and 85% of the surviving first-year males became helpers. This implies a strongly male-biased sex ratio, which is also evident from museum specimens (of both species) collected outside the breeding season (Norris, 1958), but remains unexplained. The benefits of helping are not well understood, since helpers do not copulate, their presence does not increase the total provisioning rate (since helped pairs reduce their own effort), and the contribution to breeding success is meagre: pairs with at least one helper raised more young in one out of three years only, and additional helpers had no effect (Sydeman, 1989).

The winter social system of Pygmies in Arizona is based upon stable, territorial groups whose members feed and roost together and defend a communal territory (Güntert, 1986, Güntert et al., 1988). These groups can split up temporarily or join others for foraging or roosting (see also below), particularly in cold weather. Groups are formed at the end of summer by the aggregation of families, though some mixing occurs during the autumn dispersal period. The basic groups in Arizona contained 5–16 birds, but temporary flocks numbered 2–44 individuals. Winter flocks in Colorado averaged 13 birds (McEllin, 1979). Little is known about within-group relationships. Captive flocks of four to six birds had a clearcut dominance hierarchy, males always dominating females, while in larger flocks the hierarchy was less well defined (Hay, 1983). Winter behaviour in Brown-headed Nuthatches has received hardly any study. Bent (1948) mentions groups of 6–24 birds and Audubon (1844) saw a congregation of 'fifty or more' in Florida, but in Louisiana mean group size (in mixed flocks) was only five birds (Morse, 1970). A remarkable element in the Brown-head's social behaviour is reciprocal preening, apparently not in direct connection to courtship. This behaviour was first mentioned – in passing – by Norris (1958), and later in more detail by Barbour and Degange (1982) and mentioned by Yaukey (1997) as well.

Dwarf nuthatches associate with the familiar mixed flocks of tits, other nuthatches, treecreepers and kinglets, but are often found on their own as well (c. 60% of observations; Morse, 1967; Manolis, 1977). At the height of the cone crop season, Brown-heads often remain in a small area for a long time and so become separated from mixed flocks (Morse, 1967, 1970; see also Yaukey, 1995, 1996). At artificial feeding sites, Pygmies dominate Mountain Chickadees as well as White-breasted Nuthatches, and were able gradually to monopolize the feeders (Bock, 1969). Brown-heads are often involved in conflicts with Carolina Chickadees and Red-breasted Nuthatches but may also dominate larger birds (Yaukey, 1997).

Vocalizations

As in most other nuthatches, song and call notes grade into one another 'both phonetically and functionally' (Norris, 1958). Brown-heads are somewhat less vocal and have a less varied repertoire (Löhrl, 1988) and their voice is generally lower-pitched, harsher and less piping than the Pygmy's (Fig. 92; see also Withgott and Smith, 1998). The song (*sensu* Norris, 1958; but see Löhrl, 1988) is a high-pitched staccato repetition of disyllabic notes with 40–80 double notes per minute, and may be continued for 2 minutes or more (Fig. 92). Pygmies sometimes engage in duetting, where the male and female alternate in giving double notes at high speed and in perfect synchrony. The partners may audibly differ in pitch, described as 'poo toot/pee tit/poo toot/pee tit' by Harrap and Quinn (1996). Löhrl (1988) noticed differences in vocalizations between *pygmaea* in San Francisco and *melanotis* in Arizona, and found that Arizona birds responded less strongly to the other race's vocalizations.

The disyllabic note is also used as a call note throughout the year, in shorter and less stereotyped series (described as the 'rubber ducky' call in Brown-heads, Withgott and Smith, 1998; see Fig. 92). Monosyllabic notes are also given singly or in repetition. A trilled or rolling note was heard in both species in the breeding season (particularly from females), and also used by Brown-headed Nuthatches in flight (Norris, 1958). Juvenile Pygmies produce a 'juvenile cadence' (Norris, 1958), a distinctive series of notes starting with a rapid, rather slurred series of notes that diminish in volume, followed by a louder, more rhythmic 'swee-swee swee-swee-...'. Norris suspected that this call, and perhaps some vocalizations of young Brown-heads as well, might be remnants of more complex vocalizations lost in the nuthatch lineage.

Breeding biology

Both species usually excavate their own nest holes in dead pine trees and snags, but may occasionally use natural cavities under the bark or unfinished woodpecker holes (Bent, 1948). Estimates of the frequency of excavation are considerably higher for Brown-heads (*c.* 90%; McNair, 1984) than for Pygmies (*c.* 40%; Martin, 1993). Both sexes excavate, and sometimes the male does most of the work (Norris, 1958). Up to five birds – including helpers – may work at a single cavity. As in the Red-breasted Nuthatch, several nests may be started before the final one is selected (Bent, 1948). Finished cavities are similar in both species, except that Brown-head nests are typically 1 or 2 m above the ground whereas Pygmy nests are usually at 5 m or more (Norris, 1958; McNair, 1984). Both breed in boxes, Pygmies in particular (up to 100% in thinned ponderosa pine; Brawn, 1988). There are many records of aggressiveness to other hole-nesting birds, of which bluebirds are the main competitors (Norris, 1958; Haney, 1981; Löhrl, 1988). On the other hand, there is one record of a Pygmy Nuthatch and Western Bluebird sharing a cavity and brooding side by side (Ramsay, cited by Bailey, 1939). The nest lining material includes wood chips, seed wings, bark shreds, feathers, moss, cotton, hair, etc. (Bent, 1948; Norris, 1958). Brown-head nests in particular may be largely constructed of pine seed wings. Pygmies appear to use more insulating material, perhaps because they live in cooler environments (Norris, 1958). Both species are known to fill up cracks in the nest chamber with nest material, a behaviour that might be ancestral to plastering (Löhrl, 1988). The pair roosts together in the finished cavity throughout the breeding season.

There are few descriptions of copulation behaviour. Female Pygmies solicit copulations by wing-shivering with somewhat erect tail and swaying from side to side. The male responds with a comparable, Pendulum-like display (Löhrl, 1988). Brown-heads seem to start laying several weeks before Pygmies, even at comparable latitudes and altitudes (Norris, 1958; McNair, 1984). Their clutches are found from March to mid-May (overall median 7 April), occasionally up to mid-July, and about 18 days later in North Carolina than in more southern states. However, excavation can start as early as December (Yaukey, 1997), and nestlings have been found in mid-March (Oberholser, 1974). Pygmy clutches are reported from late April to mid-June, with almost 3 weeks' difference between coastal and montane populations (medians 9 and 28 May), and considerable annual variation (24 days over only 4 years, Sydeman, 1989).

Complete clutches contain 5–11 eggs (mean = 7.0) in the Pygmy, and 3–9 (mean = 5.1) in the Brown-headed Nuthatch (Norris, 1958; McNair, 1984; see also Withgott and Smith, 1998). Brown-head clutches are even smaller in Florida (mean = 4.5), and decrease in the course of the season. Incubation lasts about 14 days in the Brown-head and 16 days (one record only) in the Pygmy Nuthatch. Small nestlings were brooded more often by Pygmies than Brown-heads, and also more often than Eurasian Nuthatches in Germany, probably because of the cooler climate (Norris, 1958). Male Pygmies take a greater share in food provisioning than females or helpers (Sydeman, 1989). Wheelock's (1905, cited by Bent, 1948) remarkable observation that young Pygmies were fed by regurgitation has never been confirmed. Brown-heads develop slightly faster and are ready to fledge at 18–19 days compared to 20–22 days for Pygmies, but the latter appear better able to fly (Norris, 1958; but see also McNair, 1984). In the day or two before fledging, young Pygmies may briefly venture out of the cavity, usually for less than a minute, a behaviour unknown from other nuthatches. Also, they may return to the nest to roost after fledging, unlike Brown-heads (Norris, 1958). More details of incubation, hatching, nestling care and development are given by Norris (1958).

Three studies in Arizona and Colorado found fairly high nest success (82%, Brawn, 1987; 89%, Sydeman *et al.*, 1988; 88%; Martin and Li, 1992). In Arizona, 5.5 and 4.4 young fledged per pair in logged and unlogged pine forest, respectively (Sydeman *et al.*, 1988). The main cause of nest loss, and of variation in breeding success, was predation by chipmunks or woodpeckers. Nest success appears to be lower in Brown-heads (52–58%; various sources in Withgott and Smith, 1998). Major causes of failure appear to be predation and windstorms. Known predators include various snakes, Red-bellied Woodpecker, flying squirrels and feral cats. Two attempts at nest parasitism by Brown-headed Cowbirds have been reported as well (Withgott and Smith, 1998). In both species there are some repeat nests after failure and occasional second broods (Norris, 1958; McNair, 1984; Sydeman *et al.*, 1988), which may be more frequent than previously thought (Withgott and Smith, 1998). The dwarf nuthatches are the only members of the genus where a detailed account of parasites is available (Norris, 1958), including blood Protozoa, Trematoda, Cestoda, Nematoda, mites, Mallophaga, hippoboscid flies and *Protocalliphora*. The limited overlap between the two species' parasite faunas has been interpreted as yet another indication of the distinctness of the two species (Norris, 1958).

Communal roosting

Pygmy Nuthatches commonly gather in large numbers to roost in a single tree-cavity, a rare behaviour among birds. Two environmental factors appear to be crucial for communal roosting behaviour: the thermal advantages gained by body contact and increased cavity temperature, and the maintenance of sufficient air flow for respiration. The behaviour and physiology of roosting nuthatches – mainly in captive flocks – have been studied in detail by Hay (1983), on which the following account is largely based.

Even in the breeding season Pygmy pairs roost together in the nest cavity, sometimes with helpers. After fledging, the family may return to roost in the nest for a few days, but then switches to a summer roost cavity (Hay and Güntert, 1983) or perhaps to roost in the open, but this is not certain. In summer, families may merge into larger roosting assemblies, which becomes normal practice in autumn and winter (see above). During heavy snowfall in winter, groups may stay in a roost cavity for up to 40 hours. Each Pygmy Nuthatch group has several roosting cavities in its home-range, and often a number of them are inspected before going to roost. Different cavity types are preferred in different seasons. Winter cavities appear to be selected for their insulation, and are typically in large trunks with thick walls and small openings. Summer cavities are more often in branches or smaller trunks and have larger openings (sometimes more than one), which is probably necessary for sufficient air flow in the absence of a strong thermal gradient between the inside and outside. Some cavities have an entrance at the bottom which even further enhances the air flow. Spring and autumn roost-sites are more difficult to characterize (Hay and Güntert, 1983). At least in captivity, small flocks prefer smaller cavities. Nest sites are generally lower in the tree than roost-sites and have smaller entrances. Because of these different requirements, the number and type of cavities inside a territory may represent a limiting resource, and this may explain why more atypical roost sites are used in heavily logged areas (Hay and Güntert, 1983).

Inside the roosting hole, individuals typically maintain close body contact with all heads pointing in the same direction. They either cling to the wall with head upward, typically in a wedge-shaped formation (Fig. 99), or lie stacked at the bottom. Often the birds cling to the walls early in the night and then move down, particularly in cold and windy conditions. They

Fig. 99 *Schematic view of a group of Pygmy Nuthatches roosting in a cavity, clinging to the wall below the entrance. Redrawn after Hay (1983).*

appear to wait for a temperature gradient to build up in the cavity, which creates an additional air flux bringing oxygen-rich air to the cavity bottom. Cavities with smooth walls, which make clinging difficult, are avoided. The typical arrangement on the bottom is stacks of four birds with up to five layers on top of one another. In this arrangement 9–49% of each individual's surface touches a neighbour. Dominant birds tend to occupy the best isolated spots at the bottom, with an estimated energy saving five times that of the top birds, based on estimates of the reduction in exposed surface area. Usually all birds face the wall opposite the entrance, except when the wind blows directly into the entrance. This orientation is supposed to minimize the air current over face and bill where heat losses are potentially largest.

Single-roosting Pygmy Nuthatches have a remarkably low basal metabolism compared to similar-sized passerines (34–42% lower) and also a low thermal conductance. Body temperature in captivity dropped as low as 35°C, particularly in larger groups. Hay (1983) concluded that body contact is a stimulus for lowering the body temperature and thereby saving energy. At 0°C, birds in pairs saved 8% and groups of four saved 18% energy compared to single birds. Birds roosting in the open in an outside aviary lowered their body temperature to 30°C while still remaining responsive and capable of flight. In natural roosts, however, birds may be found in a highly lethargic state, suggesting even further decreases in body temperature. Metabolic rates are also reduced in response to low barometric pressure, probably to conserve energy for periods of impending bad weather.

While Hay's studies in captivity suggest an optimal group size of 10–12 birds for energy saving, much larger aggregations have been reported in the wild, the advantage of which is not clearly understood. In one particular winter a single cavity was attended by 27–167 nuthatches per night, particularly when there was snow (Sydeman and Güntert, 1983). The roosting birds belonged to 5–10 distinct social groups, some of them travelling up to 1.7 km to the roost-site. The same cavity had been used by only a few groups in the previous winters. Such mass aggregations may be risky, as illustrated by reports of up to 13 dead nuthatches found inside a cavity (Knorr 1957), but no such casualties were reported in the Arizona study. An unresolved question is why birds roost communally in summer as well, when energy savings are small in relation to the respiratory problems caused by communal roosting.

Little is known about communal roosting in Brown-headed Nuthatches but it appears much less pronounced (Norris, 1958). At least in spring and summer they roost in the open more frequently than Pygmies, and it is not clear whether families roost in cavities after fledging. Even during the breeding season, pair members often roost singly and never with helpers. In fact, the only observation on communal roosting involved four individuals found in a nestbox on an unusually cold winter night (Fleetwood, 1946). The differences between the two species are probably explained by the colder climate experienced by Pygmies, with winter temperatures −40°C or below (Hay, 1983). They are in agreement with other factors such as the Pygmies' more frequent brooding of small young and use of insulating nest material (Norris, 1958).

Movements and survival

The Pygmy Nuthatch is mainly a permanent resident with some vertical migration (Bent, 1948). In the southern Rocky Mountains, for instance, while Mountain Chickadees migrate in large numbers to the lowlands when cone crops fail, Pygmy Nuthatches stay (D.A. McCallum, pers. comm.). Despite statements to the contrary in the older literature, Brown-heads are probably fully resident (Norris, 1958). Norris estimated annual survival rates (based on adult/juvenile ratios) of 42% in Pygmy and 54% in Brown-headed Nuthatches. Sydeman (1989) found annual survivorship from 41% to 70% in colour-ringed Pygmies, with higher survival in males than females (64% and 49%) but no effect of the presence of helpers.

Scientific and Common Names of Nuthatches

The scientific and common names of nuthatches as used in this book, with alternative common names in parentheses, as well as subgenus and superspecies names (in square brackets) (N = Nuthatch)

Subgenus *Callisitta*
Sitta formosa Blyth, 1843 — Beautiful Nuthatch

Subgenus *Poecilositta*
Sitta azurea Lesson, 1830 — Blue Nuthatch
(Azure N)

Subgenus *Oenositta*
Sitta [frontalis] solangiae (Delacour and Jabouille), 1930 — Yellow-billed Nuthatch (Lilac N)
Sitta [frontalis] frontalis Swainson, 1820 — Velvet-fronted Nuthatch
Sitta [frontalis] oenochlamys (Sharpe), 1877 — Sulphur-billed Nuthatch

Subgenus *Sitta*
Sitta magna Ramsay, 1876 — Giant Nuthatch
Sitta tephronota Sharpe, 1872 — Eastern Rock Nuthatch
(Rock N, Greater Rock N, Great Rock N, Great N, Persian N)

Sitta neumayer Michahelles, 1830 — Western Rock Nuthatch
(Rock N, Neumayer's Rock N, Lesser Rock N, Syrian N)

Sitta [europaea] europaea L., 1758 — Eurasian Nuthatch
(Nuthatch, European N, Common N, Wood N)

Sitta [europaea] nagaensis Godwin-Austin, 1874 — Chestnut-vented Nuthatch
(Naga Hills N, Naga N, Austen's N)

Sitta [europaea] cashmirensis Brooks, 1871 — Kashmir Nuthatch
(Brooks's N)

Sitta [europaea] castanea Lesson, 1830	Chestnut-bellied Nuthatch (Chestnut-breasted N)

Subgenus *Mesositta*

Sitta himalayensis Jardine and Shelby, 1835	White-tailed Nuthatch (Himalayan N)
Sitta victoriae Rippon, 1904	White-browed Nuthatch (Victoria N, Chin Hills N)
Sitta [pusilla] pygmaea Vigors, 1839	Pygmy Nuthatch (Californian N)
Sitta [pusilla] pusilla Latham, 1790	Brown-headed Nuthatch

Subgenus *Micrositta*

Sitta [canadensis] canadensis L., 1766	Red-breasted Nuthatch (Red-bellied N)

Sitta [canadensis] villosa Verreaux, 1865	Chinese Nuthatch
	(Snowy-browed N,
	    Black-headed N, Chinese
	    Grey N)

Sitta [canadensis] yunnanensis Ogilvie-Grant, 1900	Yunnan Nuthatch (Black-masked N)
Sitta [canadensis] krueperi Pelzeln, 1863	Krueper's Nuthatch
Sitta [canadensis] ledanti Vielliard, 1976	Algerian Nuthatch (Kabylie N, Kabylian N)
Sitta [canadensis] whiteheadi Sharpe, 1884	Corsican Nuthatch (Whitehead's N)

Subgenus *Leptositta*

Sitta [carolinensis] leucopsis Gould, 1850	White-cheeked Nuthatch (Przevalski's N)
Sitta [carolinensis] carolinensis Latham, 1790	White-breasted Nuthatch (White-bellied N)

Notes

1. Scientific and English species names follow Harrap and Quinn (1996). Superspecies (in square brackets) follow Sibley and Monroe (1990). Sources for alternative English names are Audubon (1844), Sanderson (1898), de Schauensee (1929, 1984), Wilder and Hubbard (1938), Paludan (1959), Lack (1971), Voous (1977), Howard and Moore (1980), Walters (1980), Harrison (1982), Tso-Hsin (1987), Sibley and Monroe (1990) and Cramp and Perrins (1993). Subgenera and ordering follow Wolters (1975–1982), except that *Callisitta, Poecilositta* and *Oenositta* are treated as subgenera rather than genera, and Pygmy and Brown-headed Nuthatch – not assigned to a subgenus by Wolters – are included here in *Mesositta* (see note 3).

2. 'Chinese' and 'Yunnan Nuthatch' have been used as common names for the subspecies *montium* and *nebulosa* of the Eurasian Nuthatch (Stanford and Ticehurst, 1938; Ali and Ripley, 1973). Many other common names have been given to subspecies but they are rarely used (e.g. Bent, 1948; Dement'ev and Gladkov, 1954; Ali and Ripley, 1973).

3. Glutz von Blotzheim (1993) mentions four 'Mega-superspecies', one containing the three Mediterranean nuthatches (Corsican, Algerian, Krueper's), another the three remaining *Micrositta* species, a third corresponding to Wolters' *Mesositta* plus the two dwarf nuthatches, and the fourth corresponding to the *europaea* superspecies. The same author mentions an additional superspecies *himalayensis* (White-tailed and White-browed).

Diagnostic Traits of the 24 Nuthatch Species

The following table presents an overview of the major diagnostic traits of the 24 Nuthatch species: col = basic colour of the underparts (O = ochraceous, including cinnamon, orange-brown, rufous, B = buff, W = white or whitish grey, G = grey, D = dark brown, L = lilac); sup = presence of a black (b) or white (w) supercilium (in parentheses = weakly developed); eys = dark eye-stripe; eyr = light eye-ring; cap = black cap (M = males only, fc = forecrown only); nap = light-coloured nape; bic = bi-coloured undertail-coverts; tsp = subterminal tail-spots; tpa = white tail patch. The last column lists traits that are unique to one species.

Species	col	sup	eys	eyr	cap	nap	bic	tsp	tpa	Unique traits
Beautiful	O	(w)	+	–	–	–	–	+	–	Lilac streaks, white wing bars
Blue	W	–	–	+	+	–	–	–	–	Blackish lower abdomen
Yellow-billed	L	b	–	+	–	–	–	+	–	
Velvet-fronted	L	b	–	+	–	–	–	+	–	Red bill
Sulphur-billed	L	b	–	+	–	–	–	–	–	
Giant	G	–	+	–	–	–	+	+	–	Black streaks on crown
Eastern Rock	B	–	+	–	–	–	–	–	–	
Western Rock	B	–	+	–	–	–	–	–	–	
Eurasian	~OBW	(w)	+	–	–	–	+	+	–	
Chestnut-vented	G	–	+	–	–	+	+	+	–	
Kashmir	O	–	+	–	–	+	–	+	–	
Chestnut-bellied	*OD	–	+	–	–	–	+	+	–	
White-tailed	O	(w)	+	–	–	+	–	+	+	
White-browed	O	w	+	–	–	+	–	+	+	
Brown-headed	W	–	+	–	–	+	–	+	+	Brown cap
Pygmy	W	–	+	–	–	+	–	+	+	Brownish-olive cap
Red-breasted	O	w	+	–	M	–	–	+	–	
Chinese	B	w	+	–	M	–	–	+	–	
Yunnan	G	w	+	–	–	+	–	+	–	
Krueper's	B	w	+	–	fc	–	+	+	–	Breast-patch
Algerian	B	w	+	–	fc	–	–	+	–	
Corsican	W	w	+	–	M	–	–	+	–	
White-cheeked	O	–	–	–	+	–	–	+	–	
White-breasted	W	–	–	–	+	–	–	+	–	

NOTES:

~ varying between subspecies groups.

* varying with sex.

Sex- and Age-related Morphological Variation in Eurasian Nuthatches

The following data are based on 1041 captures of over 300 different individuals in the Peerdsbos study area from 1982 to 1987. The large number of recaptures allowed me to document seasonal and age-related variation by means of within-individual comparisons. Since August was the month with the largest number of individuals trapped, this month was often used as a baseline for seasonal comparisons. Population means are calculated using means of measurements per individual.

Table 28: *Sexual differences in morphometric variables.*

Parameter	Males	Females	p
Summer weight (g)	23.4 (75)	22.1 (74)	< 0.001
Winter weight (g)	23.0 (81)	21.8 (70)	< 0.001
First-year wing (mm)	86.4 (96)	83.4 (96)	< 0.001
Adult wing (mm)	87.7 (32)	83.9 (36)	< 0.001
Tarsus length (mm)	18.0 (116)	17.8 (117)	< 0.001
Bill length (mm)	13.8 (49)	13.5 (50)	0.06
Bill depth (mm)	4.8 (88)	4.8 (90)	0.09

NOTES:

1. Values are means with number of individuals in parentheses.

2. Summer and winter weights are based on July to August captures and September to March captures, respectively, excluding roosting birds; Bill lengths are based on July to October captures for adults and October captures for first-year birds, and bill depths on July to October captures for adults and August to October captures for first-year birds (see text for further explanation).

1. *Tarsus length.* This is the only measurement that is not subject to temporal or age variation.

2. *Wing length* (unflattened chord). Wing length may change with age when flight feathers are moulted, and in between moults by wear. In nine individuals that were caught at least three times before and after their first adult moult, wing length increased on average by 1.1 mm, and this increase was significant for each of five males and two of four females. Overall, measured wing lengths increased after the first adult moult in 16 of 19 males and 12 of 18 females. First-year and adult wing lengths of the same individuals were significantly correlated but more clearly so in males ($r = 0.84$) than females ($r = 0.65$) which implies a more consistent change after moult in males. Wing length

did not appear to increase consistently after the second adult moult (3 out of 7 birds). Seasonal variation was slight except that three birds measured in May had *c.* 1 mm shorter wings, possibly because of intense wear in the nesting season.

3. *Bill length and depth* (length measured from nostril to tip, depth at anterior margin of nostril). Newly fledged juveniles had markedly shorter bills than adults, and this difference was diagnostic for juveniles until several weeks after fledging. Juvenile bills reached full adult size only by September–October (Fig. 100). Bills varied markedly in size throughout the year, being longest in late summer and shortest in spring, and male bills were slightly larger for most of the year (see Chapter 5 for a discussion of bill length changes in relation to diet). Because of variation with age and season, mean values in Table 28 were based on late summer measurements only (see Matthysen, 1989a).

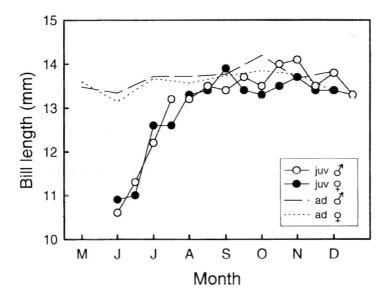

Fig. 100 *Seasonal variation in bill length of first-year and adult Nuthatches in summer and autumn. After Matthysen (1988).*

4. *Body weight.* Mean weight varied little during most of the day but increased markedly in the evening as shown by occasional roost captures in nestboxes (Fig. 101), followed by a decline during the night. Such evening fattening is characteristic of many bird species (Lehikoinen, 1987). Daytime weights were highest in summer, perhaps because of physiological requirements for moulting (Fig. 102). A similar seasonal weight peak was recorded in dwarf nuthatches by Norris (1958). Body weight increased again in early spring but reached its lowest point in June (Fig. 102), perhaps because of the energy demands of raising young, or as an adaptation to reduce energy expenditure (*cf.* Merkle and Barclay, 1996).

5. *Sexual differences* (Table 28). Males are significantly larger than females in all parameters except for bill length and depth. They are *c.* 5% heavier both in summer and winter, have 5% longer wings and *c.* 1% longer tarsi.

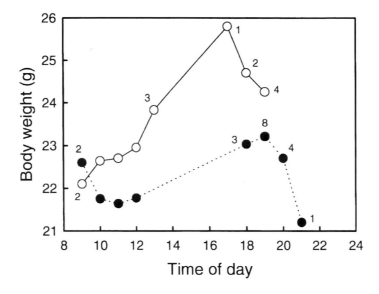

Fig. 101 *Changes in mean body weight of Nuthatches during the day in winter (November–February). Open dots = males, filled dots = females. Sample sizes below 10 are indicated.*

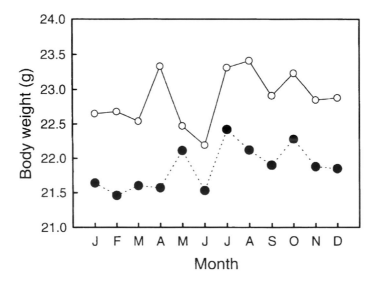

Fig. 102 *Seasonal variation in body weight of Nuthatches, excluding night captures and first-year birds in June. Open dots = males, filled dots = females.*

Population Densities of the Eurasian Nuthatch

The following table gives an overview of population density estimates of the Eurasian Nuthatch, expressed as breeding pairs per 10 ha, in various habitats across its range.

Habitat type	Density	Country	Source
~DECIDUOUS OAK AND OAK–HORNBEAM FOREST			
100-year-old oak–beech		Slovakia	A. Krištín, pers.
With nestboxes	10.0		comm.
Without nestboxes	6.2		
180-year-old oak (30 m)	7.9	Poland	Tomiałojć and Profus, 1977
130-year-old oak–hornbeam (30 m)	7.1	Poland	Tomiałojć and Profus, 1977
Oak–dominated mixed deciduous		Germany	Amann, 1993, cited
1949	2.8		by Glutz von
1992	6.9		Blotzheim, 1993
Oak–hornbeam	6.6	Germany	Pfeifer, 1955
Oak–hornbeam with lime	6.0	Poland	Ranoszek, 1969
250-year-old oak with conifers	5.7	Switzerland	Christen, 1983
70–80-year-old oak (25 m)	5.5	Poland	Tomiałojć and Profus, 1977
30–170-year-old oak	5	UK	Smith *et al.*, 1987
80–90-year-old sessile and Turkey oak	5.0	Hungary	Moskát, 1987
200-year-old oak	4.6	France	Ferry, 1974
150-year-old oak–hornbeam–lime	4.5	Poland	Głowaciński, 1975
Mature oak (25 m)	3–6	Germany	Blana, 1978
Mature oak (25 m)		Belgium	Matthysen, unpubl.
Large forests (over 100 ha)	3.4		
Forest fragments (2–30 ha)	1.8		
Japanese oak (24 m)	3.2	Japan	Fujimaki, 1988
Oak forest		France	Ferry and Frochot,
150–250 years old	3.8		1970
70–125 years old	2.1–2.7		
Coppiced with standards	1.9–5.1		
95-year-old oak–hornbeam–lime	2.8	Poland	Głowaciński, 1975
Ancient oak–hornbeam (45 m)	2.2	Poland	Wesołowski and Stawarczyk, 1991

Habitat type	Density	Country	Source
Sessile and Turkey oak	1.9–2.3	Hungary	Moskát and Fuisz, 1994
BEECH FOREST			
Mature beech (35 m)	8	Germany	Blana, 1978
Old beech	5.1	France	Blondel, 1984
Beech, 550–750 m altitude	4.9	Slovakia	A. Krištín, pers. comm.
170-year-old beech with pine	4.4	Germany	Prill, 1988
120-year-old beech with ash	4.3	Switzerland	Christen, 1983
Old beech	4.1	France	Spitz and Le Louarn, 1978, cited by Blondel, 1984
170-year-old beech	3.2	Germany	Prill, 1988
Beech, 200–400 m altitude	4.2	Germany	Zang, 1988
Beech, 450–600 m altitude	2.7	Germany	Zang, 1988
Beech with fir	3.1	Slovakia	A. Krištín, pers. comm.
Old coppice with standards	2.7	France	Ferry, 1974
100-year-old beech	2.0	Hungary	Moskát, 1987
Beech with pine	1.5–2.2	Poland	Various authors cited by Glutz von Blotzheim, 1993
160-year-old beech	1.2	Belgium	Bilcke and Joiris, 1979
Beech with 25% spruce	1.1	Luxemburg	Foyer, 1976
Beech	0.9	Japan	Nakamura, 1986
OTHER DECIDUOUS FOREST			
Woodlots with oak and beech	*c.* 4	Netherlands	van Noorden, 1986, cited by Harms and Opdam, 1989
Ash, 1992	3.2	Germany	Amann, 1993, cited by Glutz von Blotzheim, 1993
Ash, 1949	1.6	Germany	Amann, 1993, cited by Glutz von Blotzheim, 1993
Birch, ash, alder, large oaks (heavily grazed)	2.9	UK	Edington and Edington, 1972
Ancient alder–ash–spruce	2.9	Poland	Wesołowski and Stawarczyk, 1991
Oak and beech with birch	2.8	Belgium	Matthysen, 1988
Oak, magnolia, *Prunus*, sycamore	2.6	Japan	Fujimaki, 1986
Open oak and birch woodland	2.3	Sweden	Enoksson, 1990a
Birch–oak	1.9	Spain	Rivera, 1985
Alder–birch	1.8	Sweden	Nilsson, 1979
Oak and beech	1.5	Sweden	Nilsson, 1987
50-year-old alder carr	1.2	Poland	Głowaciński, 1975
†Birch–aspen patches in spruce	1.2	Sweden	Enoksson *et al.*, 1995
Birch	0.6	Japan	Nakamura, 1976

Habitat type	Density	Country	Source
EVERGREEN OAK			
Evergreen oaks	2.4	Morocco	Thévenot, 1982
Holm oak (20m)	2.1	France	Blondel and Farré, 1988
Holm oak	1.7	Spain	Herrera, 1978
*PARKLAND			
Suburban park	*c.* 8	Poland	Tomiałojć and Profus, 1977
15 ha within larger park	5–6	Germany	Löhrl, 1958
15 ha adjacent to large forest	5.2	Belgium	Matthysen, in prep
Parkland	5.0	Austria	Schneider, cited by Glutz von Blotzheim, 1993
Small parks (1–20 ha)	2.7	Belgium	Matthysen, unpubl.
IMMATURE BROADLEAVED			
Oak (10–18 m)	1.2–1.7	UK	Jones, 1972
Young broadleaved forest	0.5–1	Germany	Blana, 1978
Immature evergreen oak	1.7	Morocco	Thévenot, 1982
Immature evergreen oak (10 m)	0.2	France	Blondel and Farré, 1988
CONIFEROUS FOREST			
100-year-old spruce	5.0	Switzerland	Christen, 1983
Mature conifer	2.8	Japan	Nakamura, 1976
Spruce–fir	2.1	Germany	Löhrl, cited by Glutz von Blotzheim, 1993
Open larch forest	<2.6	Switzerland	Glutz von Blotzheim, 1993
Montane cedar and pine	1.0	Morocco	Thévenot, 1982
Spruce with oak and beech	0.9	Sweden	Nilsson, 1987
Pine with some deciduous trees	0.9	Germany	Haupt, 1992
Pine with holm oak	0.9	Spain	Obeso, 1987
100-year-old pine with oak	0.8	Poland	Głowaciński, 1975
Montane pine	0.75	Spain	Carrascal, 1984a
Spruce with pine	0.7	Sweden	Nilsson, 1979
Spruce	0.6	Slovakia	Turček, 1956
Pine	0.5–2	Germany	Blana, 1978
Bog pinewood (50 years old)	0.5	Poland	Głowaciński, 1975
Spruce	0–0.5	Germany	Blana, 1978
Cedar	0.4	France	Blondel, 1984
Pine	0.4	Germany	Haupt, 1992
Larch	0.3	Siberia	Pravosudov, 1993a
Spruce and ash (*c.* 1400 m)	0.3	Slovakia	A. Krištín, pers. comm.
Spruce	0.18	Germany	Zang, 1988
Ancient spruce–pine–birch	0.15	Poland	Wesołowski and Stawarczyk, 1991
Commercial pine	0.05–0.12	Poland	Various sources cited by Glutz von Blotzheim, 1993
Conifers	0.04	Russia	Ptushenko and Inozemtsev, 1968

NOTES:
~ unless otherwise stated, 'oak' is common oak.
* the published density estimate has been corrected for percentage forest cover.
† excluding unoccupied patches.

A Life-table for the Eurasian Nuthatch

Table 29 shows the results of an attempt to construct a life-table for the Peerdsbos population. The data in the table should be seen as gross approximations because they are based on a single study over only 5 years, and also for other reasons which are given below. The size of each age category is expressed relative to the adult population, which is assumed to remain constant between years. The starting point is the number of eggs per adult which is half the mean clutch size, and the number of fledglings which is also divided by two. It is assumed that 60% of the fledglings survive to independence (Currie and Matthysen, unpubl.). The proportion alive at the end of summer is derived from the ratio of first-year to adult birds, corrected for the mortality of adults during the summer (which is based on colour-ringed individuals, assuming no dispersal). The proportion of birds reaching their first breeding season is simply the proportion of first-year birds in spring. Survival to subsequent years is estimated from colour-ringed individuals. Survival from first to second breeding season was 41% (22 out of 54 individuals) and survival after the second breeding season 52% (9 of 17 2-year olds, 3 of 6 3-year-olds).

The proportions of first-year birds in the population in late summer and spring may have been overestimated since all summer immigrants of unknown age were considered first-year birds. Hence, survival in the first year may also be overestimated. On the other hand, a number of non-resident first-year birds may have been missed in late summer. Post-breeding survivorship is probably underestimated, particularly from first to second year, because of post-breeding dispersal (Chapters 8 and 9). Overall, the entire analysis rests on the assumption that immigration and emigration balance out within each age class, i.e. the study population is neither a source nor a sink (Pulliam, 1988).

TABLE 29: *A life-table for the Nuthatch*

Stage or age	Number per adult in previous generation	% survival to next stage	Remaining expectation of life (months)
Eggs	3.8 (100%)	78	6.5
Fledglings	2.96 (78%)	c. 60	6.8
Independence	1.8 (47%)	c. 57	10.5
Late summer	1.02 (27%)	55	15.5
1st breeding	0.56 (15%)	41	13.8
2nd breeding	0.226 (5.9%)	52	13.9
3rd breeding	0.118 (3.1%)	52	–

NOTES:
Data from the Peerdsbos population 1982–1987, except for the estimate of survival to independence (Currie and Matthysen, unpubl.). See text for further explanation.

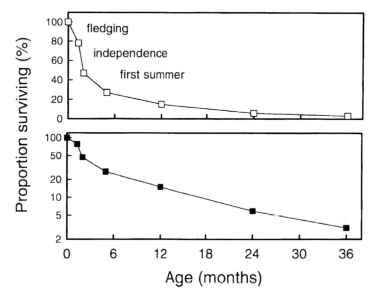

Fig. 103 *Lifetime survivorship curves of Nuthatches in the Peerdsbos study population, starting from the egg stage. Note the logarithmic axis in the lower figure.*

The calculations show that *c.* 78% of all eggs produce fledglings, 47% reach independence, and only 15% produce a breeding adult (Fig. 103). About 57% of all independent young survive their first summer, and 55% survive from late summer to the next spring. The most dramatic losses occur between egg production and recruitment into the summer population (73% loss). The bottom half of Fig. 103 shows that the survival rate remains relatively constant (i.e. a straight line on a logarithmic plot) once a first-year bird has established itself in late summer. This is also evident from the expectation of life which almost doubles after the first summer (Table 29). Based on these estimates, the breeding population has the following age-structure: 56% first-year birds, 23% second, 12% third, 6% fourth and 3% fifth-year or older birds.

Scientific Names of Species Mentioned in the Text

Birds

Australo-Papuan treecreepers	Climacteridae
Babblers	Timaliidae
Barbets	Capitonidae
Bee-eaters	*Merops*
Black Woodpecker	*Dryocopus martius*
Black-and-White Warbler	*Mniotilta varia*
Black-capped Chickadee	*Parus atricapillus*
Blue Tit	*Parus caeruleus*
Blue Jay	*Cyanocitta cristata*
Bluebirds	*Sialia*
Bowerbirds	Ptilonorhynchidae
Brown-headed Cowbird	*Molothrus ater*
Carolina Chickadee	*Parus carolinensis*
Chaffinch	*Fringilla coelebs*
Chiffchaff	*Phylloscopus collybita*
Coal Tit	*Parus ater*
Collared Flycatcher	*Ficedula albicollis*
Coral-billed Nuthatch	*Hypositta corallirostris*
Crested Black Tit	*Parus melanolophus*
Crested Tit	*Parus cristatus*
Dipper	*Cinclus cinclus*
Downy Woodpecker	*Picoides pubescens*
Elegant Tit	*Parus elegans*
Flycatchers (Old World)	Muscicapinae
Goldcrest	*Regulus regulus*
Goshawk	*Accipiter gentilis*
Great Spotted Woodpecker	*Picoides major*
Great Tit	*Parus major*
Greenfinch	*Carduelis chloris*
Green Woodpecker	*Picus viridis*
Honey Buzzard	*Pernis apivorus*
Hornbills	Bucerotidae
House Finch	*Carpodacus mexicanus*
House Sparrow	*Passer domesticus*

280

Japanese Pygmy Woodpecker	*Dendrocopos kizuki*
Jay	*Garrulus glandarius*
Kestrel	*Falco tinnunculus*
Kingfishers	Alcedinidae
Kinglets	*Regulus*
Lesser Spotted Woodpecker	*Picoides minor*
Little Owl	*Athene noctua*
Long-eared Owl	*Asio otus*
Long-tailed Tit	*Aegithalos caudatus*
Magpie	*Pica pica*
Marsh Tit	*Parus palustris*
Middle Spotted Woodpecker	*Picoides medius*
Minla	*Minla*
Mountain Chickadee	*Parus gambeli*
Nuthatch-Vanga	*Hypositta corallirostris*
Philippine creepers	Rhabdornithidae
Pied Flycatcher	*Ficedula hypoleuca*
Pine Siskin	*Carduelis pinus*
Pine Warbler	*Dendroica pinus*
Pygmy Owl	*Glaucidium passerinum*
Red-bellied Woodpecker	*Melanerpes carolinus*
Red-cockaded Woodpecker	*Picoides borealis*
Red-headed Woodpecker	*Melanerpes erythrocephalus*
Redstart	*Phoenicurus phoenicurus*
Rifleman	*Acanthisitta chloris*
Robin (European)	*Erithacus rubecula*
Rollers	*Coracias*
Sand Martin	*Riparia riparia*
Short-toed Treecreeper	*Certhia brachydactyla*
Sittellas	*Sittella*
Sparrowhawk	*Accipiter nisus*
Spotted Creeper	*Salpornis spilonotus*
Spotted Flycatcher	*Muscicapa striata*
Spotted Sandpiper	*Actitis hypoleuca*
Starling	*Sturnus vulgaris*
Tawny Owl	*Strix aluco*
Tengmalm's Owl	*Aegolius funereus*
Tit-babblers	*Alcippe*
Tit-warblers	*Leptopoecile*
Tree Sparrow	*Passer montanus*
Tufted Titmouse	*Parus bicolor*
Vangas	Vangidae
Wallcreeper	*Tichodroma muraria*
Warblers (Old World)	Sylviinae
Western Bluebird	*Sialia mexicana*
Wheatears	*Oenanthe*
White-backed Woodpecker	*Dendrocopos leucotus*
White-eyes	*Zosterops*
Willow Tit	*Parus montanus*
Wren	*Troglodytes troglodytes*
Wryneck	*Jynx torquilla*
Yellow-eyed Bulbul	*Pycnonotus penicillatus*

Other animals

Chipmunks	*Tamias*
Fallow deer	*Dama dama*
Fat dormouse	*Glis glis*
Flying squirrel	*Glaucomys volans*
Marten	*Martes*
Sock-eye salmon	*Oncorhynchus nerka*
Spruce budworm	*Choristoneura fumiferana*
Weasel	*Mustela nivalis*

Plants

Alder	*Alnus*
Algerian chestnut-leaved oak	*Quercus afares*
Algerian fir	*Abies numidica*
Almond	*Prunus*
Apricot	*Prunus*
Ash	*Fraxinus*
Aspen (Europe)	*Populus tremula*
Aspen (N America)	*Populus tremuloides*
Balsam fir	*Abies balsamea*
Beech (Europe)	*Fagus sylvatica*
Beech (N America)	*Fagus grandifolia*
Birch	*Betula*
Bishop pine	*Pinus muricata*
Black cherry	*Prunus serotina*
Caribbean pine	*Pinus caribaea*
Cedar	*Cedrus*
Cherry	*Prunus*
Cluster pine	*Pinus maritima*
Common oak	*Quercus robur*
Cork oak	*Quercus suber*
Corsican pine/Laricio	*Pinus nigra laricio*
Deodar	*Cedrus deodara*
Douglas fir	*Pseudotsuga menziesii*
Fir	*Abies*
Hawthorn	*Crataegus*
Hazel	*Coryllus avellana*
Hemlock	*Tsuga*
Hickory	*Carya*
Holm oak	*Quercus ilex*
Hornbeam	*Carpinus betulus*
Horse chestnut	*Aesculus hippocastanum*
Incense cedar	*Calocedrus decurrens*
Japanese oak	*Quercus mongolica*
Jeffrey pine	*Pinus jeffreyi*
Juniper	*Juniperus*
Khasi pine	*Pinus kesiya*
Korean pine	*Pinus koraiensis*
Larch	*Larix decidua*
Lime	*Tilia*

Loblolly pine	*Pinus taeda*
Locust	*Robinia pseudoacacia*
Lodgepole pine	*Pinus contorta*
Long-leaved Indian pine	*Pinus roxburghii*
Longleaf pine	*Pinus australis*
Magnolia	*Magnolia*
Mango	*Mangifera*
Maple	*Acer*
Monterey pine	*Pinus radiata*
Norway spruce	*Picea abies*
Oak	*Quercus*
Olive	*Olea europaea*
Pine	*Pinus*
Pistachio	*Pistacia vera*
Ponderosa pine	*Pinus ponderosa*
Poplar	*Populus*
Red oak	*Quercus rubra*
Red pine	*Pinus densiflora*
Rhododendron	*Rhododendron*
Scots pine	*Pinus sylvestris*
Scrub pine	*Pinus virginiana*
Sessile oak	*Quercus petraea*
Shortleaf pine	*Pinus echinata*
Slash pine	*Pinus ellottii*
Spruce	*Picea*
Sunflower	*Helianthus annuus*
Sweet chestnut	*Castanea sativa*
Sycamore	*Acer platanoides*
Turkey oak	*Quercus cerris*
Turkish pine	*Pinus brutia*
Walnut	*Juglans*
White fir	*Abies concolor*
White pine	*Pinus strobus*
Xcn oak/Algerian oak	*Quercus canariensis*
Yew	*Taxus baccata*

References

ADAMYAN, M. S. (1965). [The ecology of a small rock-nuthatch (*Sitta neumayer*) in Armenia]. *Ornitologiya*, 7: 157–165.

AIROLA, D. A. & BARRETT, R. H. (1985). Foraging and habitat relationships of insect-gleaning birds in a Sierra Nevada mixed-conifer forest. *Condor*, 87: 205–216.

ALERSTAM, T. & WINGE, A. (1975). [Why are the Marsh Tit and Nuthatch missing from the island of Hven? Speculations on the basis of an attempt of artificial introduction]. *Anser*, 14: 266–268.

ALEX, U. (1994). Zur geographischen Variation einiger Vogelarten Weißrußlands (Aves: Passeriformes). *Zoologische Abhandlungen (Dresden)*, 48: 139–147.

ALI, S. & RIPLEY, D. S. (1973). *Handbook of the Birds of India and Pakistan. Vol. 9. Robins to Wagtails.* Oxford University Press, Bombay.

ALLEN, A. A. (1929). Nuthatch. *Bird-Lore*, 31: 423–432.

ANDERSON, S. H. (1976). Comparative food habits of Oregon nuthatches. *Northwest Science*, 50: 213–221.

ANDRESEN, D. (1989). Über allgemeine Beziehungen zwischen Siebenschläfern und Höhlenbrütern in Nistkästen. Teil 2. *Der Falke*, 36: 156–161.

ARN, H. (1955). Mischbruten von Kohlmeisen, Blaumeisen und Kleiber. *Der Ornithologische Beobachter*, 52: 129.

ASKENMO, C., VON BRÖMSSEN, A., EKMAN, J. & JANSSON, C. (1977). Impact of some wintering birds on spider abundance in spruce. *Oikos*, 28: 90–94.

AUDUBON, J. J. B. (1844). *Birds of America.* New York.

AUSTIN, G. T. & SMITH, E. L. (1972). Winter foraging ecology of mixed insectivorous bird flocks in oak woodland in Southern Arizona. *Condor*, 74: 17–24.

AUSTIN, O. L. Jr. & KURODA, N. (1953). The birds of Japan: their status and distribution. *Bulletin of the Museum of Comparative Zoology at Harvard College*, 109: 519–520.

BAHR, P. H. (1907). The Nuthatch as a nest-builder. *British Birds*, 1: 122.

BAILEY, F. M. (1939). *Among the Birds in Grand Canyon Country.* U.S. Government Printing office, Washington, D.C.

BALL, S. C. (1947). Migration of Red-breasted Nuthatches in Gaspé. *Ecological Monographs*, 17: 501–533.

BANCROFT, J. (1987). Observations of White-breasted Nuthatch. *Blue Jay*, 45: 172–174.

BANIN, D. A., BAEME, I. R., KERIMOV, A. B. & PODDUBNAYA, N. Y. (1984). [Notes on the autumn movements of the Nuthatch *Sitta europaea amurensis* and some *Parus* species in south eastern Siberia (Primorie)]. *Ornitologiya*, 19: 191–193.

BANKS, R. C. (1970). Molt and taxonomy of Red-breasted Nuthatches. *Wilson Bulletin*, 82: 201–205.

BANKS, R. C. (1978). Prealternate molt in nuthatches. *Auk*, 95: 179–181.

BARBOUR, D. B. & DEGANGE, A. R. (1982). Reciprocal allopreening in the Brown-headed Nuthatch. *Auk*, 99: 171.

BARDIN, A. V. (1987). [Sap of trees, nectar and pollen as food sources for tits and goldcrests during early spring]. *Zoologicheskii Zhurnal,* 66: 789–790.

BAUER, H.-G. & BERTHOLD, P. (1996). *Die Brutvögel Mitteleuropas – Bestand und Gefährdung.* Aula-Verlag, Wiesbaden.

BÄUMLER, W. (1993). Zur Verbreitung und Entwicklung einiger höhlenbrütender Singvögel in Ostbayern. *Anzeiger für Schädlingskunde, Pflanzenschutz und Umweltschutz,* 66: 1–6.

BEAMAN, M. (1986). Turkey bird report 1976–1981. *Sandgrouse,* 8: 1–141.

BEAULIEU, D. (1944). *Les Oiseaux du Tranninh.* Université Indochinoise, Hanoi.

BECK, N. (1992). Conservation de la Sittelle Corse (*Sitta whiteheadi*): sa place dans les aménagements forestiers. Unpublished report, Parc Naturel Régional de la Corse, Ajaccio.

BELLATRECHE, M. (1990). Découverte d'un troisième biotope de la Sittelle Kabyle (*Sitta ledanti* Vielliard) en Algérie. *Annales de l'Institut National Agronomique El-Harrach,* 14: 13–20.

BELLATRECHE, M. (1991). Deux nouvelles localisations de la Sittelle Kabyle *Sitta ledanti* en Algérie. *L'Oiseau et la Revue Française d'Ornithologie,* 61: 269–272.

BELLATRECHE, M. & CHALABI, B. (1990). Données nouvelles sur l'aire de distribution de la Sittelle Kabyle *Sitta ledanti. Alauda,* 58: 95–97.

BENT, A. C. (1948). Life histories of North American nuthatches, wrens, thrashers and their allies. *United States National Museum Bulletin,* 195: 1–55.

BENTHAM, H. (1946). "Injury-feigning" of Nuthatch. *British Birds,* 39: 214.

BERNDT, R. & DANCKER, P. (1960). Der Kleiber *Sitta europaea* als Invasionsvogel. *Die Vogelwarte,* 20: 193–198.

BERNDT, R. & STERNBERG, H. (1968). Terms, studies and experiments on the problems of bird dispersion. *Ibis,* 110: 256–269.

BERNDT, R. & WINKEL, W. (1979). Zur Populationsentwicklung von Blaumeise, Kleiber, Gartenrotschwantz und Wendehals in Mitteleuropäischen Untersuchungsgebieten von 1927 bis 1978. *Die Vogelwelt,* 100: 55–69.

BERNER, T.O. & GRUBB, T.C. Jr. (1985). An experimental analysis of mixed-species flocking in birds of deciduous woodland. *Ecology,* 66: 1229–1236.

BEVIER, L. R. (1993). Spread-wing and tilting display of the White-breasted Nuthatch. *Connecticut Warbler,* 13: 58–62.

BIJLSMA, R. G. (1990). *Broedvogels van Roggebotzand, Reve-Abbert, Spijk-Bremerberg en Harderbos (Oostelijk Flevoland).* SOVON-rapport 90/05, SOVON, Beek-Ubbergen.

BILCKE, G. & JOIRIS, C. (1979). Récensement des oiseaux nicheurs en forêt de Soignes (Brabant); considérations critiques sur la méthode de quadrats. *Aves,* 16: 5–23.

BILCKE, G., MERTENS, R., JEURISSEN, M. & DHONDT, A. A. (1986). Influences of habitat structure and temperature on the foraging niches of the pariform guild in Belgium during winter. *Le Gerfaut,* 76: 109–129.

BIRKHEAD, T. R. (1991). *The Magpies.* T. & A.D. Poyser, London.

BLACKFORD, J. L. (1955). Pygmy Nuthatches take arboreal bath. *Condor,* 57: 304.

BLANA, H. (1978). *Die Bedeutung der Landschaftsstruktur für die Verbreitung der Vögel im Südlichen Bergischen Land.* Gesellschaft Rheinischer Ornithologen, Düsseldorf.

BLATTNER, M. & SPEISER, C. T. (1990). Schwankungen und langfristige Trends der Nistkasten-Besetzungsanteile von Singvögeln in der Region Basel und ihre Aussagekraft. *Der Ornithologische Beobachter,* 87: 223–242.

BLOMGREN, A., NORÉN, E. & STEFANSSON, O. (1979). [The invasion of Siberian Nuthatches *Sitta europaea asiatica* in 1976–77. *Vår Fagelvärld,* 38: 101–103.

BLONDEL, J. (1984). Avifaunes forestières méditerranéennes; histoire des peuplements. *Aves,* 21: 209–226.

286   *References*

BLONDEL, J. & FARRÉ, H. (1988). The convergent trajectories of bird communities along ecological successions in European forests. *Oecologia*, 75: 83–93.

BOCK, C. E. (1969). Intra- vs. interspecific aggression in Pygmy Nuthatch flocks. *Ecology*, 50: 903–905.

BOCK, C. E. & FLECK, D. C. (1995). Avian response to nest box addition in two forests of the Colorado front range. *Journal of Field Ornithology*, 66: 352–362.

BOCK, C. E. & LEPTHIEN, L. W. (1976). Synchronous eruptions of boreal seed–eating birds. *American Naturalist*, 110: 559–571.

BOHÁC, D. (1965). [Zur Brutbiologie des Kleibers *Sitta europaea*)]. *Sylvia*, 17: 244–246.

BÖHR, H.-J. (1962). Zur Kenntnis der Vogelwelt von Korfu. *Bonner Zoologische Beiträge*, 13: 50–114.

BOITEAU, M. (1991). Nidification hivernale de la Sittelle. *Nos Oiseaux*, 39: 78.

BOYLE, G. L. (1955). Nuthatch burying a nut. *British Birds*, 48: 283.

BRACKBILL, H. (1969a). White-breasted Nuthatch bill abnormality corrected by wear. *Bird-Banding*, 40: 145.

BRACKBILL, H. (1969b). Status and behavior of color-banded White-breasted Nuthatches at Baltimore. *Maryland Birdlife*, 25: 87–91.

BRAUER, P. & KIESEWETTER, K. (1983). Kleiber zieht junge Kohlmeisen auf. *Der Falke*, 30: 390–391.

BRAWN, J. D. (1987). Density effects on reproduction of cavity nesters in Northern Arizona. *Auk*, 104: 783–787.

BRAWN, J. D. (1988). Selectivity and ecological consequences of cavity nesters using natural vs. artificial nest sites. *Auk*, 105: 789–791.

BRAWN, J. D. & BALDA, R. P. (1988). Population biology of cavity nesters in Arizona: do nest sites limit breeding densities? *Condor*, 90: 61–71.

BRAZIL, M. A. (1991). *The Birds of Japan*. Christopher Helm, London.

BREHM, A. (1920). *Brehms Tierleben. Vögel, 4er Band.* Leipzig.

BRICHETTI, P. & DI CAPI, C. (1985). Distribution, population and breeding ecology of the Corsican Nuthatch *Sitta whiteheadi* Sharpe. *Rivista Italiana di ornitologia*, 55: 3–26.

BRICHETTI, P. & DI CAPI, C. (1987). Conservation of the Corsican Nuthatch *Sitta whiteheadi* Sharpe, and proposals for habitat management. *Biological Conservation*, 39: 13–21.

BROMLEY, G. F., KOSTENKO, V. A. & OKHOTINA, M. V. (1974). [The role of the Amur Nuthatch (*Sitta europaea amurensis*, Swinh.) in the revival of the Korean cedar]. In: *[Fauna and Ecology of the Terrestrial Vertebrates of the Southern Part]* (Ed. M. V. Okhotina). *Proceedings of the Institute of Biology and Pedology, Far East Science Centre*, 17: 162–166.

BUCKINGHAM, B. (1975). The winter territories of White-breasted Nuthatches. *Inland Bird Banding News*, 47: 173–178.

BUFF, U. (1987). Weniger Kleiberbruten. *Vögel der Heimat*, 57: 125.

BUNDY, G. (1971). Display of Krüper's Nuthatches. *British Birds*, 64: 461.

BURLEIGH, T. D. (1960). Three new subspecies of birds from western North America. *Auk*, 77: 210–215.

BURNIER, E. (1976). Une nouvelle espèce de l'avifaune paléarctique: la Sittelle Kabyle *Sitta ledanti*. *Nos Oiseaux*, 33: 337–340.

BUSCH, W. D. (1959). Kleiber schmarotzt bei einem Wespenbussard. *Der Falke*, 6: 178.

BUSCHE, G. (1993). Bestandsentwicklung der Waldvögel im Westen Schleswig-Holsteins 1960–1990. *Die Vogelwelt*, 114: 15–34.

BUSSE, D. & OLECH, B. (1968). On some problems of birds spending nights in nestboxes. *Acta Ornithologica Musei Zoologici Polonici*, 11: 1–26.

BUSSMANN, J. (1943). Beitrag zur Kenntnis der Brutbiologie des Kleibers (*Sitta europaea caesia*). *Der Ornithologische Beobachter*, 40: 57–67.

BUSSMAN, J. (1946). Beitrag zur Kenntnis der Brutbiologie des Kleibers, *Sitta europaea caesia*. *Der Ornithologische Beobachter*, 43: 1–5.

BUTTS, W. K. (1931). A study of the Chickadee and White-breasted Nuthatch by means of marked individuals. Part III: the White-breasted Nuthatch (*Sitta carolinensis cookei*). *Bird-Banding*, 2: 59–76.

BUXTON, P. A. (1920). (untitled). *Bulletin of the British Ornithologists' Club*, 40: 135–139.

CAMERON, A. D. & HARRISON, C. J. O. (1978). *Bird Families of the World*. H.N. Abrams Inc., New York.

CARRASCAL, L. M. (1984a). Organizacion de la comunidad de Aves de los bosques de *Pinus sylvestris* de Europa en sus limites latitudinales de distribucion. *Ardeola*, 31: 91–101.

CARRASCAL, L. M. (1984b). Cambios en el uso del espacio en un gremio de Aves durante el periodo primavero – verano. *Ardeola*, 31: 47–60.

CARRASCAL, L. M. & MORENO, E. (1992). Proximal costs and benefits of heterospecific social foraging in the Great Tit, *Parus major*. *Canadian Journal of Zoology*, 70: 1947–1952.

CARRASCAL, L. M. & MORENO, E. (1993). Food caching versus immediate consumption in the Nuthatch: the effect of social context. *Ardea*, 81: 135–141.

CARRASCAL, L. M., MORENO, E. & TELLERIA, J. (1990). Ecomorphological relationships in a group of insectivorous birds of temperate forests in winter. *Holarctic Ecology*, 13: 105–111.

CEBALLOS, P. (1969). Estudio de alimentacion del trepador azul (*Sitta europaea*) en encinares, durante los meses Marzo – Agosto. *Boll. Serv. Plagas Forest.*, 12: 89–95.

CHALABI, B. (1989). Du nouveau à propos de l'aire de distribution de la Sittelle Kabyle (*Sitta ledanti*). *Aves*, 26: 233–234.

CHAPIN, J. P. (1949). Pneumatization of the skull in birds. *Ibis*, 91: 691.

CHAPPUIS, C. (1976). Origine et évolution des vocalisations de certains oiseaux de Corse et des Baléares. *Alauda*, 44: 475–495.

CHAPPUIS, R. (1970). Nichées mixtes de mésanges. *Nos Oiseaux*, 30: 267–268.

CHASEN, F. N. & HOOGERWERF, A. (1941). The birds of the Netherlands Indian Mt Leuser expedition 1937 to North Sumatra. *Treubia* 18 (suppl.): 104.

CHRISTEN, W. (1983). Brutvogelbestände in Wäldern unterschiedlicher Baumarten- und Altersklassenzusammensetzung. *Der Ornithologische Beobachter*, 80: 281–291.

CLARK, M. & MASON, J. R. (1985). Use of nest material as insecticidal and anti-pathogenic agents by the European Starling. *Oecologia*, 67: 169–176.

CLARKE, M. (1985). Nuthatch hovering. *British Birds*, 78: 111.

COHEN, E. (1960). Little Owl taking Nuthatch. *British Birds*, 53: 574.

COLLAR, N. J., CROSBY, M. J. & STATTERSFIELD, A. J. (1994). *Birds to Watch 2: the Birdlife International World Checklist of Birds*. Birdlife Conservation Series 4. Birdlife International, Cambridge.

CRAMP, S. & PERRINS, C. M. (1993). *Handbook of the Birds of the Western Palearctic. Vol. VII: Flycatchers to Shrikes*. Oxford University Press, Oxford.

CRAWFORD, H. S., JENNINGS, D. T. & STONE, T. L. (1990). Red-Breasted Nuthatches detect early increases in Spruce Budworm populations. *Northern Journal of Applied Forestry*, 7: 81–93.

CREUTZ, G. (1953). Beeren und Früchte als Vogelnahrung. *Beiträge zur Vogelkunde*, 3: 91–103.

CUISIN, J. (1984). L'identification des crânes de petits passereaux. 5. *L'Oiseau et la Revue Francaise d'Ornithologie*, 54: 264–267.

CUYPERS, J. (1944). De Boomklever. *Wielewaal*, 11: 148–151.

DANFORD, C. G. (1878). A contribution to the ornithology of Asia Minor. *Ibis* (4th series), 11: 1–35.

DECEUNINCK, B. & BAGUETTE, M. (1991). Avifaune nicheuse de la séquence de l'Epicea (*Picea abies*) dans la région du Plateau des Tailles (Prov. du Luxembourg). *Aves*, 28: 189–207.

DEIGNAN, H. G. (1938). A new nuthatch from Yunnan. *Smithsonian Miscellaneous Collections*, 97: 1–2.

DEIGNAN, H. G. (1945). The birds of northern Thailand. *United States National Museum Bulletin*, 186: 314–319.

DE KIRILINE, L. (1952). Red-breast makes a home. *Audubon Magazine*, 54: 16–21.

DE KIRILINE, L. (1954). Irrepressible Nuthatch. *Audubon Magazine*, 56: 264–267.

DELACOUR, J. & JABOUILLE, P. (1930). Description de trente oiseaux de l'Indochine française. *L'Oiseau et la Revue Française d'Ornithologie*, 11: 393–408.

DELMÉE, E. (1948). La dispersion de la Sittelle Torche-pot, *Sitta europaea hassica* Kleinschmidt en Belgique en 1947. *Le Gerfaut*, 38: 130–146.

DELMÉE, E. (1949). Complément a l'étude de la dispersion de la Sittelle Torche-pot en Belgique. *Le Gerfaut*, 39: 115–117.

DELMÉE, E., DACHY, P. & SIMON, P. (1972). Contribution à la biologie des mésanges (Paridae) en milieu forestier. *Aves*, 9: 1–78.

DELMÉE, E., DACHY, P. & SIMON, P. (1979). Les hôtes occasionels des nichoirs à Chouettes Hulottes (*Strix aluco*). *Aves*, 16: 49–58.

DE MEERSMAN, L. (1981). Boomklever – het uitvliegen van een nest. *Wielewaal*, 47: 97–98.

DEMENT'EV, G. P. & GLADKOV, N. A. (1954). *Birds of the Soviet Union. Vol. V.* Sovetskaya Nauka, Moscow.

DE SCHAUENSEE, R. M. (1929). A further collection of birds from Siam. *Proceedings of the Academy of Natural Sciences of Philadelphia*, 81: 523–587.

DE SCHAUENSEE, R. M. (1934). Zoological results of the third De Schauensee Siamese expedition. Part II. Birds from Siam and the southern Shan states. *Proceedings of the Academy of Natural Sciences of Philadelphia*, 86: 165–280.

DE SCHAUENSEE, R. M. (1984). *The Birds of China.* Smithsonian Institution Press, Washington, DC.

DESFAYES, M. & PRAZ, J. C. (1978). Notes on habitat and distribution of montane birds in southern Iran. *Bonner Zoologische Beiträge*, 29: 18–37.

DEVILLERS, P., ROGGEMAN, W., TRICOT, J., DEL MARMOL, P., KERWIJN, C., JACOB, J.-P. & ANSELIN, A. (1988). *Atlas van de Belgische Broedvogels.* Koninklijk Belgisch Instituut voor Natuurwetenschappen, Brussel.

DE VRIES, A. (1977). De Boomklever in Nederland. *De Lepelaar*, 53: 7–9.

DE VRIES, A. (1985). *De Boomklever.* Thieme, Zutphen.

DHONDT, A. A. (1967). Snavelmisvormingen bij een Kokmeeuw, *Larus ridibundus*, en een Boomklever, *Sitta europaea*. *Le Gerfaut*, 57: 139–144.

DICKINSON, E.C., KENNEDY, R. S. & PARKES, K. C. (1991). *The Birds of the Philippines.* British Ornithologists' Union, Tring.

DIEMER, R. (1985). Rauhfußkauz *Aegolius funereus* von Kleiber in Bruthöhle eingemauert. *Anzeiger der Ornithologischen Gesellschaft Bayern*, 24: 81.

DOBINSON, H. M. & RICHARDS, A. J. (1964). The effects of the severe winter of 1962/63 on birds in Britain. *British Birds*, 57: 373–434.

DOBSON, A. P. (1987). A comparison of seasonal and annual mortality for both sexes of fifteen species of common British birds. *Ornis Scandinavica*, 18: 122–128.

DORKA, V. (1980). Insektenspeichernde Kleiber *Sitta europaea*. Zur Unterscheidung von langfristigem und kurzfristigem Nahrungsspeicher-Verhalten. *Ökologie der Vögel*, 2: 145–150.

DOUMANDJI, S. & KISSERLI, O. (1993). Paramètres écologiques de la Sittelle Kabyle *Sitta ledanti*, en chênaie mixte dans le Parc National de Taza. *Alauda*, 61: 264–265.

DUNAJEWSKI, A. (1934). Die eurasiatischen Formen der Gattung *Sitta* Linn. *Acta Ornithologica Musei Zoologici Polonici*, 1: 181–251.

DUNNING, J. B. & BOWERS, R. K. (1984). Local movements of some Arizona montane birds. *North American Bird Bander*, 9: 7.

DURANGO, A. M. & DURANGO, S. (1942). [Breeding of the Nuthatch (*Sitta e. europaea* L.)]. *Vår Fagelvärld*, 1: 33–44.

DUYCK, B. E. & McNAIR, D. B. (1991). Notes on egg-laying, incubation, and nestling periods and of food brought to the nest by four species of cavity-nesting birds. *The Chat*, 55: 21–29.

DUYCK, B. E., McNAIR, D. B. & NICHOLSON, C. P. (1991). Dirt-storing behavior by White-Breasted Nuthatches. *Wilson Bulletin*, 103: 308–309.

EAMES, J. C. , ROBSON, C. R. , CU, N. & VAN LA, T. (1992). *Vietnam Forest Project – Forest Bird Surveys 1991*. ICBP Study Report No. 51.

EDINGTON, J. M. & EDINGTON, M. A. (1972). Spatial patterns and habitat partitioning in the breeding birds of an upland wood. *Journal of Animal Ecology*, 41: 331–358.

EGGERS, J. (1977). Weiteres zum Vorkommen des Klippenkleibers (*Sitta tephronota*) in der Südosttürkei. *Die Vogelwelt*, 98: 25–27.

EDEN, S. F. (1987). Natal philopatry of the Magpie *Pica pica*. *Ibis*, 129. 477–490.

EKMAN, J. & ASKENMO, C. (1984). Social rank and habitat use in Willow Tit groups. *Animal Behaviour*, 32: 508–514.

EMLEN, J. T. (1957). Defended area? A critique of the territory concept and of conventional thinking. *Ibis*, 99: 352.

EMLEN, J. T. (1981). Divergence in the foraging responses of birds on two Bahama Islands. *Ecology* 62: 289–295.

ENGEL, R. (1942). Ein Kleiber trommelt. *Beiträge zur Fortpflanzungsbiologie der Vögel*, 18: 174.

ENOKSSON, B. (1987). Local movements in the Nuthatch (*Sitta europaea*). *Acta Regalia Societatis Scientiae Litterarum Gothoburgensis Zoologica*, 14: 36–47.

ENOKSSON, B. (1988a). Age and sex differences in dominance and foraging behaviour of Nuthatches (*Sitta europaea*). *Animal Behaviour*, 36: 231–238.

ENOKSSON, B. (1988b). Prospective resource defense and its consequences in the Nuthatch *Sitta europaea* L. Ph.D. thesis, Uppsala University.

ENOKSSON, B. (1990a). Autumn territories and population regulation in the Nuthatch *Sitta europaea*: an experimental study. *Journal of Animal Ecology*, 59: 1047–1062.

ENOKSSON, B. (1990b). Time budgets of Nuthatches *Sitta europaea* with supplementary food. *Ibis*, 132: 575–583.

ENOKSSON, B. (1993). Effects of female age on reproductive success in European Nuthatches breeding in natural cavities. *Auk*, 110: 215–221.

ENOKSSON, B., ANGELSTAM, P. & LARSSON, K. (1995). Deciduous forest and resident birds: the problem of fragmentation within a coniferous forest landscape. *Landscape Ecology*, 10: 267–275.

ENOKSSON, B. & NILSSON, S. G. (1983). Territory size and population density in relation to food supply in the Nuthatch *Sitta europaea* (Aves). *Journal of Animal Ecology*, 52: 927–935.

ERHARD, R. & WINK, M. (1987). Veränderungen des Brutvogelbestandes im Großraum Bonn: Analyse der Rasterkartierung 1975 und 1985. *Journal für Ornithologie*, 128: 477–484.

ERIKSSON, K. (1970). The invasion of *Sitta europaea asiatica* Gould into Fennoscandia in the winters of 1962/63 and 1963/64. *Annales Zoologici Fennici*, 7: 121–140.

ETCHÉCOPAR, R. D. & HUË, F. (1983). *Les Oiseaux de Chine. Passereaux.* Sociétée Nouvelle des Éditions Boubée, Paris.

FERRY, C. (1974). Comparison between breeding bird communities in an oak forest and a beech forest, censused by the IPA method. *Acta Ornithologica*, 14: 303–309.

FERRY, C. & FROCHOT, B. (1970). L'avifaune nidificatrice d'une forêt de chênes pedonculés en Bourgogne: étude de deux successions écologiques. *La Terre et la Vie*, 24: 153–250.

FIEBIG, J. (1992). Beobachtungen am Chinesenkleiber (*Sitta villosa* Verreaux) in Nordkorea und Nordostchina. *Mitteilungen des Zoologisches Museums Berlin*, 68 Suppl: *Annales Ornithologici*, 16: 135–155.

FLEETWOOD, R. J. (1946). Unusual matings of a Brown-headed Nuthatch. *Bird-Banding*, 17: 75–76.

FLOUSEK, J., HUDEC, K. & GLUTZ VON BLOTZHEIM, U. N. (1993). Immissionsbedingte Waldschäden und ihr Einfluß auf die Vogelwelt Mitteleuropas. In: *Handbuch der Vögel Mitteleuropas. Bd. 13. Passeriformes (4. Teil). Teil 1. Muscicapidae – Paridae* (Ed. U. N. Glutz von Blotzheim): pp. 11-30. Aula-Verlag, Wiesbaden.

FORBUSH, E. H. (1929). *Birds of Massachusetts and Other New England States. III.* Massachusetts Department of Agriculture.

FORD, H. A. (1987). Bird communities on habitat islands in England. *Bird Study*, 34: 205–218.

FOSSE, A. & VAILLANT, G. (1982). A propos de la couleur de la calotte chez la Sittelle Kabyle (*Sitta ledanti*). *Alauda*, 50: 228.

FOYER, H. (1976). Die Siedlungsdichte der beide Baumläuferarten und des Kleibers in einem Waldgebiet des Luxemburger Sandsteins. *Regulus*, 12: 9–18.

FRANKIS, M. P. (1991). Krüper's Nuthatch *Sitta krueperi* and Turkish Pine *Pinus brutia*: an evolving association? *Sandgrouse*, 13: 92–97.

FRELIN, C. (1975). Comportement invasionnel des Mésanges Noires (*Parus ater*) et espèces apparentées, au col de la Golèze en 1972. *L'Oiseau et la Revue Francaise d'Ornithologie*, 45: 41–64.

FUJIMAKI, Y. (1986). Breeding bird community in a deciduous broad-leaved forest in southern Hokkaido, Japan. *Japanese Journal of Ornithology*, 35: 15–23.

FUJIMAKI, Y. (1988). Breeding bird community of a *Quercus mongolica* forest in eastern Hokkaido, Japan. *Japanese Journal of Ornithology*, 37: 69–75.

GABRIELS, J. (1985). *Atlas van de Broedvogels in Limburg.* BNVR and Lisec, Hasselt.

GABRIELS, J., STEVENS, J. & VAN SANDEN, P. (1994). *Broedvogelatlas van Limburg: Veranderingen in Aantallen en Verspreiding na 1985.* Provincie Limburg, Hasselt.

GAO, W. (1978). [On the breeding behaviour and feeding habits of the Black-headed Nuthatch]. *Acta Zoologica Sinica*, 24: 261–268.

GATTER, W. (1974). Beobachtungen an Invasionsvögeln des Kleibers (*Sitta europaea caesia*) am Randecker Maar, Schwäbische Alb. *Die Vogelwarte*, 27: 203–209.

GATTER, W. & MATTES, H. (1979). Zur Populationsgröße und Ökologie des neuentdeckten Kabylenkleibers *Sitta ledanti* Vielliard 1976. *Journal für Ornithologie*, 120: 390–405.

GÉROUDET, P. (1963). *Les Passereaux. II.* Delachaux & Niestlé, Neuchatel.

GÉROUDET, P. (1991). Chronique ornithologique romande: le printemps et la nidification en 1990. *Nos Oiseaux*, 41: 137–140.

GHALAMBOR, C. & MARTIN, T. E. (MS). Ecological and evolutionary determinants of avian incubation strategies: testing alternatives in three sympatric nuthatches.

GIBBONS, D. W., REID, J. B. & CHAPMAN, R. A. (1993). *The New Atlas of Breeding Birds in Britain and Ireland, 1988–1991.* T. & A.D. Poyser, London.

GINN, H. B. & MELVILLE, D. S. (1983). *Moult in Birds.* BTO Guide no. 19, British Trust for Ornithology, Tring.

GŁOWACIŃSKI, Z. (1975). Succession of bird communities in the Niepolomice forest (southern Poland). *Ekologia Polska*, 23: 231–263.

GLUTZ VON BLOTZHEIM, U. N. (1962). *Die Brutvögel der Schweiz.* Aargauer Tagblatt AG, Aargau.

GLUTZ VON BLOTZHEIM, U. N. (1985). *Handbuch der Vögel Mitteleuropas. Bd. 10/I. Passeriformes (1. Teil).* Aula-Verlag, Wiesbaden.

GLUTZ VON BLOTZHEIM, U. N. (1993). *Handbuch der Vögel Mitteleuropas. Bd. 13 Passeriformes (4. Teil). Teil 2: Sittidae – Laniidae.* Aula-Verlag, Wiesbaden.

GOGEL, R. (1973). Trois cavités occupées dans le meme tronc. *Nos Oiseaux*, 32: 25–26.

GONZALES, R. B. & ALCALA, A. C. (1969). The foraging deployment of Velvet-fronted Nuthatches and Elegant Titmice. *Silliman Journal*, 16: 402–408.

GOOCH, S. , BAILLIE, S. R. & BIRKHEAD, T. R. (1991). Magpie *Pica pica* and songbird populations: retrospective investigation of trends in population density and breeding success. *Journal of Applied Ecology*, 28: 1068–1086.

GOODWIN, D. (1991). Great Spotted Woodpecker robbing Nuthatch. *British Birds*, 84: 196.

GOSLER, A. (1987). Sexual dimorphism in the summer bill length of the Great Tit. *Ardea*, 75: 91–98.

GRANT, P. R. (1975). The classical case of character displacement. *Evolutionary Biology*, 8: 237–337.

GREENWAY, J. C. Jr. (1967). Family Sittidae. In: *Check-list of the Birds of the World, vol. 12* (Ed. R. A. Paynter): pp. 124–149, Museum of Comparative Zoology, Cambridge, Massachusetts.

GREENWOOD, P. J. (1980). Mating systems, philopatry and dispersal in birds and mammals. *Animal Behaviour*, 28: 1140–1162.

GREENWOOD, P. J. & HARVEY, P. H. (1982). The natal and breeding dispersal of birds. *Annual Review of Ecology and Systematics*, 13: 1–21.

GREENWOOD, P. J., HARVEY, P. H. & PERRINS, C. M. (1979). The role of dispersal in the Great Tit (*Parus major*): the causes, consequences and heritability of natal dispersal. *Journal of Animal Ecology*, 48: 123–142.

GRIFFIN, D. M., HARRISON, C. J. O. & SWALES, M. K. (1984). A review of ornithological observations at Lista, south Norway. *Sterna*, 23: 3.

GRUBB, T. C. Jr. (1982). On sex-specific foraging behavior in the White-breasted Nuthatch. *Journal of Field Ornithology*, 53: 305–314.

GRUBB, T. C. Jr. & CIMPRICH, D. A. (1990). Supplementary food improves the nutritional condition of wintering woodland birds: evidence from ptilochronology. *Ornis Scandinavica*, 21: 277–281.

GRUBB, T. C. Jr. & WAITE, T. A. (1987). Caching by Red-breasted Nuthatches. *Wilson Bulletin*, 99: 696–699.

GUILLOU, J. J. (1964). Observations fait en Corse, particulièrement au Cap Corse. *Alauda*, 32: 196–225.

GÜNTERT, M. (1986). Gruppenbildung beim Zwergkleiber *Sitta pygmaea* – eine Strategie zum Überleben kalter Winter. *Der Ornithologische Beobachter*, 83: 275–280.

GÜNTERT, M., HAY, D. B. & BALDA, R. P. (1988). Communal roosting in the Pygmy Nuthatch: a winter survival strategy. *Proceedings International Ornithological Congress*, 19: 1964–1972.

HACHISUKA, M. (1930). Contributions to the birds of the Philippines No. 2. *Ornithological Society of Japan Supplementary Publications*, 14: 141–222.

HAFTORN, S. (1971). *[The Birds of Norway]*. Universitetsforlaget, Oslo.

HAFTORN, S. & SLAGSVOLD, T. (1995). Egg covering in birds: description of the behaviour in tits (*Parus* spp.) and a test of hypotheses of its function. *Fauna Norvegica Series C, Cinclus*, 18: 85–106.

HAGEMEIJER, W. J. M. & BLAIR, M. J. (1997). *The EBCC Atlas of European Breeding Birds*. T & A.D. Poyser, London.

HAILA, Y. (1983). Land birds on northern islands: a sampling metaphor for insular colonization. *Oikos*, 41: 334–351.

HAILMAN, J. P. (1979). Environmental light and conspicuous colors. In: *The Behavioral Significance of Color* (Ed. E. H. Burtt Jr.): pp. 289–354, Garland STPM Press, New York.

HANEDA, K. & ROKUGAWA, M. (1972). [Breeding biology of *Sitta europaea*]. *Bulletin of the Institute for Natural Education, Shiga Heights*, 11: 7–18.

HANEY, J. C. (1981). The distribution and life history of the Brown-headed Nuthatch (*Sitta pusilla*) in Tennessee. *The Migrant*, 52: 97–86.

HÄRDLING, R., KÄLLANDER, H. & NILSSON, J.-Å. (1995). Experimental evidence for low intra-pair cache pilfering rates in European Nuthatches. *Proceedings of the Royal Society of London B*, 260: 127–130.

HÄRDLING, R., KÄLLANDER, H. & NILSSON, J.-Å (1997). Memory for hoarded food: an aviary study of the European Nuthatch. *Condor*, 99: 526–529.

HARDY, J. W. (1965). A spectacular case of cnemidocoptiasis (scaly-leg) in the White-breasted Nuthatch. *Condor*, 67: 264–265.

HARINGTON, H. H. (1914). Notes on the nidification of some birds from Burma. *Ibis* series X, 11: 15–16.

HARMS, W. B. & OPDAM, P. (1989). Woods as habitat patches for birds: application in landscape planning in the Netherlands. In: *Changing Landscapes: an Ecological Perspective* (Ed. I. S. Zonneveld & R. T. T. Forman): pp. 73–97, Springer-Verlag, New York.

HARRAP, S. (1991). The Hainan Nuthatch. *Oriental Bird Club Bulletin*, 13: 35–36.

HARRAP, S. & QUINN, D. (1996). *Tits, Nuthatches and Treecreepers*. Christopher Helm, London.

HARRISON, C. (1982). *An Atlas of the Birds of the Western Palaearctic*. Collins, London.

HARRISON, J. M. (1955). On the occurrence of the Nuthatch, *Sitta europaea* Linnaeus, in Iraq. *Sitta europaea davidi* ssp. nov. *Bulletin of the British Ornithologists' Club*, 75: 59–60.

HARRISON, S. (1991). Local extinction in a metapopulation context – an empirical evaluation. *Biological Journal of the Linnean Society*, 42: 73–88.

HARTERT, E. (1910–1922). *Die Vögel der Paläarktischen Fauna*. Friedländer, Berlin.

HAUPT, H. (1992). Zur Brutbiologie und Ortstreue des Kleibers, *Sitta europaea*. *Der Falke*, 39: 375–381.

HAY, D. B. (1983). Physiological and behavioral ecology of communally roosting Pygmy Nuthatches (*Sitta pygmaea*). Ph.D. thesis, Northern Arizona University.

HAY, D. B. & GÜNTERT, M. (1983). Seasonal selection of tree cavities by Pygmy

Nuthatches based on cavity characteristics. *USDA Forest Service General Technical Report RM*, 99: 117–120.

HEIM DE BALSAC, H. & MAYAUD, N. (1962). *Les Oiseaux du Nord-ouest de l'Afrique.* Paul Lecheval, Paris.

HEINRICH, B., JOERG, C. C., MADDEN, S. S. & SANDERS, E. W. Jr. (1997). Black-capped Chickadees and Red-breasted Nuthatches "weigh" sunflower seeds. *Auk*, 114: 298–299.

HEINROTH, O. & HEINROTH, M. (1926). *Die Vögel Mitteleuropas.* Berlin.

HEINTZELMAN, D. S. & MacCLAY, R. (1971). An extraordinary autumn migration of White-breasted Nuthatches. *Wilson Bulletin*, 83: 129–131.

HELLEBREKERS, W. P. J. (1970). Oölogische en nidologische mededelingen in 1970 (en eerder). *Limosa*, 43: 152–155.

HELLMICH, W. (1968). *Khumbu Himal. Ergebnisse des forschungsunternehmens Nepal Himalaya. Band II.* Innsbruck.

HENDRICKS, P. (1995). Ground-caching and covering of food by a Red-Breasted Nuthatch. *Journal of Field Ornithology*, 66: 370–372.

HENZE, O. (1940). Zur Brutbiologie des Kleibers. *Beiträge zur Fortpflanzungsbiologie der Vögel*, 16: 23–27.

HERBERT, G. B., PETERLE, T. J. & GRUBB, T. C. Jr. (1989). Chronic dose effects of methyl parathion on nuthatches: cholinesterase and ptilochronology. *Bulletin of Environmental Contamination and Toxicology*, 42: 471–475.

HERRERA, C. M. (1978). Ecological correlates of residence and non-residence in a Mediterranean passerine bird community. *Journal of Animal Ecology*, 47: 871–890.

HERRERA, C. M. (1979). Ecological aspects of heterospecific flocks formation in a Mediterranean passerine bird community. *Oikos*, 33: 85–96.

HERTER, W. R. (1940). Über das "Putter" einiger Meisen-Arten. *Ornithologische Monatsberichte*, 48: 104–109.

HERTZER, H. (1972). Kleiber (*Sitta europaea*) nächtigt und brütet in Holzbeton-Nisthöhle mit nur 27 mm Fluglochdurchmesser. *Die Vogelwelt*, 93: 75.

HEYWOOD, V. H. & WATSON, R. T. (Eds.) (1995). *Global Biodiversity Assessment.* Cambridge University Press, Cambridge.

HILDÉN, O. (1977). [Occurrence of irregular migrants in Finland in 1976]. *Ornis Fennica*, 54: 170–179.

HILDÉN, O. & SAUROLA, P. (1985). [Report of the winter bird censuses in 1983/84]. *Lintumies*, 20: 218–227.

HOLMES, R. T., BONNEY, R. E. & PACALA, S. W. (1979). Guild structure of the Hubbard Brook bird community: a multivariate approach. *Ecology*, 60: 512–520.

HOOGERWERF, A. (1949). Bijdrage tot de oölogie van Java. *Limosa*, 22: 1–277.

HOPKIN, P. J. (1989). Beautiful Nuthatch (*Sitta formosa*), a species new to Thailand. *Natural History Bulletin of the Siam Society*, 37: 105–107.

HORVÁTH, L. (1961). The evolutional significance of the aberrations in the plumage of the Nuthatch (*Sitta europaea* L.). *Acta Zoologica Academiae Scientiarum Hungariae*, 7: 425–431.

HOWARD, R. & MOORE, A. (1980). *A Complete Checklist of the Birds of the World.* Oxford University Press, Oxford.

INGOLD, J. L. (1977a). Behavior of adult and juvenile White-breasted Nuthatches at the time of fledging. *Passenger Pigeon*, 39: 299–300.

INGOLD, J. L. (1977b). Territory in the White-breasted Nuthatch. M.Sc. thesis, University of Wisconsin-Milwaukee.

INGOLD, J. L. (1981). Defense of breeding territories in the White-breasted Nuthatch. *Passenger Pigeon*, 43: 41–42.

INOZEMTSEV, A. A. (1965). [The importance of highly specialized trunk-foraging birds in forest communities]. *Ornitologiya*, 7: 416–436.

INSKIPP, C. & INSKIPP, T. P. (1985). *A Guide to the Birds of Nepal*. Croom Helm, London.

ITO, I. & FUJIMAKI, Y. (1990). [Birds of parks in Obihiro city, Eastern Hokkaido]. *Japanese Journal of Ornithology*, 38: 119–129.

JACKSON, J. A. (1982). Capturing woodpecker nestlings with a noose – a technique and its limitations. *North American Bird Bander*, 7: 90–93.

JACKSON, J. A. (1983). Commensal feeding of Brown-headed Nuthatches with Red-cockaded Woodpeckers. In: *Proceedings Second Red-cockaded Woodpecker Symposium* (Ed. D. A. Wood): p. 101, State of Florida Game and Fresh Water Fish Commission, Gainesville.

JACOBS, P., MAHLER, F. & OCHANDO, B. (1978). A propos de la couleur de la calotte chez la Sittelle Kabyle (*Sitta ledanti*). *Aves*, 15: 149–153.

JAMDAR, N. (1987). An interesting feeding behaviour of the Whitecheeked Nuthatch (*Sitta leucopsis*). *Journal of the Bombay Natural History Society*, 84: 443.

JENNI, L. & WINKLER, R. (1994). *Moult and Ageing of European Passerines*. Academic Press, London.

JOHANSEN, H. (1944). Die Vogelfauna Westsibiriens. II. Teil, 1. Fortsetzung. *Journal für Ornithologie*, 92: 145–204.

JONES, P. H. (1972). Succession in breeding bird populations of sample Welsh oakwoods. *British Birds*, 65: 291–299.

JONKERS, D. A. (1990). *Monitoring-onderzoek aan Broedvogels in de Periode 1969–1985*. Rijksinstituut voor Natuurbeheer, Leersum.

JOURDAIN, F. C. R. & WITHERBY, H. C. (1918). The effect of the severe winter of 1916/1917 on British birds. *British Birds*, 12: 28.

KACZMAREK, W., SIERAKOWSKI, K. & WASILEWSKI, A. (1981). Food preference of insectivorous birds in forest ecosystems of the Kampinos national park. *Ekologia Polska*, 29: 499–518.

KÄLLANDER, H. (1983). [Great Spotted Woodpecker *Dendrocopos major* robbing Nuthatch *Sitta europaea* of hazelnuts]. *Anser*, 22: 241–242.

KÄLLANDER, H. (1993). Food caching in the European Nuthatch *Sitta europaea*. *Ornis Svecica*, 3: 49–58.

KÄLLANDER, H. (in press). The Nuthatch population of Dalby Söderskog during 15 years: trend and fluctuations. *Ornis Svecica*.

KÄLLANDER, H. & KARLSSON, J. (1981). Population fluctuations of some North European bird species in relation to winter temperatures. *Proceedings Second Nordic Congress of Ornithology*: 111–117.

KÄLLANDER, H., NILSSON, S. G. & SVENSSON, S. (1978). The Fieldfare *Turdus pilaris*, Pine Grosbeak *Pinicola enucleator*, and Nuthatch *Sitta europaea* in the winter 1976/77 – report from the Swedish winter bird census. *Vår Fagelvärld*, 37: 37–46.

KARR, J. R, NICHOLS, J. D., KLIMKIEWICZ, M. K. & BRAWN, J. D. (1990). Survival rates of birds of tropical and temperate forests: will the dogma survive? *American Naturalist*, 136: 277–291.

KASPAREK, M. (1988). *Der Bafasee: Natur und Geschichte in der Türkische Ägäis*. Max Kasparek Verlag, Heidelberg.

KENDEIGH, S. C. (1944). Measurement of bird populations. *Ecological Monographs*, 14: 67–106.

KENNEDY, C. & SOUTHWOOD, T. R. E. (1984). The number of species of insects associated with British trees: re-analysis. *Journal of Animal Ecology*, 53: 455–478.

KENWARD, R.E. (1990). *Ranges IV. Software for Analysing Animal Location Data*. Institute of Terrestrial Ecology, Wareham, UK.

KHARITONOV, I. A. (1983). Effect of habitat factors on the distribution of the Amur Nuthatch and the Marsh Tit in forests of the central Sikhote-Alin. *Soviet Journal of Ecology*, 13: 188–195.

KILHAM, L. (1968). Reproductive behavior of White-breasted Nuthatches. I. Distraction display, bill-sweeping, and nest hole defense. *Auk*, 85: 477–492.

KILHAM, L. (1971a). Roosting habits of White-breasted Nuthatches. *Condor*, 73: 113–114.

KILHAM, L. (1971b). Use of blister beetle in bill-sweeping by White-breasted Nuthatch. *Auk*, 81: 175–176.

KILHAM, L. (1972a). Reproductive behavior of White-breasted Nuthatches. II. Courtship. *Auk*, 89: 115–129.

KILHAM, L. (1972b). Death of Red-breasted Nuthatch from pitch around nest hole. *Auk*, 89: 451–452.

KILHAM, L. (1973). Reproductive behavior of the Red-breasted Nuthatch. I. Courtship. *Auk*, 90: 597–609.

KILHAM, L. (1975a). Association of Red-breasted Nuthatches with chickadees in a hemlock cone year. *Auk*, 92: 162–164.

KILHAM, L. (1975b). Breeding of Red-breasted Nuthatches in captivity. *Avicultural Magazine*, 81: 144–147.

KILHAM, L. (1981). Agonistic behavior of the White-breasted Nuthatch. *Wilson Bulletin*, 93: 271–274.

KIPP, F. A. (1965). Beobachtungen an dem Felsenkleiber *Sitta neumayer*. *Die Vogelwarte*, 23: 19–24.

KIZIROGLU, I. (1984). Untersuchungen über die Brutbiologie der höhlenbrütenden Vogelarten im Ebersberger Forst. *Anzeiger der Ornithologische Gesellschaft Bayern*, 23: 205–214.

KLEINSCHMIDT, O. (1928). Die Realgattung Kleiber, *Sitta Auto-Sitta* (Kl.). In: *Berajah, Zoographia infinita*. Kommissionsverlag von Gebauer-Schwetske Druckerei und Verlag A. G., Halle an der Saale.

KLEINSCHMIDT, O. (1933). Fremde Formenkreise des Namenkreises Spechtmeise (*Sitta*). In: *Berajah, Zoographia infinita*. Kommissionsverlag von Gebauer-Schwetske Druckerei und Verlag A. G., Halle an der Saale.

KLEJNOTOWSKI, Z. (1967). [Materials for recognizing nourishment composition of Nuthatch (*Sitta europaea* L.) nestlings]. *Roczniki Wyzszej Szkoly Rolniczej w Poznaniu*, 38: 105–107.

KLINTEROTH, L. (1978). [Differences between the nuthatches *Sitta europaea europaea* and *Sitta e. asiatica*]. *Vår Fagelvärld*, 37: 313–318.

KNEIS, P. & GÖRNER, M. (1986). Kleinvögel nutzen Asungsstellen des Schalenwildes unter dem Schnee. *Acta Ornithoecologica*, 1: 195–196.

KNORR, O. A. (1957). Communal roosting of the Pygmy Nuthatch. *Condor*, 59: 398.

KOHLER, H. (1987). Weniger Kleiberbruten. *Vögel der Heimat*, 57: 125.

KÖNIG, C. (1968). Kleiber (*Sitta europaea*) mauert Rauhfußkauz (*Aegolius funereus*) in Bruthöhle ein. *Ornithologische Mitteilungen*, 20: 110.

KÖNIG, C., KAISER, H. & MÖRIKE, D. (1995). Zur Ökologie und Bestandsentwicklung des Sperlingskauzes (*Glaucidium passerinum*) im Schwarzwald. *Jahreshefte der Gesellschafts für Naturkunde in Württemberg*, 151: 457–500.

KOOIKER, G. (1994). Influence of Magpie *Pica pica* on urban bird populations in the city of Osnabrück, northwest Germany. *Die Vogelwelt*, 115: 39–44.

KOZLOVA, E. V. (1933). The birds of South-West Transbaikalia, Northern Mongolia, and Central Gobi – Part V. *Ibis*, 75: 301–332.

KRIŠTÍN, A. (1990). Zur Kenntnis der Nahrung und Nahrungskonkurrenz des

Kleibers *Sitta europaea* und Waldbaumläufers (*Certhia familiaris*). *Beiträge zur Vogelkunde*, 36: 257–268.

KRIŠTÍN, A. (1992a). Die Nestlingsnahrung des Kleibers (*Sitta europaea* L.) in Buchenwäldern. *Acta Ornithoecologica*, 7: 341–349.

KRIŠTÍN, A. (1992b). Trophische Beziehungen zwischen Singvögeln und Wirbellosen im Eichen-Buchenwald zur Brutzeit. *Der Ornithologische Beobachter*, 89: 157–169.

KRIŠTÍN, A. (1994). Food variability of nuthatch nestlings (*Sitta europaea*) in mixed beech forests: where are limits of its polyphagy? *Biologia Bratislava*, 49: 773–779.

KRÜPER, T. (1875). Beitrag zur Ornithologie Kleinasiens. *Journal für Ornithologie*, 23: 258–285.

KUMERLOEVE, H. (1958). Sur la présence en Asie Mineure de la Sittelle naine de Krüper (*Sitta canadensis krüperi* Pelzeln). *Alauda*, 26: 81–85.

KUMERLOEVE, H. (1961). Zur Kenntnis der Avifauna Kleinasiens. *Bonnerische Zoologische Beiträge* 12, Supplement: 76–79.

LACK, D. (1971). *Ecological Isolation in Birds*. Blackwell, Oxford.

LACK, D. & LACK, E. (1953). Visible migration through the Pyrenees: an autumnal reconnaissance. *Ibis*, 95: 271–309.

LADYGIN, A. V. (1991). [Wintering birds in Lake Kuriliskoe Basin, South Kamchatka, and their relationships with salmon spawning]. *Byulleten' Moskovskogo Obshchestva Ispytatelei Prirody Otdel Biologicheskii*, 96: 17–22.

LAMBRECHTS, M. & DHONDT, A. A. (1988). The anti-exhaustion hypothesis: a new hypothesis to explain song performance and song switching in the Great Tit. *Animal Behaviour*, 36: 327–334.

LARSON, D. L. & BOCK, C. E. (1986). Eruptions of some North American boreal seed-eating birds, 1901–1980. *Ibis*, 128: 137–140.

LARSON, J. H. (1979). A winter territory study of the White-breasted Nuthatch in central Minnesota. *The Loon*, 51: 79–81.

LÁSZLÓ, V. (1988). The study of bird species foraging on the bark. *Aquila*, 95: 83–93.

LAW, J. E. (1929). Down-tree progress of *Sitta pygmaea*. *Condor*, 31: 45–51.

LEBRETON, P., BROYER, J. & CHOISY, J.-P. (1991). Relations entre activités humaines et faune sauvage en milieu forestier: impact du surpaturage en chênaie de Dombes. *Bièvre*, 12: 57–68.

LEDANT, J. P. (1978). Données comparées sur la Sittelle Corse (*Sitta whiteheadi*) et sur la Sittelle Kabyle (*Sitta ledanti*). *Aves*, 15: 154–157.

LEDANT, J. P. & JACOBS, P. (1977). La Sittelle Kabyle (*Sitta ledanti*): données nouvelles sur sa biologie. *Aves*, 14: 233–242.

LEDANT, J., JACOBS, P., OCHANDO, B. & RENAULT, J. (1985). Dynamique de la Forêt du Mont Babor et préférences écologiques de la Sittelle Kabyle *Sitta ledanti*. *Biological Conservation*, 32: 231–254.

LEGGE, V. (1983). *A History of the Birds of Ceylon*, 2nd edn., Vol. III. Tisara Prakasakayo Ltd, Dehiwala, Sri Lanka.

LEHIKOINEN, E. (1987). Seasonality of the daily weight cycle in wintering passerines and its consequences. *Ornis Scandinavica*, 18: 216–226.

LEONOVICH, V. V. & VEPRINTSEV, B. N. (1986). [New faunistic records in Sakhalin]. *Ornitologiya*, 21: 137.

LEONOVICH, V. V., DEMINA, G. V. & VEPRINTSEVA, O. D. (1996a). [The Nuthatch (*Sitta europaea* L.) and the "small" nuthatches (*S. villosa* Verreaux, *S. krueperi* Pelzeln): problems of taxonomy and. phylogeny. *Byulleten' Moskovskogo Obshchestva Ispytatelei Prirody Otdel Biologicheskii*, 101: 37–49.

LEONOVICH, V. V., DEMINA, G. V. & VEPRINTSEVA, O. D. (1996b). [On the taxonomy and phylogeny of the rock nuthatches: *Sitta neumayer rupicola* Blanford, *Sitta (tephronota) obscura* Zarudny and Loudon, *Sitta tephronota iranica* Buturlin and *Sitta tephronota tephronota* Sharpe. *Byulleten' Moskovskogo Obshchestva Ispytatelei Prirody Otdel Biologicheskii*, 101: 29-38.

LESSELLS, C. M., MATEMAN, A. C. & VISSER, J. (1996). Great Tit hatchling sex ratios. *Journal of Avian Biology*, 27: 135–142.

LI, Q. & BIAN, X. (1988). Studies on the karyotypes of birds. II. The 19 species of 12 families of passerine birds (Passeriformes, Aves). *Zoological Research*, 9: 321–326.

LIENHARDT, P. (1987). Überraschungen bei der Nistkastenkontrolle. *Vögel der Heimat*, 57: 78.

LIKNES, E. T. (1994). Seasonal variation in cold tolerance and maximal capacity for thermogenesis in White-breasted Nuthatches (Passeriformes: *Sitta carolinensis*) and Downy Woodpeckers (Passeriformes: *Picoides pubescens*). M.Sc. thesis, University of South Dakota.

LIKNES, E. T. & SWANSON, D.L. (1996). Seasonal variation in cold tolerance, basal metabolic rate, and maximal capacity for thermogenesis in White-breasted Nuthatches *Sitta carolinensis* and Downy Woodpeckers *Picoides pubescens*, two unrelated arboreal temperate residents. *Journal of Avian Biology*, 72: 279–288.

LINNAEUS, C. (1758). *Systema Naturae. Vol. I, 10th revised edn.* Facsimile reproduction, British Museum, London, 1956.

LIVESEY, T. R. (1933). Nidification of the Giant Nuthatch (*Sitta magna* Wardl. – Ramsey). *Journal of the Bombay Natural History Society*, 36: 1001–1002.

LÖHRL, H. (1956). Der Star als Bruthöhlenkonkurrent. *Die Vogelwelt*, 77: 47–50.

LÖHRL, H. (1958). Das Verhalten des Kleibers (*Sitta europaea caesia* Wolf). *Zeitschrift für Tierpsychologie*, 15: 191–252.

LÖHRL, H. (1959). Beitrag zur Avifauna Korsikas. *Journal für Ornithologie*, 100: 79–83.

LÖHRL, H. (1960). Vergleichende Studien über Brutbiologie und Verhalten der Kleiber *Sitta whiteheadi* Sharpe und *Sitta canadensis* L. *Journal für Ornithologie*, 101: 245–264.

LÖHRL, H. (1961). Vergleichende Studien über Brutbiologie und Verhalten der Kleiber *Sitta whiteheadi* Sharpe und *Sitta canadensis* L. II. *Sitta canadensis*, verglichen mit *Sitta whiteheadi*. *Journal für Ornithologie*, 102: 111–132.

LÖHRL, H. (1962). Artkennzeichen von *Sitta krüperi*. *Journal für Ornithologie*, 103: 418–419.

LÖHRL, H. (1964). Verhaltensmerkmale der Gattungen *Parus* (Meisen), *Aegithalos* (Schwanzmeisen), *Sitta* (Kleiber), *Tichodroma* (Mauerläufer) und *Certhia* (Baumläufer). *Journal für Ornithologie*, 105: 153–181.

LÖHRL, H. (1965a). Zur Vogelwelt der griechischen Insel Lesbos (Mytilene). *Die Vogelwelt*, 86: 105–112.

LÖHRL, H. (1965b). Dauernesten beim Felsenkleiber, *Sitta neumayer*. *Journal für Ornithologie*, 106: 459.

LÖHRL, H. (1966). Einige Zahlen zur Brutbiologie des Kleibers (*Sitta europaea*). *Anzeiger der Ornithologische Gesellschaft Bayern*, 7: 717–722.

LÖHRL, H. (1967). *Die Kleiber Europas: Kleiber, Felsenkleiber, Korsischer Kleiber (2nd edn.).* Die Neue Brehm Bücherei 196, A. Ziemsen Verlag, Wittenberg Lutherstadt.

LÖHRL, H. (1977). Nistökologische und ethologische Anpassungserscheinungen bei Höhlenbrütern. *Die Vogelwarte*, 29: 92–101.

LÖHRL, H. (1978). Das "Nachstürzen" – eine reflexartige Reaktion, entfallene Beute wieder zu erlangen. *Journal für Ornithologie*, 119: 325–329.

LÖHRL, H. (1982a). Zur Vogelwelt im Zedernwaldgebiet des Marokkanischen Rifs. *Die Vogelwelt*, 103: 68–71.

LÖHRL, H. (1982b). Das ausmeißeln der Bruthöhle durch einheimische Meisen (*Parus* spp.) und den Kleiber (*Sitta europaea*). *Die Vogelwelt*, 103: 121–129.

LÖHRL, H. (1987a). Versuche zur Wahl der Bruthöhle und Nisthöhe am Baum durch den Kleiber (*Sitta europaea*). *Ökologie der Vögel*, 9: 65–68.

LÖHRL, H. (1987b). Einfluß der Tiefe der Bruthöhle auf die Eizahl beim Kleiber (*Sitta europaea*). *Ökologie der Vögel*, 9: 69–70.

LÖHRL, H. (1987c). Der Bruterfolg des Kleibers (*Sitta europaea*) in Beziehung zu Brutraumgröße und Habitat. *Ökologie der Vögel*, 9: 53–63.

LÖHRL, H. (1988). Etho-ökologische Untersuchungen an verschiedenen Kleiberarten (Sittidae). Eine vergleichende Zusammenstellung. *Bonner Zoologische Monographien*, 26: 1–208.

LÖHRL, H. & THIELCKE, G. (1969). Zur Brutbiologie, Ökologie und Systematik einiger Waldvögel Afghanistans. *Bonner Zoologische Beiträge*, 20: 85–98.

LONG, C. A. (1982). Comparison of the nest-site distraction displays of Black-capped Chickadee and White-breasted Nuthatch. *Wilson Bulletin*, 94: 216–218.

LØPPENTHIN, B. (1932). Die Farbenvariation der Europäischen Baumkleiber mit besonderer Berücksichtigung der Skandinavischen Populationen und einigen Bemerkungen über ihre Verbreitung. *Vidensk. Medd. fra Dansk naturh. Foren.*, 94: 10–187.

LOSKOT, W. M., SOKOLOW, E. P. & WUNDERLICH, K. (1991). *Sitta tephronota* Sharpe–Klippenkleiber. In: *Atlas der Verbreitung Paläarktischer Vögel. 17. Lieferung.* (Ed. H. Dathe & W. M. Loskot), Academie-Verlag, Berlin.

LUDLOW, F. (1944). The birds of south-eastern Tibet. *Ibis*, 86: 43–86.

LUDLOW, F. (1951). The birds of Kongbo and Pome, South-east Tibet. *Ibis*, 93: 547–578.

LUDWIG, H. (1978). Kleiber vertreibt Buntspecht aus seiner Bruthöhle. *Der Falke*, 25: 354.

MacARTHUR, R. (1958). Population ecology of some warblers of northeastern coniferous forests. *Ecology*, 39: 599–619.

MacDONALD, D. W. & HENDERSON, D. G. (1977). Aspects of the behaviour and ecology of mixed-species bird flocks in Kashmir. *Ibis*, 119: 481–491.

MacKINNON, J. & PHILLIPS, K. (1993). *A Field Guide to the Birds of Borneo, Sumatra, Java and Bali.* Oxford University Press, Oxford.

MAKATSCH, W. (1976). *Die Eier der Vögel Europas, Band 2.* Neumann Verlag, Leipzig.

MANOLIS, T. (1977). Foraging relationships of Mountain Chickadees and Pygmy Nuthatches. *Western Birds*, 8: 13–20.

MARCHANT, J. H., HUDSON, R., CARTER, S. P. & WHITTINGTON, P. (1990). *Population Trends in British Breeding Birds.* British Trust for Ornithology, Tring.

MARÉCHAL, P. (1992). De Boomklever *Sitta europaea* mag niet aan onze aandacht ontsnappen. *Het Vogeljaar*, 40: 64–71.

MARSHALL, J. (1996). Backyard Nuthatch Active. In: *Switched-on Gutenberg: a Global Poetry Journal* (Ed. J. Harris), Vol. 1 (2). Published at http://weber.u.washington.edu/~jnh/vol1no2/mars.html (last accessed 30 January 1998).

MARTIN, T. E. (1993). Evolutionary determinants of clutch size in cavity-nesting birds: nest predation or limited breeding opportunities? *American Naturalist*, 142: 937–946.

MARTIN, T. E. (1995). Avian life history evolution in relation to nest sites, nest predation, and food. *Ecological Monographs*, 65: 101-127.

MARTIN, T. E. & LI, P. (1992). Life-history traits of open- vs. cavity-nesting birds. *Ecology*, 73: 579–592.

MATTHYSEN, E. (1986a). Postnuptial moult in a Belgian population of Nuthatches *Sitta europaea. Bird Study*, 33: 206–213.

MATTHYSEN, E. (1986b). Some observations on sex-specific territoriality in the Nuthatch. *Ardea*, 74: 177–183.

MATTHYSEN, E. (1987). Territory establishment of juvenile Nuthatches *Sitta europaea* after fledging. *Ardea*, 75: 53–57.

MATTHYSEN, E. (1988). Populatiedynamiek, sociale organisatie en habitat-kwaliteit bij de Boomklever *Sitta europaea* L. Ph.D. thesis, University of Antwerp.

MATTHYSEN, E. (1989a). Seasonal variation in bill morphology of Nuthatches *Sitta europaea*: dietary adaptations or consequences? *Ardea*, 77: 117–125.

MATTHYSEN, E. (1989b). Fledging dates of Nuthatches *Sitta europaea* in relation to age, territory and individual variation. *Bird Study*, 36: 134–140.

MATTHYSEN, E. (1989c). Territorial and non-territorial settling in juvenile Eurasian Nuthatches (*Sitta europaea* L.) in summer. *Auk*, 106: 560–567.

MATTHYSEN, E. (1989d). Nuthatch *Sitta europaea* demography, beech mast, and territoriality. *Ornis Scandinavica*, 20: 278–282.

MATTHYSEN, E. (1990a). Upward and downward movements by bark-foraging birds: the importance of habitat structure. *Ibis*, 132: 128–129.

MATTHYSEN, E. (1990b). Behavioral and ecological correlates of territory quality in the Eurasian Nuthatch (*Sitta europaea*). *Auk*, 107: 86–95.

MATTHYSEN, E. (1993). Nonbreeding social organization in migratory and resident birds. In: *Current Ornithology Vol. 11* (Ed. D. M. Power): pp. 93–141, Plenum Press, New York.

MATTHYSEN, E. (1994). Dispersal as the key process in fragmented landscapes. In: *Bird Numbers 1992. Distribution, Monitoring and Ecological Aspects. Proceedings of the 12th International Conference of IBCC and EOAC* (Ed. E.J.M. Hagemeijer & T.J. Verstrael): pp. 111–116, Statistics Netherlands, Voorburg/Heerlen & SOVON, Beek-Ubbergen.

MATTHYSEN, E. (1997). Geographical variation in the occurrence of song types in Nuthatch *Sitta europaea* populations. *Ibis*, 139: 102–106.

MATTHYSEN, E. & ADRIAENSEN, F. (1989a). Observations on the foraging behaviour of the Corsican Nuthatch *Sitta whiteheadi* in winter. *Sitta*, 3: 21–25.

MATTHYSEN, E. & ADRIAENSEN, F. (1989b). Directional dispersal by juveniles in a resident population of Nuthatches. *Ringing and Migration*, 10: 119–123.

MATTHYSEN, E. & ADRIAENSEN, F. (1989c). Notes on winter territoriality and social behaviour in the Corsican Nuthatch *Sitta whiteheadi* Sharpe. *Alauda*, 57: 155–168.

MATTHYSEN, E. & CURRIE, D. (1996). Habitat fragmentation reduces disperser success in juvenile Nuthatches *Sitta europaea*: evidence from patterns of territory establishment. *Ecography*, 19: 67–72.

MATTHYSEN, E. & DHONDT, A. A. (1983). Die Ansiedlung junger Kleiber *Sitta europaea* im Spätsommer und Herbst. *Journal für Ornithologie*, 124: 281–290.

MATTHYSEN, E. & DHONDT, A. A. (1988). Behaviour of birds after release from capture as an indicator of social status. *Ibis*, 130: 69–72.

MATTHYSEN, E. & SCHMIDT, K.-H. (1987). Natal dispersal in the Nuthatch. *Ornis Scandinavica*, 18: 313–316.

MATTHYSEN, E., GAUNT, S. L. L. & McCALLUM, D. A. (1991). A note on the vocalizations of the Chinese Nuthatch. *Wilson Bulletin*, 103: 706–710.

MATTHYSEN, E., CIMPRICH, D. & GRUBB, T. C. Jr. (1992). Is social organization in winter determined by short- or long-term benefits? A case study on migrant Red-breasted Nuthatches *Sitta canadensis*. *Ornis Scandinavica*, 23: 43–48.

MATTHYSEN, E., ADRIAENSEN, F. & DHONDT, A. A. (1995). Dispersal distances of Nuthatches, *Sitta europaea*, in a highly fragmented forest habitat. *Oikos*, 72: 375–381.

MAUERSBERGER, G. (1989). Zur Ernährungsweise des Chinesenkleibers, *Sitta villosa* Verreaux. *Acta Ornithoecologica*, 2: 79–86.

MAYR, E. & AMADON, D. (1951). A classification of recent birds. *American Museum Novitates*, 1496: 1–42.

MAYR, E. & SHORT, L. L. (1970). *Species Taxa of North American Birds. A Contribution to Comparative Systematics*. Nuttall Ornithological Club, Cambridge, Massachusetts.

McELLIN, S. M. (1979). Population demographies, spacing, and foraging behaviours of White-breasted and Pygmy Nuthatches in Ponderosa Pine habitat. In: *The Role of Insectivorous Birds in Forest Ecosystems* (Ed. J. G. Dickson *et al.*), pp. 301–330, Academic Press, London.

McGREGOR, R. C. (1920). Some features of the Philippine Ornis. *Philippine Journal of Science*, 16: 361–437.

McNAIR, D. B. (1983). Brown-headed Nuthatches store pine seeds. *Chat*, 47: 47–48.

McNAIR, D. B. (1984). Clutch-size and nest placement in the Brown-headed Nuthatch. *Wilson Bulletin*, 96: 296–301.

MEAD, C. J. & CLARK, J. A. (1991). Report on bird ringing for Britain and Ireland for 1990. *Ringing and Migration*, 12: 139–176.

MEADEN, F. (1970). Notes on breeding the Common Whitethroat, Nuthatch, Willow Warbler and Waxwing. *Avicultural Magazine*, 76: 9–15.

MELCHIOR, E., MENTGEN, E., PELTZER, R., SCHMITT, R. & WEISS, J. (1987). *Atlas der Brutvögel Luxemburgs*. Letzebuerger Natur- a Vulleschutzliga, Luxemburg.

MENGEL, R. A. (1965). *The Birds of Kentucky*. Ornithological Monographs No. 3, American Ornithologists' Union.

MENSCHAERT, L. (1991). *Zijn er Nog Vogels: Waarnemingen van Vogels in het Zuidwesten van Oost-Vlaanderen*. Wielewaal Schelde-Leie.

MERKLE, M. S. & BARCLAY, R. M. R. (1996). Body mass variation in breeding Mountain Bluebirds *Sialia currucoides*: evidence of stress or adaptation for flight? *Journal of Animal Ecology*, 65: 401–413.

MITCHELL, T. L. (1993). Tool use by a White-breasted Nuthatch. *Bulletin of the Oklahoma Ornithological Society*, 26: 6–7.

MÖCKEL, R. (1992). Häufigkeitsveränderungen höhlenbrütender Singvögel des Fichtenwaldes während des "Waldsterbens" im Westerzgebirge. *Zoologisches Jahrbuch für Systematik*, 119: 437–493.

MÖLLERSTEN, B. (1985). Nuthatches, *Sitta europaea*, catching flying insects. *Vår Fågelvärld*, 44: 226.

MONTIER, D. J. (1977). *Atlas of Breeding Birds of the London Area*. Batsford, London.

MOORE, H. J. & BOSWELL, C. (1956). Field observations on the birds of Iraq, Part II. Pteroclidae – Timaliidae. *Iraq Natural History Museum Publications*, 10: 111–213.

MOORE, N. W. & HOOPER, M. D. (1975). On the number of bird species in British woods. *Biological Conservation*, 8: 239–250.

MORENO, E. & CARRASCAL, L.M. (1995). Hoarding Nuthatches spend more time hiding a husked seed than an unhusked seed. *Ardea*, 83: 391–395.

MORENO, J. (1981). Feeding niches of woodland birds in a montane coniferous forest in central Spain during winter. *Ornis Scandinavica*, 12: 148–159.

MORENO, J., LUNDBERG, A. & CARLSON, A. (1981). Hoarding of individual Nuthatches *Sitta europaea* and Marsh Tits *Parus palustris*. *Holarctic Ecology*, 4: 263–269.

MÖRIKE, K. (1964). Vergleich der Stimmen der beiden Felsenkleiber. *Jahreshefte der Vereins Vaterländische Naturkunde Württembergs*, 119: 397–398.

MORRISON, M. L., TIMOSSI, I. C., WITH, K. A. & MANLEY, P. N. (1985). Use of tree species by forest birds during winter and summer. *Journal of Wildlife Management*, 49: 1098–1102.

MORRISON, M. L., WITH, K. A., TIMOSSI, I. C., BLOCK, W. M. & MILNE, K. A. (1987). Foraging behavior of bark-foraging birds in the Sierra Nevada. *Condor*, 89: 201–204.

MORSE, D. H. (1967). Foraging relationships of Brown-headed Nuthatches and Pine Warblers. *Ecology*, 48: 94–103.

MORSE, D. H. (1968). The use of tools by Brown-headed Nuthatches. *Wilson Bulletin*, 80: 220–224.

MORSE, D. H. (1970). Ecological aspects of some mixed-species foraging flocks of birds. *Ecological Monographs*, 40: 119–168.

MORSE, D. H. (1978). Structure and foraging patterns of flocks of tits and associated species in an English woodland during the winter. *Ibis*, 120: 298–312.

MOSKÁT, C. (1987). Estimating bird densities during the breeding season in Hungarian deciduous forests. *Acta Regalia Societatis Scientiae Litterarum Gothoburgensis Zoologica*, 14: 153–161.

MOSKÁT, C. & FUISZ, T. (1994). Forest management and bird communities in the Beech and Oak forests of the Hungarian mountains. In: *Bird Numbers 1992. Distribution, Monitoring and Ecological Aspects. Proceedings of the 12th International Conference of IBCC and EOAC* (Ed. E.J.M. Hagemeijer & T.J. Verstrael): pp. 29–38, Statistics Netherlands, Voorburg/Heerlen & SOVON, Beek-Ubbergen,.

MUGAAS, J. N. & TEMPLETON, J. R. (1970). Thermoregulation in the Red breasted Nuthatch (*Sitta canadensis*). *Condor*, 72: 125–132.

MURIN, B., KRISTIN, A., DAROLOVA, A., DANKO, S. & KROPIL, R. (1994). [Breeding bird population sizes in Slovakia]. *Sylvia*, 30: 97–105.

MURRAY, B. G. Jr. (1966). Migration of age and sex classes of passerines on the Atlantic coast in autumn. *Auk*, 83: 352–360.

MURRAY, R. D. (1991). The first successful breeding of Nuthatch in Scotland. *Scottish Bird Report (1989)*, 22: 51–55.

MYLNE, C. K. (1959). Birds drinking the sap of a birch tree. *British Birds*, 52: 426–427.

NAKAMURA, H. (1986). Ecological studies of the beech forest bird communities on Kayanodaira Heights. *Bulletin of the Institute for Natural Education, Shiga Heights*, 23: 9–20.

NAKAMURA, T. (1976). Ecological grade of birds' community from coniferous to deciduous woods. III. Annual change 1973 to 1976. *Bulletin of the Institute for Natural Education, Shiga Heights*, 15: 31–42.

NESBITT, S. A. & HETRICK, W. M. (1976). Foods of the Pine Warbler and Brown-headed Nuthatch. *Florida Field Naturalist*, 4: 28–33.

NEUFELDT, I. A. & WUNDERLICH, K. (1984). *Sitta krüperi* Pelzeln – Türkenkleiber, Rotbrustkleiber. In: *Atlas der Verbreitung Palaearktischer Vögel, 12. Lieferung* (Ed. H. Dathe & I. A. Neufeldt), Academie-Verlag, Berlin.

NEUMANN, J. (1961). Kleiberbrut in Mauer. *Der Ornithologische Beobachter*, 58: 143.

NICOLAI, V. (1986). The bark of trees: thermal properties, microclimate and fauna. *Oecologia*, 69: 148–160.

NILSSON, J.-Å. (1994). Energetic stress and the degree of fluctuating asymmetry: implications for a long-lasting, honest signal. *Evolutionary Ecology*, 8: 248–255.

NILSSON, J.-Å., KÄLLANDER, H. & PERSSON, O. (1993). A prudent hoarder:

effects of long-term hoarding in the European Nuthatch, *Sitta europaea. Behavioral Ecology*, 4: 369–373.

NILSSON, S. G. (1976). Habitat, territory size, and reproductive success in the Nuthatch *Sitta europaea. Ornis Scandinavica*, 7: 179–184.

NILSSON, S. G. (1979). Seed density, cover, predation and the distribution of birds in a beech wood in southern Sweden. *Ibis*, 121: 177–185.

NILSSON, S. G. (1982). Seasonal variation in the survival rate of adult Nuthatches *Sitta europaea* in Sweden. *Ibis*, 124: 96–100.

NILSSON, S. G. (1984). The evolution of nest-site selection among hole-nesting birds : the importance of nest predation and competition. *Ornis Scandinavica*, 15: 167–175.

NILSSON, S. G. (1987). Limitation and regulation of population density in the Nuthatch *Sitta europaea* (Aves) breeding in natural cavities. *Journal of Animal Ecology*, 56: 921–938.

NIXON, J. & NIXON, S. (1985). Nuthatch, *Sitta europaea. Peregrine*, 5: 167.

NORRIS, R. A. (1958). Comparative biosystematics and life history of the nuthatches *Sitta pygmaea* and *Sitta pusilla. University of California Publications in Zoology*, 56: 119–300.

NOSKE, R. A. (1985). Huddle-roosting behaviour of the Varied Sittella *Daphoenositta chrysoptera* in relation to social status. *Emu*, 85: 188–194.

NOSKE, R. A. (1986). Intersexual niche segregation among three bark-foraging birds of eucalypt forests. *Australian Journal of Ecology*, 11: 255–267.

NOTHDURFT, W. (1978). Ungewöhnliche Verharzungen bei einem Kleiber *Sitta europaea. Anzeiger der Ornithologische Gesellschaft Bayern*, 17: 183.

NOUR, N. (1997). Ecological and behavioural effects of habitat fragmentation on forest birds. Ph.D. thesis, University of Antwerp.

NOWAK, E. (1965). Vögel mit mißgebildeten Schnäbeln. *Der Falke*, 12: 122–130.

OBERHOLSER, H. C. (1974). *The Bird Life of Texas, Vol. 2.* University of Texas Press, Austin.

OBESO, J. R. (1987). Comunidades de passeriformes en bosques mixtos de altitudes medias de la Sierra de Cazorla. *Ardeola*, 34: 37–59.

OBESO, J. R. (1988). Alimentacion de *Sitta europaea* en pinares de la Sierra de Cazorla, SE Espana, durante el verano y el otoño. *Ardeola*, 35: 45–50.

O'HALLORAN, K. A. & CONNER, R. N. (1987). Habitat used by Brown-headed Nuthatches. *Bulletin of the Texas Ornithological Society*, 20: 7–13.

OLSSON, V. (1957). [Unusual nesting-places for Wheatear (*Oenanthe oenanthe*), Jay (*Garrulus glandarius*), and Nuthatch (*Sitta europaea*)]. *Vår Fagelvärld*, 16: 43–48.

OPDAM, P. (1991). Metapopulation theory and habitat fragmentation: a review of holarctic breeding bird studies. *Landscape Ecology*, 5: 93–106.

OPDAM, P., VAN DORP, D. & TER BRAAK, C. J. F. (1984). The effect of isolation on the number of woodland birds in small woods in the Netherlands. *Journal of Biogeography*, 11: 473–478.

OPDAM, P., RIJSDIJK, G. & HUSTINGS, F. (1985). Bird communities in small woods in an agricultural landscape: effects of area and isolation. *Biological Conservation*, 34: 333–352.

OPDAM, P., VAN APELDOORN, R., SCHOTMAN, A. & KALKHOVEN, J. (1993). Population responses to landscape fragmentation. In: *Landscape Ecology of a Stressed Environment* (Ed. C. C. Vos & P. Opdam): pp. 147–171, Chapman & Hall, London.

OSBORNE, P. (1982). Some effects of Dutch elm disease on nesting farmland birds. *Bird Study*, 29: 2–16.

PALUDAN, K. (1959). On the birds of Afghanistan. *Vidensk. Medd. Dansk. Naturh. For.*, 122: 249–252.

PANOV, E. (1989). [*Natural Hybridization and Ethological Isolation in Birds*]. Academy of Sciences of the USSR, Moscow.

PARSLOW, J. (1973). *Breeding Birds of Britain and Ireland: a Historical Survey.* T. & A.D. Poyser, Berkhamsted.

PARTRIDGE, L. & ASHCROFT, R. (1976). Mixed-species flocks of birds in Hill Forest in Ceylon. *Condor*, 78: 449–453.

PASQUET, E. (1998). Phylogeny of the nuthatches of the *Sitta canadensis* group and its evolutionary and biogeographical implications. *Ibis*, 140: 150–156.

PAYN, W. A. (1927). Some notes on the birds of Corsica. *Ibis*, 69: 74–81.

PAYN, W. A. (1931). Further notes on the birds of Corsica. *Ibis*, 73: 14–18.

PAZ, U. (1987). *The Birds of Israel.* Christopher Helm, London.

PETERS, W. D. & GRUBB, T. C. (1983). An experimental analysis of sex-specific foraging in the Downy Woodpecker, *Picoides pubescens*. *Ecology*, 64: 1437–1443.

PETIT, D. R., PETIT, L. J. & PETIT, K. E. (1989). Winter caching ecology of deciduous woodland birds and adaptations for protection of stored food. *Condor*, 91: 766–776.

PEUS, F. (1954). Zur Kenntnis der Brutvögel Griechenlands. *Bonner Zoologische Beiträge*, Sonderheft I: 1–50.

PFEIFER, S. (1955). Ergebniße zweier Versuche zur Steigerung der Siedlungsdichte der Vögel auf forstlicher Kleinfläche und benachbarter Großfläche. *Waldhygiene*, 1: 76–78.

PFEIFER, S. & KEIL, W. (1959). Siebenjährige Untersuchunge zur Ernährungsbiologie nestjunger Singvögel. *Luscinia*, 32: 13–18.

PHILLIPS, A. R. (1986). *The Known Birds of North and Middle America. Part I. Hirundidae to Mimidae; Certhiidae.* Denver Museum of Natural History, Denver.

PHILLIPS, W. W. A. (1939). Nests and eggs of Ceylon birds. *Ceylon Journal of Science (B)*, 21: 113–137.

PIERCE, V. & GRUBB, T. C. Jr. (1981). Laboratory studies of foraging in four bird species of deciduous woodland. *Auk*, 98: 307–320.

PLATTNER, J. & SUTTER, E. (1947). Ergebnisse der Meisen- und Kleiberberingung in der Schweiz (1920–1941) (II. Teil). *Der Ornithologische Beobachter*, 44: 1–35.

POLIVANOV, V. M. (1981). [*Ecology of Hole-nesting Birds in Primor'e*]. Academy of Sciences of the USSR, Moscow.

POLIVANOV, V. M. & POLIVANOVA, N. N. (1986). [Ecology of woodland birds on the northern macro-slopes of the north-west Caucasus]. *Trudy Teberdinsk. gos. zapoved*, 10: 11–164.

POLIVANOVA, N. N. (1985). [The food of nestlings of woodland birds in Teberda Natural Reserve]. In: *Ptitsy Severo-zapadnogo Kavkaza* (Ed. A. M. Amirkhanov): pp. 101–124, "Nauka", Moscow.

POST, F. & ONGENAE, J. P. (1990). Over de Boomklever en andere zware jongens. In: *Vogels in Midden-Brabant* (Eds. F. Post, A. Braam & R. Buskens): pp. 34–40, Werkgroep voor Vogel- en Natuurbescherming Midden-Brabant, Oisterwijk.

PRANTY, B. (1995). Tool use by Brown-headed Nuthatches in two Florida slash pine forests. *Florida Field Naturalist*, 23: 33–34.

PRAVOSUDOV, V. V. (1993a). Social organization of the Nuthatch *Sitta europaea asiatica*. *Ornis Scandinavica*, 24: 290–296.

PRAVOSUDOV, V. V. (1993b). Breeding biology of the Eurasian Nuthatch in northeastern Siberia. *Wilson Bulletin*, 105: 475–482.

PRAVOSUDOV, V. V. (1995). Clutch size and fledging rate in the Eurasian

Nuthatch breeding in natural cavities is unrelated to nest cavity size. *Journal of Field Ornithology*, 66: 231–235.

PRAVOSUDOV, V. V. & GRUBB, T. C. Jr. (1993). White-breasted Nuthatch (*Sitta carolinensis*). In: *The Birds of North America* No. 54 (Ed. A. Poole & F. Gill): pp. 1–16, Academy of Natural Sciences, Philadelphia; American Ornithologists' Union, Washington.

PRAVOSUDOV, V. V., PRAVOSUDOVA, E. V. & ZIMIREVA, E. Y. (1996). The diet of nestling Eurasian Nuthatches. *Journal of Field Ornithology*, 67: 114–118.

PRILL, H. (1988). Siedlungsdichte und Nistökologie des Kleibers im Naturschutzgebiet Serrahn. *Ornithologisches Rundbrief Mecklenburgs*, 31: 61–69.

PTUSHENKO, E. S. & INOZEMTSEV, A. A. (1968). [*Biology and Economic Significance of Birds in the Moscow Region and Surrounding Areas*]. Academy of Sciences of the USSR, Moscow.

PULLIAM, H. R. (1988). Sources, sinks and population regulation. *American Naturalist*, 132: 652–661.

RADFORD, M. C. (1954). Notes on the winter roosting and behaviour of a pair of Nuthatches. *British Birds*, 47: 166–168.

RADFORD, M. C. (1955). Nuthatch roosting times in relation to light as measured with a photometer. *British Birds*, 48: 71–74.

RADFORD, M. C. (1957). Observations on broods of Nuthatches leaving the nest. *British Birds*, 50: 526–528.

RAND, A. L. (1959). The pitch-plastering of the Red-breasted Nuthatch. *Audubon Magazine*, 61: 270–272.

RAND, A. L. & FLEMING, R. L. (1957). Birds from Nepal. *Fieldiana: Zoology*, 41: 118–119.

RAND, A. L. & RABOR, D. S. (1967). New birds from Luzon, Philippine Islands. *Fieldiana: Zoology*, 51: 89.

RANOSZEK, E. (1969). [Quantitative observations on breeding populations of birds in *Querceto-carpinetum* forest at the Odra River]. *Notatki Ornitologiczne*, 10: 10–14.

REISER, O. (1905). *Materialien zu einer Ornis Balcanica. III. Griechenland und die Griechischen Inseln*. Bosnisch-Herzegowinischen Landesmuseum, Wien.

RICHARDS, T. J. (1949). Concealment of food by Nuthatch, Coal Tit and Marsh Tit. *British Birds*, 42: 360–361.

RICHARDSON, F. (1942). Adaptive modifications for tree-trunk foraging in birds. *University of California Publications in Zoology*, 46: 317–368.

RIPLEY, S. D. (1959). Character displacement in Indian nuthatches (*Sitta*). *Postilla*, 42: 1–11.

RISBERG, L. (1977). Fagelrapport för 1976. *Vår Fagelvärld*, 36: 266–285.

RITCHISON, G. (1981). Breeding biology of the White-breasted Nuthatch. *The Loon*, 53: 184–187.

RITCHISON, G. (1983). Vocalisations of the White-breasted Nuthatch. *Wilson Bulletin*, 95: 440–451.

RIVERA, J. G. (1985). Comportamiento de alimencion del Trepador Azul, *Sitta europaea* (L.) a lo largo del ano, en un bosque de la Cordillera Cantabrica Occidental (Galicia). *Cyanopica*, 3: 463–470.

ROBERTS, T. J. (1992). *The Birds of Pakistan. Vol. 2. Passeriformes*. Oxford University Press, Oxford.

ROBINSON, H. C. (1927). *The Birds of the Malay Peninsula. Vol. 1. The Commoner Birds*. Witherby, London.

RODGERS, J. A. Jr., KALE, H. W. II & SMITH, H. T. (1995). *Rare and Endangered Biota of Florida. Vol. V. Birds*. University Press of Florida, Gainesville.

ROHÁCEK, F. (1919). Beiträge zur Biologie der *Sitta neumayer* Mich. *Ornithologisches Jahrbuch,* 29: 130–136.

ROKITANSKY, G. (1962). Über das Nest des Felsenkleibers (*Sitta neumayer*). *Die Vogelwelt,* 82: 28–29.

ROOT, T. L. (1988). *Atlas of Wintering North American Birds. An Analysis of Christmas Bird Count Data.* University of Chicago Press, Chicago.

ROUND, P. D. (1983). Some recent bird records from Northern Thailand. *Natural History Bulletin of the Siam Society,* 31: 123–138.

ROUND, P. D. (1984). The status and conservation of the bird community in Doi Suthep-Pui national park, North-west Thailand. *Natural History Bulletin of the Siam Society,* 32: 21–46.

ROUND, P. D. (1988). *Resident Forest Birds in Thailand: Their Status and Conservation.* ICBP Monograph No. 2, ICBP, Cambridge, UK.

SANDERSON, E. D. (1898). The economic value of the White-bellied Nuthatch and Black-capped Chickadee. *Auk,* 15: 144–155.

SARUDNY, N. & HÄRMS, M. (1923). Bemerkungen über einige Vögel Persiens. III. Gattung *Sitta* L. *Journal für Ornithologie,* 71: 398–421.

SAUER, J. R., HINES, J. E., GOUGH, G., THOMAS, I. & PETERJOHN, B. G. (1997). *The North American Breeding Bird Survey: Results and Analysis. Version 96.3.* Patuxent Wildlife Research Center, Laurel, Maryland.(http://www.mbr.nbs.gov/bbs/bbs.html, last accessed 30 September 1997).

SCHÄFER, E. (1938). Ornithologische Ergebnisse zweier Forschungsreisen nach Tibet. *Journal für Ornithologie* Suppl., 38: 1–349.

SCHÄFER, E. & DE SCHAUENSEE, R. M. (1938). Zoological results of the second Dolan expedition to western China and eastern Tibet, 1934–1936. Part II: Birds. *Proceedings of the Academy of Natural Sciences of Philadelphia,* 90: 185–260.

SCHANDY, T. (1981). Green Woodpeckers throw out Nuthatches. *Vår Fuglefauna,* 4: 171.

SCHERNER, E. R. (1983). Kleiber (*Sitta europaea*) mit weißem Brustgefieder in Niedersachsen. *Beiträge zur Vogelkunde,* 29: 119.

SCHMIDT, K.-H. (1984). Frühjarstemperaturen und Legebeginn bei Meisen (*Parus*). *Journal für Ornithologie,* 125: 321–331.

SCHMIDT, K.-H. & HAMANN, H.-J. (1983). Unterbrechung der Legefolge bei Höhlenbrütern. *Journal für Ornithologie,* 124: 163–176.

SCHMIDT, K.-H., MÄRZ, M. & MATTHYSEN, E. (1992). Breeding success and laying date of Nuthatches *Sitta europaea* in relation to habitat, weather and breeding density. *Bird Study,* 39: 23–30.

SCHMIDT, R. (1979). Bioakustische Untersuchungen zur Frage der Korrelation zwischen Lautäußerungen und Erregungsgrad des Kleibers (*Sitta europaea*). Diplomarbeit, University of Kaiserslautern.

SCHOEVAART, S. (1981). De Boomklever in Groot-Amsterdam: verleden, heden, toekomst. *De Gierzwaluw,* 19: 140–142.

SCHÖNFELD, M. & BRAUER, R. (1972). Ergebnisse der 8jährigen Untersuchungen an der Höhlenbrüterpopulation eines Eichen-Hainbuchen-Linden-Waldes in der "Alten Göhle" bei Freyburg/Unstrut. *Hercynia,* 9: 40–68.

SCHOTMAN, A. G. M. & MEEUWSEN, H. A. M. (1994). *Voorspelling van het Effect van Ontsnipperende Maatregelen in het Gebied Rolde/Gieten.* IBN-rapport 118, Instituut voor Bos- en Natuuronderzoek, Wageningen.

SCHÜZ, E. (1957). Ein Vergleich der Vogelwelt von Elbursgebirge und Alpen. *Der Ornithologische Beobachter,* 54: 9–33.

SCHUSTER, L. (1930). Über die Beerennahrung der Vögel. *Journal für Ornithologie,* 78: 273–301.

SEALY, S. G. (1984). Capture and caching of flying carpenter ants by Pygmy Nuthatches. *Murrelet*, 65: 49–51.

SIBLEY, C. G. & AHLQUIST, J. E. (1982). The relationships of the Australo-Papuan Sittellas *Daphoenositta* as indicated by DNA–DNA hybridization. *Emu*, 82: 173–176.

SIBLEY, C. G. & AHLQUIST, J. E. (1990). *Phylogeny and Classification of Birds*. Yale University Press, New Haven.

SIBLEY, C. G. & MONROE, B. L. Jr. (1990). *Distribution and Taxonomy of Birds of the World*. Yale University Press, New Haven.

SIKORA, S. (1975a). [Occurrence and composition of food of Nuthatch (*Sitta europaea* L.) in forest biotopes]. *Roczniki Akademii Rolniczej w Poznanin*, 87: 193–210.

SIKORA, S. (1975b). [Investigations on the biology of Nuthatch (*Sitta europaea* L.)]. *Roczniki Akademii Rolniczej w Poznanin*, 87: 171–191.

SIMPSON, M. B. Jr. (1976). Breeding season habitat and distribution of the Red-breasted Nuthatch in the southern Blue Ridge Mountain Province. *The Chat*, 40: 23–25.

SLESSERS, M. (1970). Bathing behavior of land birds. *Auk*, 87: 95–99.

SMALL, R.J., HOLZWART, J.C. & RUSCH, D.H. (1993). Are Ruffed Grouse more vulnerable to mortality during dispersal? *Ecology*, 74: 2020–2026.

SMITH, H. C., GARTHWAITE, P. F., SMYTHIES, B. E. & TICEHURST, C. B. (1940). Notes on the birds of Nattaung, Karenni. *Journal of the Bombay Natural History Society*, 41: 577–593.

SMITH, K. W., AVERIS, B. & MARTIN, J. (1987). The breeding bird community of oak plantations in the Forest of Dean, Southern England. *Acta Oecologica-Oecologia Generalis*, 8: 209–217.

SMITH, P. W. & SMITH, S. A. (1994). A preliminary assessment of the Brown-headed Nuthatch in the Bahamas. *Bahamas Journal of Science*, 1: 22–26.

SMYTHIES, B. E. (1960). *The Birds of Borneo*. The Sabah Society & The Malayan Nature Society, Kuala Lumpur.

SMYTHIES, B. E. (1986). *The Birds of Burma*. Nimrod Press, Liss, UK.

SNOW, D. W. (1954). Trends in geographical variation in Palaearctic members of the genus *Parus*. *Evolution*, 8: 19–28.

SPEIRS, J. M. (1985). *Birds of Ontario*. Natural Heritage/Natural History Inc., Toronto.

STALLCUP, P. L. (1968). Spatio-temporal relationships of nuthatches and wood-peckers in ponderosa pine forests of Colorado. *Ecology*, 49: 831–843.

STANFORD, J. K. & MAYR, E. (1941). The Vernay–Cutting expedition to Northern Burma. Part II. *Ibis*, 83: 56–105.

STANFORD, J. K. & TICEHURST, C. B. (1938). On the birds of Northern Burma. Part I. *Ibis*, 80: 65–102.

STAUFFER, D. F. & BEST, L. B. (1982). Nest-site selection by cavity-nesting birds of riparian habitats in Iowa. *Wilson Bulletin*, 94: 329–337.

STECHOW, J. (1937). Ein Beitrag zur Frage des Vertriebenwerdens der erwachse-nen Jungvögel durch die Alten. *Beiträge zur Fortpflanzungsbiologie der Vögel*, 13: 54–55.

STEINFATT, O. (1938). Das Brutleben des Kleibers, *Sitta europaea* (*homeyeri* Hartert). *Mitteilungen des Vereins sächsischer Ornithologen*, 5: 167–180.

STRESEMANN, E. (1919). *Sitta europaea homeyeri*, eine reine Rasse oder eine Mischrasse? *Verhandlungen der Ornithologische Gesellschaft Bayern*, 14: 139–147.

STRESEMANN, E. (1925). Zur Systematik der Felsenkleiber. *Ornithologische Monatsberichte*, 33: 106–109.

STRESEMANN, E. & HEINRICH, G. (1940). Die Vögel des Mount Victoria. *Mitteilungen des Zoologisches Museums Berlin*, 24: 151–264.

STRESEMANN, E., MEISE, W. & SCHÖNWETTER, M. (1937). Aves Beickianae. Beiträge zur Ornithologie von Nordwest-Kansu nach den Forschungen von Walter Beick in den Jahren 1926–1933. *Journal für Ornithologie*, 85: 375–576.

SUTTON, G. M. (1967). *Oklahoma Birds*. Oklahoma University Press, Norman.

SVÄRDSON, G. (1955). [The Nuthatch (*Sitta europaea*) as an irruption-bird in Sweden]. *Vår Fagelvärld*, 14: 235–240.

SVENSSON, S. (1975). [Chaffinch stealing food from Nuthatch]. *Anser*, 14: 265–266.

SVENSSON, S. (1981). Population fluctuations in tits *Parus*, Nuthatch *Sitta europaea* and Treecreeper *Certhia familiaris* in South Sweden. *Proceedings Second Nordic Congress of Ornithology* (1979): 9–18.

SYDEMAN, W. J. (1989). Effects of helpers on nestling care and breeder survival in Pygmy Nuthatches. *Condor*, 91: 147–155.

SYDEMAN, W. J. (1991). Facultative helping by Pygmy Nuthatches. *Auk*, 108: 173–175.

SYDEMAN, W. J. & GÜNTERT, M. (1983). Winter communal roosting in the Pygmy Nuthatch. *USDA Forest Service General Technical Report RM*, 99: 121–124.

SYDEMAN, W. J., GÜNTERT, M. & BALDA, R. P. (1988). Annual reproductive yield in the cooperative Pygmy Nuthatch (*Sitta pygmaea*). *Auk*, 105: 70–77.

SZARO, R. C. & BALDA, R. P. (1979). Bird community dynamics in a ponderosa pine forest. *Studies in Avian Biology*, 3: 1–66.

SZÉKELY, T. (1987). Foraging behaviour of woodpeckers (*Dendrocopos* spp.), Nuthatch (*Sitta europaea*) and treecreeper (*Certhia* sp.) in winter and in spring. *Ekologia Polska*, 35: 101–114.

SZÉKELY, T. & JUHÁSZ, T. (1993). Flocking behaviour of tits (*Parus* spp.) and associated species: the effect of habitat. *Ornis Hungarica*, 3: 1–6.

SZÉKELY, T., SZÉP, T. & JUHÁSZ, T. (1989). Mixed species flocking of tits (*Parus* spp.): a field experiment. *Oecologia*, 78: 490–495.

TACZANOWSKI, M. L. (1882). Notice sur la Sittelle d'Europe (*Sitta europaea Linn.*). *Bulletin de la Société Zoologique de France*, 1882: 1–5.

TELLERÍA, J. L. & SANTOS, T. (1995). Effects of forest fragmentation on a guild of wintering passerines: the role of habitat selection. *Biological Conservation*, 71: 61–67.

THÉVENOT, M. (1982). Contribution à l'étude écologique des Passereaux forestiers du Plateau Central et de la corniche du Moyen Atlas (Maroc). *L'Oiseau et la Revue Francaise d'Ornithologie*, 52: 21–86.

THIBAULT, J.-C. (1983). *Les Oiseaux de la Corse. Histoire et Répartition aux XIXe et XXe Siècles*. Parc Naturel Régional de la Corse, Ajaccio.

THIJSSE, J.P. (1903). *Het Vogeljaar*. A.G. Van Schoonderbeek, Laren.

THOMPSON, D. W. (1936). *A Glossary of Greek Birds*. Oxford University Press, London.

THOMPSON, P. S., GREENWOOD, J. J. D. & GREENWAY, K. (1993). Birds in European gardens in the winter and spring of 1988–89. *Bird Study*, 40: 120–134.

TINBERGEN, J. M., VAN BALEN, H. & VAN ECK, H. M. (1985). Density dependent survival in an isolated Great Tit population: Kluyvers data reanalysed. *Ardea*, 73: 38–48.

TOMIAŁOJĆ, L. & PROFUS, P. (1977). Comparative analysis of breeding bird communities in two parks of Wrocław and in an adjacent *Querco-Carpinetum* forest. *Acta Ornithologica*, 16: 117–177.

TOOK, G. E. (1946). "Injury-feigning" of Nuthatch. *British Birds*, 39: 117.

TRAMER, E. J. (1994). Feeder access: deceptive use of alarm calls by a White-breasted Nuthatch. *Wilson Bulletin*, 106: 573.

TSO-HSIN, C. (1987). *A Synopsis of the Avifauna of China.* Paul Parey, Hamburg.

TSO-HSIN, C., WEN-NING, T. & TZE-YU, W. (1964). [A new subspecies of the Velvet-fronted Nuthatch from Hainan – *Sitta frontalis chienfengensis*, subsp. nov.]. *Acta Zootaxonomica Sinica*, 1: 1–5.

TUCKER, G. M. & HEATH, M. F. (1994). *Birds in Europe: their Conservation Status.* BirdLife International (BirdLife Conservation Series No. 3), Cambridge, UK.

TURČEK, F. J. (1956). On the bird population of the spruce forest community in Slovakia. *Ibis*, 98: 24–33.

TURČEK, F. J. (1961). *Ökologische Beziehungen der Vögel und Gehölze.* Verlag der Slowakischen Akademie der Wissenschaften, Bratislava.

TYLER, W. H. (1916). A study of a White-breasted Nuthatch. *Wilson Bulletin*, 28: 18–25.

UTLEY, J. P. (1944). Entrance to Nuthatch's hole enlarged and no mud used. *British Birds*, 37: 95.

UTTENDÖRFER, O. (1952). *Neue Ergebnisse über die Ernährung der Greifvögel und Eulen.* Eugen Ulmer, Stuttgart.

VAN BALEN, J. H., BOOY, C. J. H., VAN FRANEKER, J. A. & OSIECK, E. R. (1982). Studies on hole-nesting birds in natural nest sites. 1. Availability and occupation of natural nest sites. *Ardea*, 70: 1–24.

VAN DE CASTEELE, T. (1994). Invloed van habitaatfragmentatie op het sociaal gedrag van de Boomklever (*Sitta europaea*). Licentiaatsthesis, University of Antwerp.

VAN DEN BRINK, J. N. (1951). Late trek of omzwervingen van de Boomklever, *Sitta europaea* subsp. *Limosa*, 24: 67.

VAN DORP, D. & OPDAM, P. F. M. (1987). Effects of patch size, isolation and regional abundance on forest bird communities. *Landscape Ecology*, 1: 59–73.

VAN MARLE, J. G. & VOOUS, K. H. (1988). *The Birds of Sumatra. An Annotated Check-list.* British Ornithologists' Union, Tring.

VAN NOORDEN, B., OPDAM, P. & SCHOTMAN, A. (1988). Dichtheid van bosvogels in geïsoleerde loofbosjes. *Limosa*, 61: 19–25.

VAN STYVENDAELE, B. (1963). Vergelijkend onderzoek naar de broedvogelstand van een bosgemeenschap. *Giervalk*, 2: 197–224.

VAURIE, C. (1950). Notes on some Asiatic nuthatches and creepers. *American Museum Novitates*, 1472: 1–39.

VAURIE, C. (1951). Adaptive differences between two sympatric species of nuthatches (*Sitta*). *Proceedings International Ornithological Congress*, 10: 163–166.

VAURIE, C. (1957). Systematic notes on Palearctic Birds. No. 29. The subfamilies Tichodromadinae and Sittinae. *American Museum Novitates*, 1854: 1–26.

VAURIE, C. (1959). *The Birds of the Palearctic Fauna: Passeriformes.* London, Witherby.

VENABLES, L. S. V. (1938). Nesting of the Nuthatch. *British Birds*, 32: 26–33.

VERBOOM, J. & SCHOTMAN, A. (1994). Responses of metapopulations to a changing landscape. In: *Bird Numbers 1992. Distribution, Monitoring and Ecological Aspects. Proceedings of the 12th International Conference of IBCC and EOAC* (Ed. E.J.M. Hagemeijer & T.J. Verstrael): pp. 117–121, Statistics Netherlands, Voorburg/Heerlen & SOVON, Beek-Ubbergen.

VERBOOM, J., SCHOTMAN, A., OPDAM, P. & METZ, J. A. J. (1991a). European Nuthatch metapopulations in a fragmented agricultural landscape. *Oikos*, 61: 149–156.

VERBOOM, J., OPDAM, P. & SCHOTMAN, A. (1991b). Kerngebieden en kleinschalig landschap: een benadering met een metapopulatiemodel. *Landschap*, 8: 3–14.

VERBOOM, J., METZ, J. A. J. & MEELIS, E. (1993). Metapopulation models for impact assessment of fragmentation. In: *Landscape Ecology of a Stressed Environment* (Ed. C. C. Vos & P. Opdam): pp. 172–191, Chapman and Hall, London.

VERMEULEN, N. (1991). De boomklever blijft graag plakken. *Grasduinen*, 1991 (1): 28–31.

VIELLIARD, J. (1976). La Sittelle Kabyle. *Alauda*, 44: 351–352.

VIELLIARD, J. (1978). Le Djebel Babor et sa Sittelle, *Sitta ledanti* Vielliard 1976. *Alauda*, 46: 1–42.

VIELLIARD, J. (1980). Remarques complémentaires sur la Sittelle Kabyle *Sitta ledanti* Vielliard 1976. *Alauda*, 48: 139–150.

VIEWEG, A. (1989). Kleiber beteiligt sich an der Fütterung einer Starenbrut. *Der Falke*, 36: 164–165.

VILKS, E. K. (1966). [Migrations and territorial behaviour of Latvian tits and nuthatches according to ringing records. In: *Migrations of Birds of the Latvian SSR. (Ornithological investigations No. 4)]*, pp. 69–88.

VILKS, K. A. & VILKA, E. K. (1961). Seasonal distribution of tits and nuthatches in Latvia and their supplementary feeding in winter. In: *Birds of the Baltic Region: Ecology and Migrations, Proceedings of the 4th Baltic Ornithological Conference* (Y.Y. Lusis, Z. D. Spurin, E. J. Taurins & E. K. Vilka): pp. 134–144, Academy of Sciences of the Latvian S.S.R., Riga.

VITTERY, A. & SQUIRE, J. E. (1972). *Ornithological Society of Turkey Bird Report 1968–1969.* Ornithological Society of Turkey, Sandy, UK.

VON JORDANS, A. (1923). Über seltenere und über fragliche Vogelformen meiner Sammlung. *Falco* Sunderheft, 19: 8–26.

VON KNORRE, D., GRÜN, G., GÜNTHER, R. & SCHMIDT, K. (1986). *Die Vogelwelt Thüringens.* Gustav Fischer Verlag, Jena.

VOOUS, K. H. (1977). List of recent Holarctic bird species: Passerines. *Ibis*, 119: 376–408.

VOOUS, K. H. & VAN MARLE, J. G. (1953). The distributional history of the Nuthatch *Sitta europaea. Ardea*, 41 (supplement): 1–68.

VUILLEUMIER, F. & MAYR, E. (1987). New species of birds described from 1976 to 1980. *Journal für Ornithologie*, 128: 137–150.

WAHLSTEDT, J. (1965). [Siberian Nuthatches (*Sitta europaea asiatica*) in Norrbotten, Sweden 1962–64]. *Vår Fagelvärld*, 24: 172–182.

WAITE, T. A. (1987). Vigilance in the White-breasted Nuthatches: effects of dominance and sociality. *Auk*, 104: 429–434.

WAITE, T. A. & GRUBB, T. C. Jr. (1988a). Diurnal caching rhythm in captive White-breasted Nuthatches *Sitta carolinensis. Ornis Scandinavica*, 19: 68–70.

WAITE, T. A. & GRUBB, T. C. Jr. (1988b). Copying of foraging locations in mixed-species flocks of temperate-deciduous woodland birds: an experimental study. *Condor*, 90: 132–140.

WALPOLE-BOND, J. (1931). Notes on the songs and cries of the British Nuthatch. *British Birds*, 25: 70–71.

WALTERS, M. (1980). *The Complete Birds of the World.* David & Charles, London.

WARD, I. (1982). Nuthatch hovering. *British Birds*, 75: 537.

WATERS, J. R., NOON, B. R. & VERNER, J. (1990). Lack of nest site limitation in a cavity-nesting bird community. *Journal of Wildlife Management*, 54: 239–245.

WELLER, L. G. (1949). Nuthatches walling up abandoned nesting-hole. *British Birds*, 42: 56–57.

WESOŁOWSKI, T. (1989). Nest-sites of hole-nesters in a primaeval temperate forest (Bialowieza National Park, Poland). *Acta Ornithologica*, 25: 321–351.

WESOŁOWSKI, T. (1994). Variation in the numbers of resident birds in a primaeval temperate forest: are winter weather, seed crop, caterpillars and interspecific competition involved? In: *Bird Numbers 1992. Distribution, Monitoring and Ecological Aspects. Proceedings of the 12th International Conference of IBCC and EOAC* (Ed. E.J.M. Hagemeijer & T.J. Verstrael): pp. 203–211, Statistics Netherlands, Voorburg/Heerlen & SOVON, Beek-Ubbergen.

WESOŁOWSKI, T. & STAWARCZYK, T. (1991). Survival and population dynamics of Nuthatches *Sitta europaea* breeding in natural cavities in a primeval temperate forest. *Ornis Scandinavica*, 22: 143–154.

WHISTLER, H. (1963). *Popular Handbook of Indian Birds* (4th edn. revised and enlarged by N. B. Kinnear). Oliver & Boyd, Edinburgh.

WHITTLE, H. G. (1926). Recent history of a pair of White-breasted Nuthatches, nos. 117455 and 117456. *Bulletin of the Northeastern Bird-Banding Association*, 2: 72–74.

WIDMER, R. (1987). Weniger Kleiberbruten. *Vögel der Heimat*, 57: 125.

WIDRLECHNER, M. P. & DRAGULA, S. K. (1984). Relation of cone-crop size to irruptions of four seed-eating birds in California. *American Birds*, 38: 840–844.

WILDE, N. A. J. (1973). Nuthatch assuming camouflage posture. *British Birds*, 66: 230–231.

WILDER, G. D. & HUBBARD, G. D. (1938). *Birds of Northeastern China.* Handbook Nr 6. Peking Natural History Bulletin, Peking.

WILLIAMS, J. B. & BATZLI, G. O. (1979a). Competition among bark-foraging birds in Central Illinois: experimental evidence. *Condor*, 81: 122–132.

WILLIAMS, J. B. & BATZLI, G. O. (1979b). Interference competition and niche shifts in the bark-foraging guild in central Illinois. *Wilson Bulletin*, 91: 400–411.

WILLIAMS, J. B. & BATZLI, G. O. (1979c). Winter diet of a bark-foraging guild of birds. *Wilson Bulletin*, 91: 126–131.

WILLSON, M. F. (1970). Foraging behavior of some winter birds of deciduous woods. *Condor*, 72: 169–174.

WILSON, E. O. & WILLIS, E. O. (1975). Applied biogeography. In: *Ecology and Evolution of Communities* (Ed. M.L. Cody & J.M. Diamond): pp. 522–534, Belknap Press, Cambridge, Massachusetts.

WINKEL, W. (1970). Hinweise zur Art- und Alterbestimmung von Nestlingen höhlenbrütenden Vogelarten anhand ihrer Körperentwicklung. *Die Vogelwelt*, 91: 52–59.

WINKEL, W. (1989). Zum Dispersionsverhalten und Lebensalter des Kleibers (*Sitta europaea caesia*). *Die Vogelwarte*, 35: 37–48.

WINKEL, W. (1996). Zum primären Geschlechterverhältnis des Kleibers (*Sitta europaea*). *Die Vogelwarte*, 38: 194–196.

WINKEL, W. & HUDDE, H. (1988). Über das Nächtigen von Vögeln in künstlichen Nisthöhlen während des Winters. *Die Vogelwarte*, 34: 174–188.

WINKLER, H. & BOCK, W. J. (1976). Analyse der Kräfteverhältnisse bei Klettervögeln. *Journal für Ornithologie*, 117: 397–418.

WINKLER, R. (1979). Zur pneumatisation des Schädeldachs der Vögel. *Der Ornithologische Beobachter*, 76: 49–118.

WITHERBY, H. F., JOURDAIN, F. C. R., TICEHURST, N. F. & TUCKER, B. W. (1940). *The Handbook of British Birds, Vol. I.* H. F. & G. Witherby, London.

WITHGOTT, J. H. & SMITH, K. G. (1998). Brown-headed Nuthatch (*Sitta pusilla*). In: *The Birds of North America* No. 349 (Ed. A. Poole & F. Gill): pp. 1–24, Academy of Natural Sciences, Philadelphia; American Ornithologists' Union, Washington.

WITVLIET, W. (1987). Weniger Kleiberbruten. *Vögel der Heimat*, 57: 124.
WOLTERS, H. E. (1975–1982). *Die Vogelarten der Erde*. Paul Parey, Hamburg.
WOOD, D. S. (1992). Color and size variation in eastern White-breasted Nuthatches. *Wilson Bulletin*, 104: 599–611.
WOODREY, M. S. (1990). Economics of caching versus immediate consumption by White-Breasted Nuthatches – the effect of handling time. *Condor*, 92: 621–624.
WOODREY, M. S. (1991). Caching behavior in free-ranging White-Breasted Nuthatches – the effects of social dominance. *Ornis Scandinavica*, 22: 160–166.
WOODS, P. E. & BENS, C. M. (1981). Interspecific aggressive display of the White-breasted Nuthatch at a feeder. *Inland Bird Banding News*, 53: 17–19.
WUNDERLICH, K. (1986). *Sitta neumayer* Michahelles – Felsenkleiber. In: *Atlas der Verbreitung Paläarktischer Vögel. 13. Lieferung* (Ed. H. Dathe & I.A. Neufeldt), Academie-Verlag, Berlin.
WUNDERLICH, K. (1988). *Sitta leucopsis* Gould – Weißwangenkleiber. In: *Atlas der Verbreitung Paläarktischer Vögel. 15. Lieferung* (Ed. H. Dathe & I.A. Neufeldt), Academic-Verlag, Berlin.
WYDLER, A. (1973). Vom Kleiber eingemauerte Feldsperlingsbrut. *Der Ornithologische Beobachter*, 70: 184.
YANAGAWA, H. & SHIBUYA, T. (1996). Causes of wild bird mortality in eastern Hokkaido. II. *Research Bulletin of Obihiro University Natural Science*, 19: 251–258.
YATES, G. M. (1983). "Body-brushing" by Nuthatches. *British Birds*, 76: 142–143.
YAUKEY, P. H. (1995). Effects of food supplementation and predator simulation on nuthatches and parids within mixed-species flocks. *Wilson Bulletin*, 107: 542–547.
YAUKEY, P. H. (1996). Patterns of avian population density, habitat use, and flocking behavior in urban and rural habitats during winter. *Professional Geographer*, 48: 70–81.
YAUKEY, P. H. (1997). Multiscale patterns of flocking and activities of Brown-headed Nuthatches. *Physical Geography*, 18: 88–100.
YUNICK, R. P. (1980). Timing of completion of skull pneumatization of the Black-capped Chickadee and the Red-breasted Nuthatch. *North American Bird Bander*, 5: 43–46.
YUNICK, R. P. (1988). An assessment of the White-breasted Nuthatch and Red-breasted Nuthatch on recent New York State Christmas counts. *Kingbird*, 38: 95–104.
ZANG, H. (1980). Zum Geschlechterverhältnis beim Kleiber (*Sitta europaea*). *Vogelkundige Berichte Niedersächsens* Sonderheft, 12: 52–55.
ZANG, H. (1988). Der Einfluß der Höhenlage auf die Biologie des Kleibers (*Sitta europaea*) im Harz. *Journal für Ornithologie*, 129: 161–174.
ZINK, G (1981). *Der Zug Europäischer Singvögel – ein Atlas der Wiederfunde beringter Vögel. 3. Lieferung*. Vogelwarte Radolfzell, Möggingen.
ZIPPELIUS, H.-M. (1973). Das Kopfabwärtsklettern des Kleibers (*Sitta europaea*). *Bonner Zoologische Beiträge*, 24: 48–50.
ZUBERBIER, G. M. & GRUBB, T. C. Jr. (1992). Ptilochronology: wind and cold temperatures fail to slow induced feather growth in captive White-breasted Nuthatches *Sitta carolinensis* maintained on ad libitum food. *Ornis Scandinavica*, 23: 139–142.
ZUKAL, J. (1992). Mammals in the food of the Long-eared Owl, *Asio otus* L. *Lynx*, 26: 21–26.

Index

Note: Bold page numbers indicate a section specific to a nuthatch species.

Algerian Nuthatch *see* Mediterranean
   nuthatches

Beautiful Nuthatch 6, 7, 9, 10, 13, 17,
   219, **232**, 269, 271
beech mast 125, 152, 154
bill morphology 52
Blue Nuthatch 6, 7, 9, 10, 11, 219,
   232–234, 237, 269, 271
breeding success 103–109, 170
breeding biology 86–112

calls 82–85
causes of mortality 148
Chestnut-bellied Nuthatch 6, 11, 12,
   17, 25, 30, 31, 38, 111, 214, 219,
   223–226, 228, 234, 270, 271
Chestnut-vented Nuthatch 6, 9, 17, 25,
   30, 31, 219, 223, 224, **227–228**, 229,
   230, 269, 271
Chinese Nuthatch 5, 6, 9, 16, 17, 24,
   30, 187–189, 197, **219–221**, 270,
   271
climbing technique 14, 47
clutch size 99
comfort behaviour 14
competition for nests 93
Corsican Nuthatch *see* Mediterranean
   nuthatches
courtship 94

diet 47–51
disease 148
dispersal 131–142
   breeding dispersal 142, 149, 173
   distance 132–135, 171
   timing 135–138, 141, 172
displays 70
distribution 24–29
dwarf nuthatches 4–7, 9, 11, 16–18, 20,
   74, 228, 238, 239, 242–247, 252,
   257–268, 270, 271

breeding biology 264
communal roosting 266
foraging and food 51, 56, 260
habitat and densities 259
morphology and taxonomy 258
movements and survival 268
social behaviour 262
vocalizations 264

Eastern Rock Nuthatch *see* rock
   nuthatches
egg-laying 96
evolution 5

fighting 69
fly-catching 46
food handling 12
foraging sites 53–57
foraging behaviour 46
forest fragmentation 163–180
forest degradation 159, 161

Giant Nuthatch 4, 6, 9, 13, 17, 219,
   230–232, 269, 271

habitat 36–44
hoarding 12, 57–61, 193, 209, 224, 227,
   235, 243, 252, 262
home-range 65

incubation 99
irruption 138, 143–146

Kashmir Nuthatch 6, 17, 25, 30, 31,
   111, 198, 201, 214, 219, 222, 223,
   226–227, 228, 269, 271
Krueper's Nuthatch *see* Mediterranean
   nuthatches

landscape planning 180
life-table 278–279

Mediterranean nuthatches 4, 5, 6, 7, 9,
    10, 12, 14, **17**, 20, 24, **183–199**, 201,
    214, 220, 221, 258, 270, 271
  breeding biology 194
  distribution and habitat 190–193
  foraging and food 56, 193
  morphology 186
  movements 198
  phylogeny 5, 188–190
  population status 198
  social behaviour 196
  vocalizations 197
metapopulation 175
mixed flocks 75
morphology 7, 24, 31–34
moult 11, 34–35

nest sites 87–91
nest defence 94
nest-building 16, 90–93
nestboxes 45, 90
non-territorial residents 117, 122–128

pair-bond 71, 128
Peerdsbos study 63–64
Pendulum display 13, 94, 211, 220,
    226, 227, 247, 254, 256, 265
phylogeny 5–6, 188
plastering 16, 17, 91–93, 209, 220, 225,
    229, 232, 235
plumage variation 11, 31
population density 37–44, 164
population models 176–180
population trends 157
population size 153–156, 162
population threats 12, 198, 230–232
post-fledging period 110
predation 148, 110
ptilochronology 61, 249

Red-breasted Nuthatch 5–7, 9, 11,
    16–18, 33, 39, 189, 193, 198, 221,
    238, 239, 240, 243, 246, 248,
    250–257, 270, 271
  breeding biology 255
  foraging and food 51, 56, 251
  habitat and densities 251
  movements 256
  social behaviour and roosting 253
  vocalizations 254
rock nuthatches 4, 6, 7, 9, 12, 14, 16,
    17, 20, 24, 198, 225–228, 269, 271

affinities 201
behaviour and displays 212
breeding biology 211
character displacement 200, 205
distribution 202
foraging and food 208
habitat 205
morphology 201, 205
movements 216
nest 209
social organization 216
vocalizations 214
roosting 76

second broods 111
sex ratio 109, 150
sexing and ageing 31–32
size variation 9, 33, 52, 274
song duetting 16, 197, 215, 229
song 76–82
source-sink 174
studies on nuthatches 18–20
study areas 63, 166
Sulphur-billed Nuthatch 6, 9, **236–237**,
    269, 271
survival rates 149, 170, 278

taxonomy 24–30
territory 65–69, 113–130
  establishment 113–119, 122–128
  quality 119
  shifts 115, 173
  size 66–68
time budget 73
tool use 241, 262
trios 71

Velvet-fronted Nuthatch 6, 7, 9, 10, 11,
    12, 17, 214, 219, 224, 233, **234–236**,
    269, 271

Wallcreeper 5, 6
Western Rock Nuthatch *see* rock
    nuthatches
White-breasted Nuthatch 5–7, 9, 14,
    16, 17, 20, 53, 54, 56, 58, 61, 75, 76,
    100, 222, 223, 228, 233, **240–250,
    252, 254, 255, 270, 271**
  breeding biology 246
  ecology and movements 249
  foraging and food 51, 56, 241–243
  habitat and densities 240

social behaviour and roosting 243
vocalizations 245
White-browed Nuthatch 6, 9, 12, 219,
  228, **230**, 258, 270, 271
White-cheeked Nuthatch 5, 6, 8, 9, 16,
  17, 30, 220, 221, **222–223**, 226, 227,
  233, 239, 270, 271
White-tailed Nuthatch 6, 7, 9, 16, 219,
  221, 222, 224, 226, **228–230**, 258,
  270, 271

winter cold 152, 157
world distribution 7

Yellow-billed Nuthatch 6, 9, 12, **236**,
  269, 271
Yunnan Nuthatch 5, 6, 9, 188, 189,
  219, **221–222**, 228, 270, 271